CAMBRIDGE MONOGRAPHS ON MECHANICS AND APPLIED MATHEMATICS

GENERAL EDITORS

G. K. BATCHELOR, PH.D., F.R.S.

Professor of Applied Mathematics in the University of Cambridge

J. W. MILES, PH.D.

Professor of Applied Mathematics, University of California, La Jolla

MAGNETIC FIELD GENERATION IN ELECTRICALLY CONDUCTING FLUIDS

MAGNETIC FIELD GENERATION IN ELECTRICALLY CONDUCTING FLUIDS

H. K. MOFFATT

PROFESSOR OF APPLIED MATHEMATICS, UNIVERSITY OF BRISTOL

CAMBRIDGE UNIVERSITY PRESS

CAMBRIDGE

LONDON · NEW YORK · MELBOURNE

Published by the Syndics of the Cambridge University Press
The Pitt Building, Trumpington Street, Cambridge CB2 1RP
Bentley House, 200 Euston Road, London NW1 2DB
32 East 57th Street, New York, NY 10022, USA
296 Beaconsfield Parade, Middle Park, Melbourne 3206, Australia

First published 1978

Printed in Great Britain at the University Press, Cambridge

Library of Congress Cataloguing in Publication Data

Moffatt, Henry Keith, 1935–

Magnetic field generation in electrically conducting fluids.

(Cambridge monographs on mechanics and applied mathematics)

Bibliography: p. 325

1. Dynamo theory (Cosmic physics) I. Title.
QC809.M25M63 538 77–4398
ISBN 0 521 21640 0

CONTENTS

PREFACE

Understanding of the process of magnetic field generation by self-inductive action in electrically conducting fluids (or 'dynamo theory' as the subject is commonly called) has advanced dramatically over the last decade. The subject divides naturally into its kinematic and dynamic aspects, neither of which were at all well understood prior to about 1960. The situation has been transformed by the development of the two-scale approach advocated by M. Steenbeck, F. Krause and K.-H. Rädler in 1966, an approach that provides essential insights into the effects of fluid motion having either a random ingredient, or a space-periodic ingredient, over which spatial averages may usefully be defined. Largely as a result of this development, the kinematic aspect of dynamo theory is now broadly understood, and recent inroads have been made on the much more difficult dynamic aspects also.

Although a number of specialised reviews have appeared treating dynamo theory in both solar and terrestrial contexts, this monograph provides, I believe, the first self-contained treatment of the subject in book form. I have tried to focus attention on the more fundamental aspects of the subject, and to this end have included in the early chapters a treatment of those basic results of magnetohydrodynamics that underly the theory. I have also however included two brief chapters concerning the magnetic fields of the Earth and the Sun, and the relevant physical properties of these bodies, and I have made frequent reference in later chapters to specific applications of the theory in terrestrial and solar contexts. Thus, although written from the point of view of a theoretically oriented fluid dynamicist, I hope that the book will be found useful by graduate students and researchers in geophysics and astrophysics, particularly those whose main concern is geomagnetism or solar magnetism.

My treatment of the subject is based upon a course of lectures given in various forms over a number of years to graduate students reading Part III of the Mathematical Tripos at Cambridge University. I was also privileged to present the course to students of the 3me Cycle in Theoretical Mechanics at the Université Pierre et Marie Curie, Paris, during the academic year 1975–6. The material of all the chapters, except the difficult chapter 8 on the theory of S. I. Braginskii, has been subjected in this way to student criticism, and has greatly benefited in the process.

The single idea which recurs throughout and which I hope gives some unity to the treatment is the idea of 'lack of reflexional symmetry' of a fluid flow, the simplest measure of which is its 'helicity'. In a sense, this is a book about helicity; the invariance and topological interpretation of this pseudo-scalar quantity are discussed at an early stage (chapter 2) and the central importance of helicity in the dynamo context is emphasised in chapters 7 and 8. Helicity is also the theme of chapter 10 (on helical wave motions) and of chapter 11, in which its influence in turbulent flows with and without magnetic fields is discussed. A preliminary and much abbreviated account of some of these topics has already appeared in my review article (Moffatt, 1976) in volume 16 of Advances in Applied Mechanics.

It is a pleasure to record my gratitude to many colleagues with whom I have enjoyed discussions and correspondence on dynamo theory; in particular to Willi Deinzer, David Gubbins, Uriel Frisch, Robert Kraichnan, Fritz Krause, Willem Malkus, Karl-Heinz Rädler, Paul Roberts, Michael Stix and Nigel Weiss; also to Glyn Roberts, Andrew Soward and Michael Proctor whose initial research it was my privilege to supervise, and who have since made striking contributions to the subject; and finally to George Batchelor who as Editor of this series, has given constant encouragement and advice. To those who have criticised the manuscript and helped eliminate errors in it, I offer warm thanks, while retaining full responsibility for any errors, omissions and obscurities that may remain.

I completed the writing during the year 1975–6 spent at the Université Pierre et Marie Curie, and am grateful to M. Paul Germain and M. Henri Cabannes and their colleagues of the Laboratoire de Mécanique Théorique for inviting me to work in such a stimulating and agreeable environment.

CHAPTER I

INTRODUCTION AND HISTORICAL BACKGROUND

'How could a rotating body such as the Sun become a magnet?' This was the title of a famous 'brief communication' from Sir Joseph Larmor in 1919 to the British Association for the Advancement of Science; and the question was certainly a natural one since the origin of the magnetic field of the Sun was at that time a total mystery. Curiously, the magnetic field of the Earth did not then excite similar inquiry, because it was still believed that this could be explained in terms of permanent magnetisation. However it has now long been known that the temperature of the Earth's interior is well above the critical temperature (the Curie point) at which ferromagnetic materials lose their 'permanent' magnetisation and that some other explanation for the Earth's magnetic field must be found. The fact that a large fraction of the Earth's interior is now known (by inference from seismological observations) to be in a liquid state is profoundly relevant to the problem.

Not only the Earth and the Sun; it is probably safe to state that a magnetic field is a normal accompaniment of any cosmic body that is both fluid (wholly or in part) and rotating. There appears to be a sort of universal validity about this statement which applies quite irrespective of the length-scales considered. For example, on the planetary length-scale, Jupiter shares with the Earth the property of strong rotation (its rotation period being approximately 10 hours) and it is believed to have a fluid interior composed of an alloy of liquid metallic hydrogen and helium (Hide, 1974); it exhibits a surface magnetic field of order 10 gauss in magnitude[1] (as compared with the Earth's field of order 1 gauss). On the stellar length-scale magnetic fields as weak as the solar field (~ 1 gauss) cannot be detected in general; there are however numerous examples of stars

[1] The international unit of magnetic field is the tesla (T), and $1 \mathrm{T} = 1 \mathrm{Wb\,m^{-2}} = 10^4$ gauss.

which rotate with periods ranging from several days to several months, and with detectable surface magnetic fields in the range 10^2 to 3×10^4 gauss (Preston, 1967). And on the galactic length-scale, our own galaxy rotates about the normal to the plane of its disc with a period of order 3×10^8 years and exhibits a galactic-scale magnetic field roughly confined to the plane of the disc whose typical magnitude is of order 3 or 4×10^{-6} gauss.

The detailed character of these naturally occurring magnetic fields and the manner in which they evolve in time will be described in subsequent chapters; for the moment it is enough to state that it is the mere existence of these fields (irrespective of their detailed properties) which provides the initial motivation for the various investigations which will be described in this monograph.

Larmor put forward three alternative and very tentative suggestions concerning the origin of the Sun's magnetic field, only one of which has in any sense stood the test of time. This suggestion, which is fundamental to hydromagnetic dynamo theory, was that motion of the electrically conducting fluid within the rotating body, might by its inductive action in flowing across the magnetic field generate just those currents $\mathbf{J}(\mathbf{x})$ required to provide the self-same field $\mathbf{B}(\mathbf{x})$.

This type of 'bootstrap' effect is most simply illustrated with reference to a system consisting entirely of solid (rather than fluid) conductors. This is the 'homopolar' disc dynamo (Bullard, 1955) illustrated in fig. 1.1. A solid copper disc rotates about its axis with angular velocity Ω, and a current path between its rim and its axle is provided by the wire twisted as shown in a loop round the axle. This system can be unstable to the growth of magnetic perturbations. For suppose that a current $I(t)$ flows in the loop; this generates a magnetic flux Φ across the disc, and, provided the conductivity of the disc is not too high, this flux is given by $\Phi=M_0I$ where M_0 is the mutual inductance between the loop and the rim of the disc. (The proviso concerning the disc conductivity is necessary as is evident from the consideration that a superconducting disc would not allow any flux to cross its rim; a highly conducting disc in a time-dependent magnetic field tends to behave in the same way.) Rotation of the disc leads to an electromotive force $\mathscr{E}=\Omega\Phi/2\pi$ which drives the current I, and the equation for $I(t)$ is then

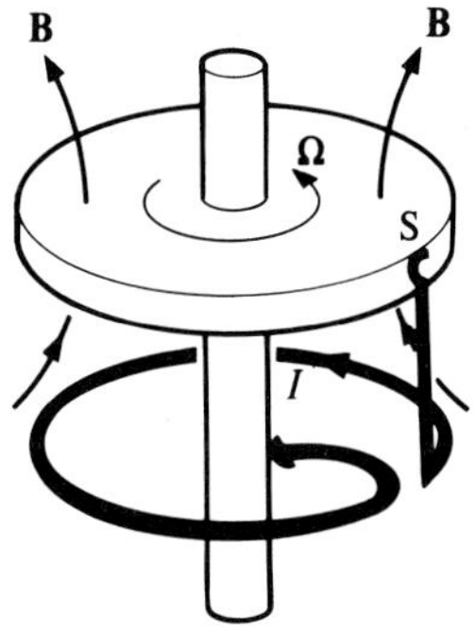

Fig. 1.1 The homopolar disc dynamo. Note that the twist in the wire which carries the current $I(t)$ must be in the same sense as the sense of rotation Ω.

$$L\frac{\mathrm{d}I}{\mathrm{d}t}+RI=\mathscr{E}=M\Omega I, \tag{1.1}$$

where $M=M_0/2\pi$ and L and R are the self-inductance and resistance of the complete current circuit. The device is evidently unstable to the growth of I (and so of Φ) from an infinitesimal level if

$$\Omega>R/M. \tag{1.2}$$

In this circumstance the current grows exponentially, as does the retarding torque associated with the Lorentz force distribution in the disc. Ultimately the disc angular velocity slows down to the critical level $\Omega_0=R/M$ at which the driving torque G just balances this retarding torque, and the current can remain steady. In this equilibrium, the power supplied $\Omega_0 G$ is equal to the rate of Joule dissipation RI^2 (assuming that frictional torques are negligible).

This type of example is certainly suggestive, but it differs from the conducting fluid situation in that the current is constrained by the twisted geometry to follow a very special path that is particularly conducive to dynamo action (i.e. to the conversion of mechanical energy into magnetic energy). No such geometrical constraints are apparent in, say, a spherical body of fluid of uniform electrical conductivity, and the question arises whether fluid motion within such a sphere, or other simply-connected region, can drive a suitably contorted current flow to provide the same sort of homopolar (i.e. self-excited) dynamo effect.

There are however two properties of the disc dynamo which reappear in some of the hydromagnetic situations to be considered later, and which deserve particular emphasis at this stage. Firstly, there is a discontinuity in angular velocity at the sliding contact S between the rotating disc and the stationary wire, i.e. the system exhibits *differential rotation*. The concentration of this differential rotation at the single point S is by no means essential for the working of the dynamo; we could in principle distribute the differential rotation arbitrarily by dividing the disc into a number of rings, each kept in electrical contact with its neighbours by means of lubricating films of, say, mercury, and by rotating the rings with different angular velocities. If the outermost ring is held fixed (so that there is no longer any sliding at the contact S), then the velocity field is entirely axisymmetric, the differential rotation being distributed across the plane of the disc. The system will still generally work as a dynamo provided the angular velocity of the inner rings is in the sense indicated in fig. 1.1 and sufficiently large.

Secondly, the device *lacks reflexional symmetry*: in fig. 1.1 the disc must rotate in the same sense as the twist in the wire if dynamo action is to occur. Indeed it is clear from (1.1) that if $\Omega < 0$ the rotation leads only to an *accelerated decay* of any current that may initially flow in the circuit. Recognition of this essential lack of reflexional symmetry provides the key to understanding the nature of dynamo action as it occurs in conducting fluids undergoing complex motions.

It was natural however for early investigators to analyse systems having a maximum degree of symmetry in order to limit the analytical difficulties of the problem. The most natural 'primitive' system to consider in the context of rotating bodies such as the Earth or the Sun is one in which both the velocity field and magnetic field are axisymmetric. Cowling (1934) considered this idealisation in an investigation of the origin of the much more local and intense magnetic fields of sunspots (see chapter 5), but concluded that a steady axisymmetric field could not be maintained by axisymmetric motions. This first 'anti-dynamo' theorem was reinforced by later investigations (Backus & Chandrasekhar, 1956; Cowling, 1957*b*) and it was finally shown by Backus (1957) that axisymmetric motions could at most extend the natural decay time of an axisymmetric field in a spherical system by a factor of about 4. In the context of the Earth's magnetic field, whose natural decay time is of the order of 10^4–10^5 years (see chapter 4), this modest delaying action is totally inadequate to explain the continued existence of the main dipole field for a period of the same order (3×10^9 years) as the age of the Earth itself (the evidence being from studies of rock magnetism), and its relative stability over periods of order 10^6 years and greater (Bullard, 1968). It was clear that non-axisymmetric configurations had to be considered if any real progress in dynamo theory were to be made. It is in fact the essentially three-dimensional character of 'the dynamo problem' (as the problem of explaining the origin of the magnetic field of the Earth or of any other cosmic body has come to be called) that provides both its particular difficulty and its peculiar fascination.

Recognition of the three-dimensional nature of the problem led Elsasser (1946) to initiate the study of the interaction of a prescribed non-axisymmetric velocity field with a general non-axisymmetric magnetic field in a conducting fluid contained within a rigid spherical boundary, the medium outside this boundary being assumed non-conducting. Elsasser advocated the technique of expansion of both fields in spherical harmonics, a technique that was greatly developed and extended in the pioneering study of Bullard & Gellman (1954). The discussion of § 7(e) of this remarkable paper shows clear recognition of the desirability of two ingredients in the velocity field for effective dynamo action: (i) a differential rotation which would draw out the lines of force of the poloidal magnetic field to generate a toroidal field (for the definition of these terms, see chapter 2), and (ii) a non-axisymmetric motion capable of distorting a toroidal line of force by an upwelling followed by a twist in such a way as to provide a feedback to the poloidal field.

Interaction of the velocity field $\mathbf{u}(\mathbf{x})$ and magnetic field $\mathbf{B}(\mathbf{x})$ (through the $\mathbf{u}\wedge\mathbf{B}$ term in Ohm's law) leads to an infinite set of coupled ordinary differential equations for the determination of the various spherical harmonic ingredients of possible steady magnetic field patterns, and numerical solution of these equations naturally involves truncation of the system and discretisation of radial derivatives. These procedures are of course legitimate in a numerical search for a solution that is known to exist, but they can lead to erroneous conclusions when the existence of an exact steady solution to the problem is in doubt. The dangers were recognised and accepted by Bullard & Gellman, but it has in fact since been demonstrated that the velocity field $\mathbf{u}(\mathbf{x})$ that they proposed most forcibly as a candidate for steady dynamo action in a sphere is a failure in this respect under the more searching scrutiny of modern high-speed computers (Gibson & Roberts, 1969).

The inadequacy of purely computational approaches to the problem intensified the need for theoretical approaches that do not, at the fundamental level, require recourse to the computer. In this respect a breakthrough in understanding was provided by Parker (1955*b*) who argued that the effect of the non-axisymmetric upwellings referred to above might be incorporated by an averaging procedure in equations for the components of the *mean magnetic field* (i.e. the field averaged over the azimuth angle φ about the axis of rotation of the system). Parker's arguments were heuristic rather than deductive, and it was perhaps for this reason that some years elapsed before the power of the approach was generally appreciated. The theory is referred to briefly in Cowling's (1957*a*) monograph 'Magnetohydrodynamics' with the following conclusions:[2] 'The argument is not altogether satisfactory; a more detailed analysis is really needed. Parker does not attempt such an analysis; his mathematical discussion is limited to elucidating the consequences if his picture of what occurs is accepted. But clearly his suggestion deserves a good deal of attention.'

This attention was not provided for some years, however, and was finally stimulated by two rather different approaches to the problem, one by Braginskii (1964*a,b*) and the other by Steenbeck, Krause & Rädler (1966). The essential idea behind Braginskii's approach was that, while steady axisymmetric solutions to the dynamo problem are ruled out by Cowling's theorem, nevertheless weak departures from axisymmetry might provide a means of regeneration of the mean magnetic field. This approach can succeed only if the fluid conductivity σ is very high (or equivalently if the magnetic diffusivity $\lambda = (\mu_0\sigma)^{-1}$ is very small), and the theory was

[2] It is only fair to note that this somewhat guarded assessment is eliminated in the more recent edition of the book (Cowling, 1975*a*).

developed in terms of power series in a small parameter proportional to $\lambda^{1/2}$. By this means, Braginskii demonstrated that, as Parker had argued, non-axisymmetric motions could indeed provide an effective mean toroidal electromotive force (emf) in the presence of a predominantly toroidal magnetic field. This emf. drives a toroidal current thus generating a poloidal field, and the dynamo cycle anticipated by Bullard & Gellman can be completed.

The approach advocated by Steenbeck, Krause & Rädler (1966) is potentially more general, and is applicable when the velocity field consists of a mean and a turbulent (or random) ingredient having widely different length-scales L and l, say ($L \gg l$). Attention is then focussed on the evolution of the mean magnetic field on scales large compared with l. The mean-field approach is of course highly developed in the theory of shear flow turbulence in non-conducting fluids (see, for example, Townsend, 1975) and it had previously been advocated in the hydromagnetic context by, for example, Kovasznay (1960). The power of the approach of Steenbeck *et al.* (1966) however lay in recognition of the fact that the turbulence can give rise to a mean electromotive force having a component parallel to the prevailing local mean magnetic field (as in Braginskii's model); and these authors succeeded in showing that this effect would certainly occur whenever the statistical properties of the background turbulence *lack reflexional symmetry*; this is the random counterpart (whose meaning will be made clear in chapter 7) of the purely geometrical property of the simple disc dynamo discussed above.

Since 1966, there has been a growing flood of papers developing different aspects of these theories and their applications to the Earth and Sun and other systems. It is the aim of this book to provide a coherent account of the most significant of these developments, and reference to specific papers published since 1966 will for the most part be delayed till the appropriate point in the text.

Several other earlier papers are, however, historical landmarks and deserve mention at this stage. The fact that turbulence could be of crucial importance for dynamo action was recognised independently by Batchelor (1950) and Schlüter & Biermann (1950), who considered the effect of a random velocity field on a random magnetic field, both having zero mean. Batchelor perceived that random stretching of magnetic lines of force would lead to exponential increase of magnetic energy in a fluid of infinite conductivity; and, on the basis of the analogy with vorticity (see § 3.5), he obtained a criterion for just how large the conductivity must be for this conclusion to remain valid, and an estimate for the ultimate equilibrium level of magnetic energy density that might be expected when Lorentz forces react back upon the velocity field. Schlüter & Biermann, by arguments based on the concept of equipartition of energy, obtained a different criterion for growth and a much greater estimate for the ultimate level of magnetic energy density. Yet a third possibility was advanced by Saffman (1963), who came to the conclusion that, although the magnetic energy might increase for a while from a very weak initial level, ultimately it would always decay to zero due to accelerated ohmic decay associated with persistent decrease in the characteristic length-scale of the magnetic field. It is now known from consideration of the effect of turbulence which lacks reflexional symmetry (see chapter 7) that none of the conclusions of the above papers can have any general validity, although the question of what happens when the turbulence is reflexionally symmetric remains to some extent open (see §§ 7.12 and 11.4).

The problem as posed by Batchelor (1950) has to some extent been bypassed through recognition of the fact that it is the ensemble-average magnetic field that is of real interest and that if this average vanishes, as in the model conceived by Batchelor, the model can have little direct relevance for the Earth and Sun, both of which certainly exhibit a non-zero dipole moment. It is fortunate that the problem has been bypassed, because in a rather pessimistic diagnosis of the various conflicting theories, Kraichnan & Nagarajan (1967) concluded that 'Equipartition arguments, the vorticity analogy, and the known turbulence approximations are all found inadequate for predicting whether the magnetic energy eventually dies away or grows exponentially. Lack of bounds on errors makes it impossible to predict reliably the sign of the eventual net growth rate of magnetic energy.' Kraichnan & Nagarajan have not yet been proved wrong as regards the basic problem with homogeneous isotropic reflexionally symmetric turbulence. Parker (1970) comments on the situation in the following terms: 'Cyclonic turbulence,[3] together with large-scale shear, generates magnetic field at a very high rate. Therefore we ask whether the possible growth of fields in random turbulence without cyclonic ordering . . . is really of paramount physical interest. We suggest that, even if random turbulence could be shown to enhance magnetic field densities, the effect in most astrophysical objects would be obscured by the more rapid generation of fields by the cyclonic turbulence and non-uniform rotation.'

The crucial importance of a lack of reflexional symmetry in fluid motions conducive to dynamo action is apparent also in the papers of Herzenberg (1958) and Backus (1958) who provided the first examples of laminar velocity fields inside a sphere which could be shown by rigorous procedures to be capable of sustained dynamo action. Herzenberg's model involved two spherical rotors rotating with angular velocities $\boldsymbol{\omega}_1$ and $\boldsymbol{\omega}_2$ and separated by vector distance $\mathbf{R}$ inside the conducting sphere. The configuration can be described as right-handed or left-handed according as the triple scalar product $[\boldsymbol{\omega}_1, \boldsymbol{\omega}_2, \mathbf{R}]$ is positive or negative. A necessary condition for dynamo action (see § 6.9) was that this triple scalar product should be non-zero, and the configuration then certainly lacks reflexional symmetry.

The Backus (1958) dynamo followed the pattern of the Bullard & Gellman dynamo, but decomposed temporally into mathematically tractable units. The velocity field considered consisted of three active phases separated by long periods of rest (or 'stasis') to allow unwanted high harmonics of the magnetic field to decay to a negligibly low level. The three phases were: (i) a vigorous differential rotation which generated strong toroidal field from pre-existing poloidal field; (ii) a non-axisymmetric poloidal convection which regenerated poloidal field from toroidal; (iii) a rigid rotation through an angle of $\pi/2$ to bring the newly generated dipole moment into alignment with the direction of the original dipole moment. The lack of reflexional symmetry lies here in the mutual relationship between the phase (i) and phase (ii) velocity fields (see § 6.12).

The viewpoint adopted in this book is that random fluctuations in

3 This is Parker's terminology for turbulence whose statistical properties lack reflexional symmetry.

the velocity field and the magnetic field are almost certainly present both in the Earth's core and in the Sun's convection zone, and that a realistic theory of dynamo action should incorporate effects of such fluctuations at the outset. Laminar theories are of course not without value, particularly for the mathematical insight that they provide; but anyone who has conscientiously worked through such papers as those of Bullard & Gellman (1954), Herzenberg (1958) and Backus (1958) will readily admit the enormous complexity of the laminar problem. It is a remarkable fact that acceptance of turbulence (or possibly random wave motions) and appropriate averaging procedures actually leads to a dramatic simplification of the problem. The reason is that the mean fields satisfy equations to which Cowling's anti-dynamo theorem does not apply, and which are therefore amenable to an axisymmetric analysis with distinctly positive and encouraging results. The equations admit both steady solutions modelling the Earth's quasi-steady dipole field, and, in other circumstances, time-periodic solutions which behave in many respects like the magnetic field of the Sun with its 22-year periodic cycle.

A further crucial advantage of an approach involving random fluctuations is that dynamic, as opposed to purely kinematic, considerations become to some extent amenable to analysis. A kinematic theory is one in which a kinematically possible velocity field $\mathbf{u}(\mathbf{x}, t)$ is assumed known, either in detail or at least statistically when random fluctuations are involved, and its effect on magnetic field evolution is studied. A dynamic theory is one in which $\mathbf{u}(\mathbf{x}, t)$ is constrained to satisfy the relevant equations of motion (generally the Navier–Stokes equations with buoyancy forces, Coriolis forces and Lorentz forces included according to the context); and again the effect of this velocity field on magnetic field evolution is studied. It is only since the advent of the 'mean-field electrodynamics' of Braginskii and of Steenbeck, Krause & Rädler that progress on the dynamic aspects of dynamo theory has become possible. As compared with the advanced state of kinematic theory, the dynamic theory is still relatively undeveloped; this situation has been changing however over the last few years and it is reasonable to anticipate that, over the next decade or so, dynamic theory will mature to the same level as kinematic theory.

The general pattern of the book will be as follows. Chapter 2 will be devoted to simple preliminaries concerning magnetic field structure and diffusion in a stationary conductor. Chapter 3 will be concerned with the interplay of convection and diffusion effects insofar as these influence magnetic field evolution in a moving fluid. In chapters 4 and 5 we shall digress from the purely mathematical development to provide a necessarily brief survey of the observed properties of the Earth's magnetic field (and other planetary fields) and of the Sun's magnetic field (and other astrophysical fields) and of the relevant physical properties of these bodies. This is designed to provide more detailed motivation for the material of subsequent chapters. Some readers may find this motivation superfluous; but it is necessary, particularly when it comes to the study of specific dynamic models, to consider limiting processes in which the various dimensionless numbers characterising the system are either very small or very large; and it is clearly desirable that such limiting processes should at the least be not in contradiction with observation in the particular sphere of relevance claimed for the theory.

In chapter 6, the kinematic dynamo problem will be defined, and various exact results in the laminar context (including the various anti-dynamo theorems) will be obtained, and the Bullard & Gellman approach will be briefly described. In chapter 7 the effects of a random velocity field having zero mean (conceived either as turbulence as traditionally understood, or as a field of random waves) on magnetic field evolution will be analysed; and we shall follow this in chapter 8 with a discussion of Braginskii's theory (as reformulated by Soward, 1972) within the general framework of mean-field electrodynamics. In chapter 9, properties of the 'dynamo equations' that emerge from the mean-field theory will be studied, and various attacks (analytical and numerical) on the solution of these equations will be described.

Chapters 10–12 will be devoted to the dynamic theory as at present understood. When a magnetic field grows as a result of the action of a random velocity field, the Lorentz force ultimately reacts back on the motion in two ways: firstly, the growing field tends to suppress the turbulent fluctuations that are partly or wholly responsible for its growth; secondly, the mean Lorentz force modifies any pre-existing mean velocity field (or generates a mean velocity field if none exists initially). These effects will be separately analysed in chapters 10, 11 and 12. Chapter 12 also includes consideration of the coupled disc dynamo model of Rikitake (1958); although this model has only three degrees of freedom (as compared with the doubly infinite freedom of the fluid conductor) it exhibits the right sort of couplings between magnetic and dynamic modes, and the behaviour of solutions of the governing equations (which are by no means trivial) is extraordinarily suggestive in the context of the problem of explaining the random reversals of the Earth's magnetic field (see chapter 4).

Given the present state of knowledge, it is inevitable that the kinematic theory will occupy a rather greater proportion of the book than it would ideally deserve. It must be remembered however that any results that can be obtained in kinematic theory on the minimal assumption that $\mathbf{u}(\mathbf{x}, t)$ is a kinematically possible but otherwise *arbitrary* velocity field will have a generality that transcends any dynamical model that is subsequently adopted for the determination of $\mathbf{u}$. It is important to seek this generality because, although there is little uncertainty regarding the equations governing magnetic field evolution (i.e. Maxwell's equations and Ohm's law), there are wide areas of grave uncertainty concerning the relevance of different dynamic models in both terrestrial and solar contexts; for example, it is not yet known what the ultimate source of energy is for core motions that drive the Earth's dynamo. In this situation, any results that do not depend on the details of the governing dynamical equations (whatever these may be) are of particular value. For this reason, the postponement of dynamical considerations to the last three chapters should perhaps be welcomed rather than lamented.

CHAPTER 2

MAGNETOKINEMATIC PRELIMINARIES

2.1. Structural properties of the B-field

In §§ 2.1–2.4 we shall be concerned with basic instantaneous properties of magnetic field distributions $\mathbf{B}(\mathbf{x})$; the time dependence of $\mathbf{B}$ is for the moment irrelevant. The first, and perhaps the most basic, of these properties is that if S is any closed surface with unit outward normal $\mathbf{n}$, then

$$\int_S \mathbf{B} \,.\, \mathbf{n} \, \mathrm{d}S = 0, \tag{2.1}$$

i.e. magnetic poles do not exist in isolation. This global statement implies the existence of a single-valued vector potential $\mathbf{A}(\mathbf{x})$ satisfying

$$\mathbf{B} = \nabla \wedge \mathbf{A}, \qquad \nabla \,.\, \mathbf{A} = 0. \tag{2.2}$$

$\mathbf{A}$ is not uniquely defined by these equations since we may add to it the gradient of any harmonic function without affecting $\mathbf{B}$; but, in problems involving an infinite domain, $\mathbf{A}$ is made unique by imposing the boundary condition

$$\mathbf{A} \to 0 \quad \text{as } |\mathbf{x}| \to \infty. \tag{2.3}$$

At points where $\mathbf{B}$ is differentiable, (2.1) implies that

$$\nabla \,.\, \mathbf{B} = 0, \tag{2.4}$$

and across any surface of discontinuity S_d of physical properties of the medium (or of other relevant fields such as the velocity field), (2.1) implies that

$$[\mathbf{n} \,.\, \mathbf{B}] = (\mathbf{n} \,.\, \mathbf{B})_+ - (\mathbf{n} \,.\, \mathbf{B})_- = 0, \tag{2.5}$$

where the + and − refer to the two sides of S_d, and $\mathbf{n}$ is now the unit normal on S_d directed from the − to the + side. We shall always use the square bracket notation to denote such surface quantities.

The lines of force of the **B**-field (or '**B**-lines') are determined as the integral curves of the differential equations

$$\mathrm{d}\mathbf{x} \wedge \mathbf{B} = 0. \tag{2.6}$$

A **B**-line may, exceptionally, close on itself. More generally it may cover a closed surface S in the sense that, if followed far enough, it passes arbitrarily near every point of S. It is also conceivable that a **B**-line may be space-filling in the sense that, if followed far enough, it passes arbitrarily near every point of a three-dimensional region V; there are no known examples of solenoidal **B**-fields, finite and differentiable everywhere, with this property, but it is nevertheless a possibility that no topological arguments have yet been able to eliminate, and indeed it seems quite likely that **B**-fields of any degree of complexity will in general be space-filling.

Now let C be any (unknotted) closed curve spanned by an open orientable surface S with normal **n**. The flux Φ of **B** across S is defined by

$$\Phi = \int_S \mathbf{B} \,.\, \mathbf{n} \,\mathrm{d}S = \oint_C \mathbf{A} \,.\, \mathrm{d}\mathbf{x}, \tag{2.7}$$

where the line integral is described in a right-handed sense relative to the normal on S. A *flux-tube* is the aggregate of **B**-lines passing through a closed curve (usually of small or infinitesimal extent). By virtue of (2.1), Φ is constant along a flux-tube.

A measure of the degree of structural complexity of a **B**-field is provided by a set of integrals of the form

$$I_m = \int_{V_m} \mathbf{A} \,.\, \mathbf{B} \,\mathrm{d}^3\mathbf{x} \quad (m = 1, 2, 3, \cdots), \tag{2.8}$$

where V_m is any volume with surface S_m on which $\mathbf{n} \,.\, \mathbf{B} = 0$. Suppose, for example, that **B** is identically zero except in two flux-tubes occupying volumes V_1 and V_2 of infinitesimal cross-section following the closed curves C_1 and C_2 (fig. 2.1(a)) and let Φ_1 and Φ_2 be the respective fluxes. Note that if the tubes are linked as in the figure and the field directions in the tubes are as indicated by the arrows, then each tube has a right-handed orientation relative to the other; if one arrow is reversed the relative orientation becomes left-handed; if both are reversed it remains right-handed. For the

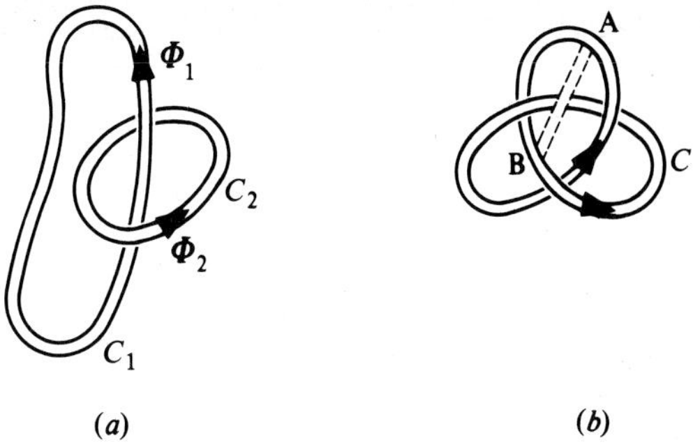

Fig. 2.1 (*a*) The two flux-tubes are linked in such a way as to give positive magnetic helicity. (*b*) A flux-tube in the form of a right-handed trefoil knot; insertion of equal and opposite flux-tube elements between the points A and B as indicated gives two tubes linked as in (*a*).

configuration as drawn, evidently $\mathbf{B}\,\mathrm{d}^3\mathbf{x}$ may be replaced by $\Phi_1\,\mathrm{d}\mathbf{x}$ on C_1 and $\Phi_2\,\mathrm{d}\mathbf{x}$ on C_2 with the result that

$$I_1 = \Phi_1 \oint_{C_1} \mathbf{A}\,.\,\mathrm{d}\mathbf{x} = \Phi_1\Phi_2, \tag{2.9}$$

and similarly

$$I_2 = \Phi_2 \oint_{C_2} \mathbf{A}\,.\,\mathrm{d}\mathbf{x} = \Phi_1\Phi_2. \tag{2.10}$$

More generally, if the tubes wind round each other N times (i.e. N is the 'winding number' of C_1 relative to C_2) then

$$I_1 = I_2 = \pm N\Phi_1\Phi_2, \tag{2.11}$$

the + or − being chosen according as the relative orientation is right- or left-handed. The integrals I_1 and I_2 are therefore intimately related to the fundamental topological invariant N of the pair of curves C_1 and C_2.

If the **B**-field has limited spatial extent (i.e. $\mathbf{B} \equiv 0$ outside some sufficiently large sphere), then one possible choice of the volume of integration is the whole three-dimensional space V_∞, the corresponding integral being denoted I_∞. For the case of the two linked flux-tubes, evidently

$$I_\infty = I_1 + I_2. \tag{2.12}$$

More generally, if the V_m can be chosen so that $V_\infty = \bigcup_{m=1}^{\infty} V_m$, then

$$I_\infty = \sum_{m=1}^{\infty} I_m. \tag{2.13}$$

If a single flux-tube (with flux Φ) winds round itself before closing (i.e. if it is knotted) then the integral I_∞ for the associated magnetic field will in general be non-zero. Fig. 2.1(*b*) shows the simplest non-trivial possibility: the curve C is a right-handed trefoil knot; insertion of the two self cancelling elements between the points A and B indicates that this is equivalent to the configuration of fig. 2.1(*a*) with $\Phi_1 = \Phi_2 = \Phi$ so that $I = 2\Phi^2$. Knotted tubes may always be decomposed in this way into two or more linked tubes.

It may of course happen that $\mathbf{A} \cdot \mathbf{B} \equiv 0$; it is well known that this is the necessary and sufficient condition for the existence of scalar functions $\varphi(\mathbf{x})$ and $\psi(\mathbf{x})$ such that

$$\mathbf{A} = \psi \nabla \varphi, \qquad \mathbf{B} = \nabla \psi \wedge \nabla \varphi. \tag{2.14}$$

In this situation, the **B**-lines are the intersections of the surfaces $\varphi = \text{cst.}$, $\psi = \text{cst.}$, and the **A**-lines are everywhere orthogonal to the surfaces $\varphi = \text{cst.}$ It is clear from the above discussion that **B**-fields having linked or knotted **B**-lines cannot admit such a representation.

The same limitation applies to the use of Clebsch variables φ, ψ, χ, defined (if they exist) by the equations

$$\mathbf{A} = \psi \nabla \varphi + \nabla \chi, \qquad \mathbf{B} = \nabla \psi \wedge \nabla \varphi. \tag{2.15}$$

For example, if **B** is a field admitting such a representation, with φ, ψ and χ single-valued differentiable functions of $\mathbf{x}$, then

$$\mathbf{A} \cdot \mathbf{B} = \nabla \chi \cdot (\nabla \psi \wedge \nabla \varphi),$$

and

$$\begin{aligned} I_m &= \int_{V_m} \nabla \chi \cdot (\nabla \psi \wedge \nabla \varphi)\, \mathrm{d}^3\mathbf{x} \\ &= \int_{V_m} \nabla \cdot (\chi \nabla \psi \wedge \nabla \varphi)\, \mathrm{d}^3\mathbf{x} \\ &= \int_{S_m} \chi \mathbf{n} \cdot \mathbf{B}\, \mathrm{d}S = 0, \end{aligned} \tag{2.16}$$

since $\mathbf{n}.\mathbf{B}=0$ on S_m. Conversely, if $I_m \neq 0$ (as will happen if the $\mathbf{B}$-lines are knotted or linked), then (2.15) is not a possible global representation for $\mathbf{A}$ and $\mathbf{B}$ (although it may be useful in a purely local analysis).

The characteristic local structure of a field for which $\mathbf{A}.\mathbf{B}$ is non-zero may be illustrated with reference to the example (in Cartesian coordinates)

$$\mathbf{B}=(B_0, 0, 0), \qquad \mathbf{A}=(A_0, -\tfrac{1}{2}B_0 z, \tfrac{1}{2}B_0 y), \tag{2.17}$$

where A_0 and B_0 are constants; note that possible Clebsch variables are

$$\varphi = z, \qquad \psi = B_0 y, \qquad \chi = A_0 x - \tfrac{1}{2}B_0 yz; \tag{2.18}$$

there is here no closed surface on which $\mathbf{n}.\mathbf{B}=0$, and the problem noted above does not arise. Clearly $\mathbf{A}.\mathbf{B}=A_0B_0$, and the $\mathbf{A}$-lines are the helices with parametric representation

$$x=(2A_0/B_0)t, \qquad y=\cos t, \qquad z=\sin t. \tag{2.19}$$

These helices are right-handed or left-handed according as A_0B_0 is positive or negative.

The quantity $\mathbf{A}.(\nabla \wedge \mathbf{A})$ for *any* vector field A($\mathbf{x}$) is called the *helicity density* of the field $\mathbf{A}$; its integral I_∞ over V_∞ is then the helicity of $\mathbf{A}$; the integrals I_m over V_m can be described as 'partial' helicities. The helicity density is a pseudo-scalar quantity, being the scalar product of a polar vector and an axial vector; its sign therefore changes under change from a right-handed to a left-handed frame of reference. A field $\mathbf{A}$ that is 'reflexionally symmetric' (i.e. invariant under the change from a right-handed to a left-handed reference system represented by the reflexion $\mathbf{x}'=-\mathbf{x}$) must therefore have zero helicity density. The converse is not true, since of course other pseudo-scalar quantities such as $(\nabla \wedge \mathbf{A}).\nabla \wedge (\nabla \wedge \mathbf{A})$ may be non-zero even if $\mathbf{A}.(\nabla \wedge \mathbf{A}) \equiv 0$.

2.2. Magnetic field representations

In a spherical geometry, the most natural coordinates to use are spherical polar coordinates (r, θ, φ) related to Cartesian coordinates

(x, y, z) by

$$x = r \sin\theta \cos\varphi, \qquad y = r \sin\theta \sin\varphi, \qquad z = r\cos\theta. \tag{2.20}$$

Let us first recall some basic results concerning the use of this coordinate system.

Let $\psi(r, \theta, \varphi)$ be any scalar function of position. Then

$$\nabla^2\psi = \frac{1}{r^2}\frac{\partial}{\partial r} r^2 \frac{\partial\psi}{\partial r} + \frac{1}{r^2} L^2\psi, \tag{2.21}$$

where

$$L^2\psi = \left(\frac{1}{\sin\theta}\frac{\partial}{\partial\theta}\sin\theta\frac{\partial}{\partial\theta} + \frac{1}{\sin^2\theta}\frac{\partial^2}{\partial\varphi^2}\right)\psi. \tag{2.22}$$

The vector identity

$$(\mathbf{x}\wedge\nabla)^2\psi \equiv r^2\nabla^2\psi - 2(\mathbf{x}\,.\,\nabla)\psi - \mathbf{x}\,.\,(\mathbf{x}\,.\,\nabla)\nabla\psi \tag{2.23}$$

leads to the identification

$$L^2\psi = (\mathbf{x}\wedge\nabla)^2\psi. \tag{2.24}$$

L^2 is the angular momentum operator of quantum mechanics. Its eigenvalues are $n(n+1)$ $(n = 0, 1, 2, \ldots)$, and the corresponding eigenfunctions are the surface harmonics

$$S_n(\theta, \varphi) = \sum_{m=0}^{n} A_n^m P_n^m(\cos\theta)\,\mathrm{e}^{im\varphi}, \tag{2.25}$$

where $P_n^m(\cos\theta)$ are associated Legendre polynomials and the A_n^m are arbitrary complex constants; i.e.

$$L^2 S_n = n(n+1)S_n. \tag{2.26}$$

Now let $f(r, \theta, \varphi)$ be any smooth function having zero average over spheres $r = \text{cst.}$, i.e.

$$4\pi r^2\langle f\rangle \equiv \int_0^{2\pi}\int_0^{\pi} f(r, \theta, \varphi)\sin\theta\,\mathrm{d}\theta\,\mathrm{d}\varphi = 0. \tag{2.27}$$

We may expand f in surface harmonics

$$f(r, \theta, \varphi) = \sum_{n=1}^{\infty} f_n(r)S_n(\theta, \varphi), \tag{2.28}$$

the term with $n = 0$ being excluded by virtue of (2.27). The functions S_n satisfy the orthogonality relation

$$\langle S_n S_{n'} \rangle = 0, \qquad (n \neq n'), \tag{2.29}$$

and so the coefficients $f_n(r)$ are given by

$$f_n(r) = \langle f S_n \rangle / \langle S_n^2 \rangle. \tag{2.30}$$

If now

$$L^2 \psi = f(r, \theta, \varphi), \tag{2.31}$$

then clearly the operator L^2 may be inverted to give

$$\psi = L^{-2} f = \sum_{n=1}^{\infty} f_n(r)[n(n+1)]^{-1} S_n(\theta, \varphi), \tag{2.32}$$

the result also satisfying $\langle \psi \rangle = 0$.

Note that any function of the form $f = \mathbf{x} . \nabla \wedge \mathbf{A}$, where $\mathbf{A}$ is an arbitrary smooth vector field, satisfies the condition (2.27); for

$$\int_{S(r)} \mathbf{x} . \nabla \wedge \mathbf{A} \, \mathrm{d}S = r \int_{S(r)} \mathbf{n} . (\nabla \wedge \mathbf{A}) \, \mathrm{d}S = r \int_{V(r)} \nabla . (\nabla \wedge \mathbf{A}) \, \mathrm{d}V = 0, \tag{2.33}$$

where $S(r)$ is the surface of the sphere of radius r and $V(r)$ its interior.

A *toroidal magnetic field* $\mathbf{B}_T$ is any field of the form

$$\mathbf{B}_T = \nabla \wedge (\mathbf{x} T(\mathbf{x})) = -\mathbf{x} \wedge \nabla T, \tag{2.34}$$

where $T(\mathbf{x})$ is any scalar function of position. Note that addition of an arbitrary function of r to T has no effect on $\mathbf{B}_T$, so that without loss of generality we may suppose that $\langle T \rangle = 0$. Note further that $\mathbf{x} . \mathbf{B}_T = 0$, so that the lines of force of $\mathbf{B}_T$ ('$\mathbf{B}_T$-lines') lie on the spherical surfaces $r = \text{cst}$.

A *poloidal magnetic field* $\mathbf{B}_P$ is any field of the form

$$\mathbf{B}_P = \nabla \wedge \nabla \wedge (\mathbf{x} P(\mathbf{x})) = -\nabla \wedge (\mathbf{x} \wedge \nabla P), \tag{2.35}$$

where $P(\mathbf{x})$ is any scalar function of position which again may be assumed to satisfy $\langle P \rangle = 0$. $\mathbf{B}_P$ does in general have a non-zero radial component.

It is clear from these definitions that the curl of a toroidal field is a poloidal field. Moreover the converse is also true; for

$$\nabla \wedge \nabla \wedge \nabla \wedge (\mathbf{x} P) = -\nabla^2 \nabla \wedge (\mathbf{x} P) = -\nabla \wedge (\mathbf{x} \nabla^2 P); \tag{2.36}$$

The latter identity can be trivially verified in Cartesian coordinates.

Now suppose that

$$\mathbf{B} = \mathbf{B}_P + \mathbf{B}_T = \nabla \wedge \nabla \wedge (\mathbf{x}P) + \nabla \wedge (\mathbf{x}T). \tag{2.37}$$

Then

$$\mathbf{x} \cdot \mathbf{B} = -(\mathbf{x} \wedge \nabla)^2 P, \qquad \mathbf{x} \cdot (\nabla \wedge \mathbf{B}) = -(\mathbf{x} \wedge \nabla)^2 T, \tag{2.38}$$

so that P and T may be obtained in the form

$$P = -L^{-2}(\mathbf{x} \cdot \mathbf{B}), \qquad T = -L^{-2}\mathbf{x} \cdot (\nabla \wedge \mathbf{B}). \tag{2.39}$$

Conversely, given any solenoidal field $\mathbf{B}$, if we define P and T by (2.39), then (2.37) is satisfied, i.e. the decomposition of $\mathbf{B}$ into poloidal and toroidal ingredients is always possible.

If we 'uncurl' (2.37), we obtain the vector potential of $\mathbf{B}$ in the form

$$\mathbf{A} = \nabla \wedge (\mathbf{x}P) + \mathbf{x}\mathrm{T} + \nabla U, \tag{2.40}$$

where U is a scalar 'function of integration'. Since $\nabla \cdot \mathbf{A} = 0$, U and T are related by

$$\nabla^2 U = -\nabla \cdot (\mathbf{x}T) \tag{2.41}$$

By virtue of this condition, $\mathbf{x}T + \nabla U$ may itself be expressed as a poloidal field:

$$\mathbf{x}T + \nabla U = \nabla \wedge \nabla \wedge (\mathbf{x}S), \qquad S = -L^{-2}(r^2 T + \mathbf{x} \cdot \nabla U). \tag{2.42}$$

The toroidal part of $\mathbf{A}$ is simply

$$\mathbf{A}_T = \nabla \wedge (\mathbf{x}P) = -\mathbf{x} \wedge \nabla P. \tag{2.43}$$

Axisymmetric fields

A $\mathbf{B}$-field is *axisymmetric* about a line Oz (the axis of symmetry) if it is invariant under rotations about Oz. In this situation, the defining scalars P and T of (2.37) are clearly independent of the azimuth angle φ, i.e. $T = T(r, \theta)$, $P = P(r, \theta)$. The toroidal field $\mathbf{B}_T = -\mathbf{x} \wedge \nabla T$ then takes the form

$$\mathbf{B}_T = (0, 0, B_\varphi), \qquad B_\varphi = -\partial T/\partial \theta, \tag{2.44}$$

and so the $\mathbf{B}_T$-lines are circles $r=\text{cst.}$, $\theta=\text{cst.}$ about Oz. Similarly,

$$\mathbf{A}_T=(0,0,A_\varphi),\qquad A_\varphi=-\partial P/\partial\theta, \tag{2.45}$$

and, correspondingly, in spherical polars,

$$\mathbf{B}_P=\nabla\wedge\mathbf{A}_T=\left(-\frac{1}{r^2\sin\theta}\frac{\partial\chi}{\partial\theta},\frac{1}{r\sin\theta}\frac{\partial\chi}{\partial r},0\right), \tag{2.46}$$

where

$$\chi=r\sin\theta,\qquad A_\varphi=-r\sin\theta\,\partial P/\partial\theta. \tag{2.47}$$

The scalar χ is the analogue of the Stokes stream function $\psi(r,\theta)$ for incompressible axisymmetric velocity fields. The $\mathbf{B}_P$-lines are given by $\chi=\text{cst.}$, and the differential

$$2\pi\,\mathrm{d}\chi=(B_\theta\,\mathrm{d}r-B_r r\,\mathrm{d}\theta)2\pi r\sin\theta \tag{2.48}$$

represents the flux across the infinitesimal annulus obtained by rotating about Oz the line element joining (r,θ) and $(r+\mathrm{d}r,\theta+\mathrm{d}\theta)$. It is therefore appropriate to describe χ as the *flux-function* of the field $\mathbf{B}_P$.

When the context allows no room for ambiguity, we shall drop the suffix φ from B_φ and A_φ, and express $\mathbf{B}$ in the simple form

$$\mathbf{B}=B\mathbf{i}_\varphi+\nabla\wedge(A\mathbf{i}_\varphi), \tag{2.49}$$

where $\mathbf{i}_\varphi$ is a unit vector in the φ-direction[1].

The two-dimensional analogue

Geometrical complications inherent in the spherical geometry frequently make it desirable to seek simpler representations. In particular, if we are concerned with processes in a spherical annulus $r_1<r<r_2$ with $r_2-r_1\ll r_1$, a local Cartesian representation $Oxyz$ is appropriate (fig. 2.2). Here Oz is now in the radial direction (i.e. the vertical direction in terrestrial and solar contexts), Ox is south and Oy is east; hence (r,θ,φ) are replaced by (z,x,y).

The field decomposition analogous to (2.37) is then

$$\mathbf{B}=\mathbf{B}_P+\mathbf{B}_T=\nabla\wedge\nabla\wedge(\mathbf{i}_zP)+\nabla\wedge(\mathbf{i}_zT), \tag{2.50}$$

[1] $\mathbf{i}_q$ will generally denote a unit vector in the direction of increasing q where q is any generalised coordinate.

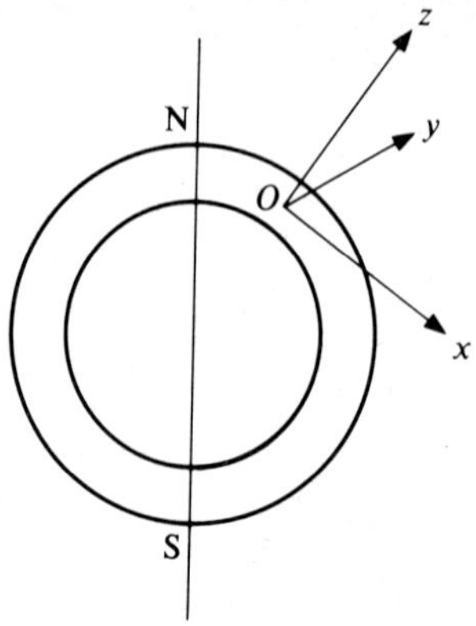

Fig. 2.2 Local Cartesian coordinate system in a spherical annulus geometry; Ox is directed south, Oy east, and Oz vertically upwards.

and P and T are given by

$$\nabla_2^2 P = -\mathbf{i}_z \, . \, \mathbf{B}, \qquad \nabla_2^2 T = -\mathbf{i}_z \, . \, (\nabla \wedge \mathbf{B}), \tag{2.51}$$

where ∇_2^2 is the two-dimensional Laplacian

$$\nabla_2^2 = (\mathbf{i}_z \wedge \nabla)^2 = \partial^2/\partial x^2 + \partial^2/\partial y^2. \tag{2.52}$$

If the field $\mathbf{B}$ is independent of the coordinate y (the analogue of axisymmetry), then $P = P(x, z)$, $T = T(x, z)$. The 'toroidal' field becomes

$$\mathbf{B}_T = B\mathbf{i}_y, \qquad B = -\partial T/\partial x, \tag{2.53}$$

and the 'poloidal' field becomes

$$\mathbf{B}_P = \nabla \wedge (A\mathbf{i}_y), \qquad A = -\partial P/\partial x. \tag{2.54}$$

The $\mathbf{B}_P$-lines are now given by $A(x, z) = \text{cst.}$, and A is the flux-function of the $\mathbf{B}_P$-field.

Quite apart from the spherical annulus context, two-dimensional fields are of independent interest, and consideration of idealised two-dimensional situations can often provide valuable insights. It is then in general perhaps more natural to regard Oz as the direction of invariance of $\mathbf{B}$ and to express $\mathbf{B}$ in the form

$$\mathbf{B} = \nabla \wedge (A(x, y)\mathbf{i}_z) + B(x, y)\mathbf{i}_z. \tag{2.55}$$

2.3. Relations between electric current and magnetic field

In a steady situation, the magnetic field $\mathbf{B}(\mathbf{x})$ is related to the electric current distribution $\mathbf{J}(\mathbf{x})$ by Ampere's law; in integral form this is

$$\oint_C \mathbf{B} \,.\, \mathrm{d}\mathbf{x} = \mu_0 \int_S \mathbf{J} \,.\, \mathbf{n} \,\mathrm{d}S, \tag{2.56}$$

where S is any open orientable surface spanning the closed curve C, and μ_0 is a constant associated with the system of units used[2]. If $\mathbf{B}$ is measured in Wb m^{-2} (1 Wb m^{-2} = 10^4 gauss) and $\mathbf{J}$ in A m^{-2}, then

$$\mu_0 = 4\pi \times 10^{-7} \text{ Wb A}^{-1} \text{ m}^{-1}. \tag{2.57}$$

It follows from (2.56) that in any region where $\mathbf{B}$ and $\mathbf{J}$ are differentiable,

$$\nabla \wedge \mathbf{B} = \mu_0 \mathbf{J}, \qquad \nabla \,.\, \mathbf{J} = 0, \tag{2.58}$$

and the corresponding jump conditions across surfaces of discontinuity are

$$[\mathbf{n} \wedge \mathbf{B}] = \mu_0 \mathbf{J}_S, \qquad [\mathbf{n} \,.\, \mathbf{J}] = 0, \tag{2.59}$$

where $\mathbf{J}_S$ (A m^{-1}) represents surface current distribution. Surface currents (like concentrated vortex sheets) can survive only if dissipative processes do not lead to diffusive spreading, i.e. only in a perfect electrical conductor. In fluids or solids of finite conductivity, we may generally assume that $\mathbf{J}_S = 0$, and then (2.59) together with (2.5) implies that all components of $\mathbf{B}$ are continuous:

$$[\mathbf{B}] = 0. \tag{2.60}$$

In an unsteady situation, (2.56) is generally modified by the inclusion of Maxwell's displacement current; it is well known however that this effect is negligible in treatment of phenomena whose time-scale is long compared with the time for electromagnetic waves to cross the region of interest. This condition is certainly satisfied in the terrestrial and solar contexts, and we shall therefore neglect displacement current throughout; this has the effect of filtering electromagnetic waves from the system of governing equa-

[2] Permeability effects are totally unimportant in the topics to be considered and may be ignored from the outset.

tions. The resulting equations are entirely classical (i.e. non-relativistic) and are sometimes described as the 'pre-Maxwell equations'.

In terms of the vector potential **A** defined by (2.2), (2.58) becomes the (vector) Poisson's equation

$$\nabla^2 \mathbf{A} = -\mu_0 \mathbf{J}. \tag{2.61}$$

Across discontinuity surfaces, **A** is in general continuous (since $\mathbf{B} = \nabla \wedge \mathbf{A}$ is in general finite) and (2.60) implies further that the normal gradient of **A** must also be continuous; i.e. in general

$$[\mathbf{A}] = 0, \qquad [(\mathbf{n} \,.\, \nabla)\mathbf{A}] = 0. \tag{2.62}$$

Multipole expansion of the magnetic field

Suppose now that S is a closed surface, with interior V and exterior $\hat{V}$, and suppose that $\mathbf{J}(\mathbf{x})$ is a current distribution entirely confined to V, i.e. $\mathbf{J} \equiv 0$ in $\hat{V}$; excluding the possibility of surface current on S, we must then also have $\mathbf{n} \,.\, \mathbf{J} = 0$ on S. From (2.58), **B** is irrotational in $\hat{V}$, and so there exists a scalar potential $\Psi(\mathbf{x})$ such that, in $\hat{V}$,

$$\mathbf{B} = -\nabla \Psi, \qquad \nabla^2 \Psi = 0. \tag{2.63}$$

Note that in general Ψ is not single-valued; it is, however, single-valued if $\hat{V}$ is simply-connected, and we shall assume this to be the case. We may further suppose that $\Psi \to 0$ as $|\mathbf{x}| \to \infty$.

Relative to an origin O in V, the general solution of (2.63) vanishing at infinity may be expressed in the form

$$\Psi(\mathbf{x}) = \sum_{n=1}^{\infty} \Psi^{(n)}(\mathbf{x}), \qquad \Psi^{(n)}(\mathbf{x}) = -\mu^{(n)}_{ij\ldots s}(r^{-1})_{,ij\ldots s}; \tag{2.64}$$

here $\mu^{(n)}_{ij\ldots s}$ is the multipole moment tensor of rank n, $r = |\mathbf{x}|$, and a suffix i after the comma indicates differentiation with respect to x_i. The term with $n = 0$ is omitted by virtue of (2.1). The terms with $n = 1, 2$ are the dipole and quadrupole terms respectively; in vector notation

$$\Psi^{(1)}(\mathbf{x}) = -\boldsymbol{\mu}^{(1)} \,.\, \nabla(r^{-1}), \qquad \Psi^{(2)}(\mathbf{x}) = -\boldsymbol{\mu}^{(2)} \colon \nabla\nabla(r^{-1}), \tag{2.65}$$

and similarly for higher terms.

The field $\mathbf{B}$ in $\hat{V}$ clearly has the expansion

$$\mathbf{B}(\mathbf{x}) = \sum_{n=1}^{\infty} \mathbf{B}^{(n)}(\mathbf{x}), \qquad B_\alpha^{(n)}(\mathbf{x}) = \mu_{ij\ldots s}^{(n)}(r^{-1})_{,ij\ldots s\alpha}, \tag{2.66}$$

and, since $\nabla^2(r^{-1}) = 0$ in $\hat{V}$, the expression for $\mathbf{B}^{(n)}$ may be expressed in the form $\mathbf{B}^{(n)} = \nabla \wedge \mathbf{A}^{(n)}$ where

$$A_q^{(n)} = -\varepsilon_{qmi}\mu_{mj\ldots s}^{(n)}(r^{-1})_{,ij\ldots s}. \tag{2.67}$$

The first two terms of the expansion for the vector potential, corresponding to (2.65), thus have the form

$$\mathbf{A}^{(1)}(\mathbf{x}) = -\boldsymbol{\mu}^{(1)} \wedge \nabla(r^{-1}), \qquad \mathbf{A}^{(2)}(\mathbf{x}) = -\boldsymbol{\mu}^{(2)} \dot{\wedge} \nabla\nabla(r^{-1}). \tag{2.68}$$

The tensors $\boldsymbol{\mu}^{(n)}$ may be determined as linear functionals of $\mathbf{J}(\mathbf{x})$ as follows. The solution of (2.61) vanishing at infinity is

$$\mathbf{A}(\mathbf{x}) = \frac{\mu_0}{4\pi}\int_V |\mathbf{x}-\mathbf{x}'|^{-1}\mathbf{J}(\mathbf{x}')\, \mathrm{d}^3\mathbf{x}', \tag{2.69}$$

(and it may be readily verified using $\nabla \cdot \mathbf{J} = 0$, $\mathbf{n} \cdot \mathbf{J} = 0$ on S, that this satisfies $\nabla \cdot \mathbf{A} = 0$). The function $|\mathbf{x}-\mathbf{x}'|^{-1}$ has the Taylor expansion

$$|\mathbf{x}-\mathbf{x}'|^{-1} = \sum_n \frac{(-1)^n}{n!} x_i' x_j' \ldots x_s' (r^{-1})_{,ij\ldots s}. \tag{2.70}$$

Substitution in (2.69) leads immediately to $\mathbf{A} = \sum \mathbf{A}^{(n)}(\mathbf{x})$ where

$$A_q^{(n)}(\mathbf{x}) = c_{ij\ldots sq}(r^{-1})_{,ij\ldots s}, \qquad c_{ij\ldots sq} = \frac{\mu_0}{4\pi}\frac{(-1)^n}{n!}\int_V x_i x_j \ldots x_s J_q(\mathbf{x})\, \mathrm{d}^3\mathbf{x}. \tag{2.71}$$

Comparison with (2.67) then gives the equivalent relations

$$c_{ij\ldots sq} = -\varepsilon_{qmi}\mu_{mj\ldots s}^{(n)}, \qquad \mu_{mj\ldots s}^{(n)} = -\tfrac{1}{2}\varepsilon_{qmi}c_{ij\ldots sq}. \tag{2.72}$$

In particular, for the terms $n = 1, 2$ we have

$$\begin{aligned} \boldsymbol{\mu}^{(1)} &= \frac{\mu_0}{8\pi}\int_V \mathbf{x} \wedge \mathbf{J}\, \mathrm{d}^3\mathbf{x}, \\ \mu_{mj}^{(2)} &= -\frac{\mu_0}{16\pi}\int_V x_j(\mathbf{x} \wedge \mathbf{J})_m\, \mathrm{d}^3\mathbf{x}. \end{aligned} \tag{2.73}$$

Axisymmetric fields

If $\mathbf{J}(\mathbf{x})$ is axisymmetric about the direction of the unit vector $\mathbf{i}_z = \mathbf{e}$, then these results can be simplified. Choosing spherical polar coordinates (r, θ, φ) based on the polar axis Oz, we have evidently

$$\boldsymbol{\mu}^{(1)} = \mu^{(1)}\mathbf{e}, \qquad \mu^{(1)} = \tfrac{1}{4}\mu_0 \iint J_\varphi(r, \theta) r^3 \sin^2\theta \, dr \, d\theta. \tag{2.74}$$

Likewise, $\mu^{(2)}_{mj}$ must be axisymmetric about Oz; and since $\mu^{(2)}_{jj} = 0$ from (2.73*b*), it must therefore take the form

$$\mu^{(2)}_{mj} = \mu^{(2)}(e_m e_j - \tfrac{1}{3}\delta_{mj}). \tag{2.75}$$

Putting $m = 3, j = 3$ in (2.73) and (2.75) then gives $\mu^{(2)}$ in the form

$$\mu^{(2)} = -\frac{3\mu_0}{16} \iint J_\varphi(r, \theta) r^4 \sin^2\theta \cos\theta \, dr \, d\theta. \tag{2.76}$$

In this axisymmetric situation, the expansion (2.64) clearly has the form

$$\Psi(\mathbf{x}) = -\sum_{n=1}^{\infty} \mu^{(n)} \frac{\partial^n}{\partial z^n}\left(\frac{1}{r}\right). \tag{2.77}$$

2.4. Force-free fields

We shall have frequent occasion to refer to magnetic fields for which $\mathbf{B}$ is everywhere parallel to $\mathbf{J} = \mu_0^{-1}\nabla \wedge \mathbf{B}$ and it will therefore be useful at this stage to gather together some properties of such fields[3], which are described as 'force-free' (Lüst & Schlüter, 1954) since the associated Lorentz force $\mathbf{J} \wedge \mathbf{B}$ is of course identically zero. For any force-free field, there exists a scalar function of position $K(\mathbf{x})$ such that

$$\nabla \wedge \mathbf{B} = K\mathbf{B}, \qquad \mathbf{B} \cdot \nabla K = 0, \tag{2.78}$$

the latter following from $\nabla \cdot \mathbf{B} = 0$. K is therefore constant on $\mathbf{B}$-lines, and if $\mathbf{B}$-lines cover surfaces then K must be constant on each such surface. A particularly simple situation is that in which K is constant everywhere; in this case, taking the curl of (2.78)

[3] In general, a vector field $\mathbf{B}(\mathbf{x})$, with the property that $\nabla \wedge \mathbf{B}$ is everywhere parallel to $\mathbf{B}$, is known as a *Beltrami field*.

immediately leads to the Helmholtz equation

$$(\nabla^2 + K^2)\mathbf{B} = 0. \tag{2.79}$$

Note however that this process cannot in general be reversed: a field **B** that satisfies (2.79) does not necessarily satisfy either of the equations $\nabla \wedge \mathbf{B} = \pm K\mathbf{B}$.

The simplest example of a force-free field, with $K = \text{cst.}$, is, in Cartesian coordinates,

$$\mathbf{B} = B_0(\sin Kz, \cos Kz, 0). \tag{2.80}$$

The property $\nabla \wedge \mathbf{B} = K\boldsymbol{B}$ is trivially verified. The **B**-lines, as indicated in fig. 2.3(*a*), lie in the x–y plane and their direction rotates

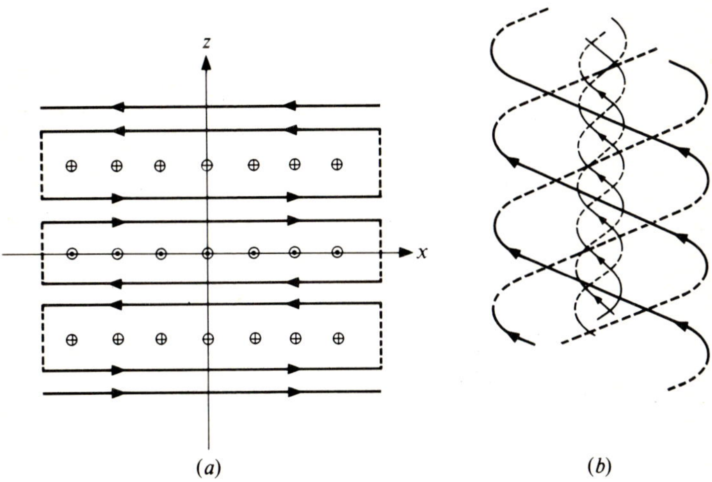

Fig. 2.3 (*a*) Lines of force of the field (2.80) (with $K > 0$). ⊙ indicates a line in the positive y-direction (i.e. into the paper), and ⊕ indicates a line in the negative y-direction; the lines of force rotate in a left-handed sense with increasing z. Closing the lines of force by means of the dashed segments leads to linkages consistent with the positive helicity of the field. (*b*) Typical helical lines of force of the field (2.82); the linkages as illustrated are negative, and therefore correspond to a negative value of K in (2.82).

with increasing z in a sense that is left-handed or right-handed according as K is positive or negative. The vector potential of **B** is simply $\mathbf{A} = K^{-1}\mathbf{B}$, so that its helicity density is uniform:

$$\mathbf{A} \cdot \mathbf{B} = K^{-1}\mathbf{B}^2 = K^{-1}B_0^2. \tag{2.81}$$

If we imagine the lines of force closed by the dashed lines as indicated in the figure, then the resulting linkages are consistent with the discussion of § 2.1.

A second example (fig. 2.3(*b*)) of a force-free field, with K again constant, is, in cylindrical polars (s, φ, z),

$$\mathbf{B} = B_0(0, J_1(Ks), J_0(Ks)), \tag{2.82}$$

where J_n is the Bessel function of order n. Here the **B**-lines are helices on the cylinders $s = \text{cst}$. Again $\mathbf{A} = K^{-1}\mathbf{B}$, and

$$\mathbf{A} \cdot \mathbf{B} = K^{-1}\boldsymbol{B}^2 = K^{-1}B_0^2[(J_1(Ks))^2 + (J_0(Ks))^2], \tag{2.83}$$

and again any simple closing of lines of force would lead to linkages (which are negative, corresponding to a negative value of K, in Fig. 2.3(*b*).

In both these examples, the **J**-field extends to infinity. There are in fact *no* force-free fields, other than $\mathbf{B} \equiv 0$, for which **J** is confined (as in § 2.3) to a finite volume V and **B** is everywhere differentiable and $O(r^{-3})$ at infinity[4]. To prove this, let T_{ij} be the Maxwell stress tensor, given by

$$T_{ij} = \mu_0^{-1}(B_iB_j - \tfrac{1}{2}\mathbf{B}^2\delta_{ij}), \tag{2.84}$$

with the properties

$$(\mathbf{J} \wedge \mathbf{B})_i = T_{ij,j}, \qquad T_{ii} = -(2\mu_0)^{-1}\mathbf{B}^2, \tag{2.85}$$

and suppose that $\mathbf{J} \wedge \mathbf{B} \equiv 0$. Then

$$0 = \int x_i T_{ij,j}\, \mathrm{d}^3\mathbf{x} = \int_{S_\infty} n_j x_i T_{ij}\, \mathrm{d}S - \int T_{ii}\, d^3\mathbf{x}, \tag{2.86}$$

the volume integrals being over all space. Now since $\mathbf{B} = O(r^{-3})$ as $r \to \infty$, $T_{ij} = O(r^{-6})$ and so the integral over S_∞ vanishes. Hence the integral of $\mathbf{B}^2$ vanishes, and so $\mathbf{B} \equiv 0$. The proof fails if surfaces of discontinuity of **B** (and so of T_{ij}) are allowed, since then further surface integrals which do not in general vanish must be included in (2.86).

[4] This condition means of course that the only source for **B** is the current distribution **J**, and there are no further 'sources at infinity'. The leading term of the expansion (2.66) is clearly $O(r^{-3})$. It is perhaps worth noting that the proof still goes through under the weaker condition $\mathbf{B} = o(r^{-3/2})$ corresponding to finiteness of the magnetic energy $\int T_{ii}\, \mathrm{d}^3\mathbf{x}$.

Force-free fields in spherical geometry

We can however construct solutions of (2.78) which are force-free in a finite region V and current-free in the exterior region $\hat{V}$ and which do *not* vanish at infinity. This can be done explicitly (Chandrasekhar, 1956) as follows, in the important case when V is the sphere $r<R$. First let

$$K=\begin{cases} K_0 & (r<R) \\ 0 & (r>R) \end{cases} \tag{2.87}$$

where K_0 is constant. We have seen in § 2.2 that, under the poloidal and toroidal decomposition

$$\mathbf{B}=\nabla\wedge\nabla\wedge\mathbf{x}P+\nabla\wedge\mathbf{x}T, \tag{2.88}$$

we have

$$\nabla\wedge\mathbf{B}=-\nabla\wedge(\mathbf{x}\nabla^2P)+\nabla\wedge\nabla\wedge(\mathbf{x}T), \tag{2.89}$$

and so (2.78) is satisfied provided

$$T=KP \quad \text{and} \quad (\nabla^2+K^2)P=0. \tag{2.90}$$

Here K is discontinuous across $r=R$; but continuity of $\mathbf{B}$ ((2.60)) requires that T, P and $\partial P/\partial r$ be continuous across $r=R$, or equivalently

$$P=0, \qquad [\partial P/\partial r]=0 \quad \text{on} \quad r=R. \tag{2.91}$$

The simplest solution of (2.90), (2.91) is given in spherical polars (r, θ, φ) by

$$P(r,\theta)=\begin{cases} Ar^{-1/2}J_{3/2}(K_0r)\cos\theta & (r<R) \\ -B_0(r-R^3/r^2)\cos\theta & (r>R) \end{cases}, \tag{2.92}$$

where

$$J_{3/2}(K_0R)=0, \quad \text{and} \quad 3B_0=-A(\mathrm{d}/\mathrm{d}R)(R^{-1/2}J_{3/2}(K_0R)), \tag{2.93}$$

the conditions (2.93) following from (2.91). The corresponding flux-function $\chi(r, \theta)$ is then given by (2.47); the $\mathbf{B}_P$-lines, given by $\chi=$ cst., are sketched in fig. 2.4(a) for the case where $K_0<0$ and $|K_0|R$ is the smallest zero of $J_{3/2}(x)$. For $r>R$, the $\mathbf{B}$-lines are identical with the streamlines in irrotational flow past a sphere, and

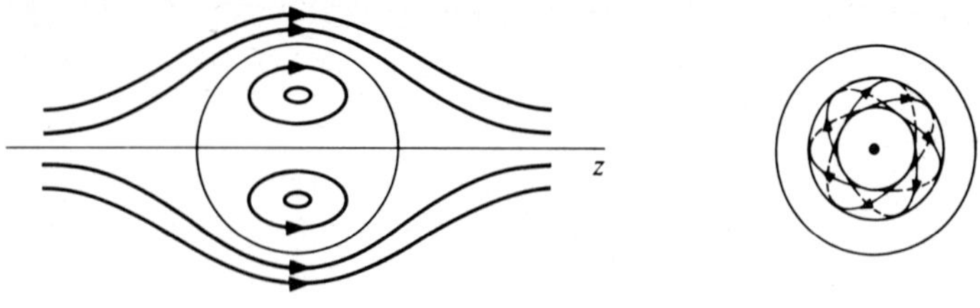

Fig. 2.4 (*a*) The $\mathbf{B}_P$-lines ($\chi = \text{cst.}$) of the force-free field given by (2.92), with K_0R the smallest zero of $J_{3/2}(x)$. (*b*) A typical **B**-line (a helix on a toroidal surface); the axis of symmetry is here perpendicular to the paper.

$\mathbf{B} \sim \mathbf{B}_0 = B_0\mathbf{i}_z$ as $r \to \infty$. For $r < R$, the **B**-lines lie on a family of nested toroidal surfaces (fig. 2.4(*b*)). The poloidal field has a neutral point on $\theta = \pi/2$ at $r = r_c$ where $(\mathrm{d}/\mathrm{d}r)(r^{1/2}J_{3/2}(K_0r)) = 0$, and the circle $r = r_c$, $\theta = \pi/2$ is the degenerate torus of the family (described as the 'magnetic axis' of the field). Each **B**-line is a helix and the pitch of the helices decreases continuously from infinity on the magnetic axis to zero on the sphere $r = R$ as we move outwards across the family of toroidal surfaces.

Note once again that $\mathbf{A} = K^{-1}\mathbf{B}$ for $r < R$, and that (if $K_0 < 0$)

$$\int_{r<R} \mathbf{A}\,.\,\mathbf{B}\,\mathrm{d}^3\mathbf{x} = K_0^{-1}\int_{r<R} \mathbf{B}^2\,\mathrm{d}^3\mathbf{x} < 0, \qquad (2.94)$$

consistent with the fact that there is an indisputable degree of linkage in the lines of force within the sphere; e.g. each line of force winds round the magnetic axis which is a particular **B**-line of the field. The **B**-lines in $r < R$ in general cover the toroidal surfaces; but if the pitch p, defined as the increase $\Delta\varphi$ in the azimuth angle as the torus is circumscribed once by the **B**-line, is $2\pi m/n$ where m and n are integers (which may be assumed to have no common factor) then the **B**-line is closed; moreover, if $m \geqslant 2$ and $n \geqslant 3$, the curve is knotted! The corresponding knot is known as the torus knot $K_{m,n}$; $K_{2,3}$ is just the trefoil knot of fig. 2.1(*b*). It is an intriguing property of this **B**-field that if we take a subset of the **B**-lines consisting of one **B**-line on each toroidal surface, then every torus knot is represented once and only once in this subset, since $p/2\pi$ passes through every rational number m/n once as it decreases continuously from infinity to zero; and yet the closed **B**-lines are exceptional in that they constitute a subset of measure zero of the set of all **B**-lines inside the sphere!

More complicated force-free fields can be constructed either by choosing higher zeros of $J_{3/2}(K_0R)$ (in which case there is more than one magnetic axis) or by replacing (2.92) by more general solutions of (2.90). The above example is however quite sufficient as a sort of prototype for spatial structures with which we shall later be concerned.

2.5. Lagrangian variables and magnetic field evolution

We must now consider magnetic field evolution in a moving fluid conductor. Let us specify the motion in terms of the displacement field $\mathbf{x}(\mathbf{a}, t)$, which represents the position at time t of the fluid particle that passes through the point $\mathbf{a}$ at a reference instant $t=0$; in particular

$$\mathbf{x}(\mathbf{a}, 0) = \mathbf{a}, \qquad (\partial x_i/\partial a_j)_{t=0} = \delta_{ij}. \tag{2.95}$$

Each particle is labelled by its initial position $\mathbf{a}$. The mapping $\mathbf{x} = \mathbf{x}(\mathbf{a}, t)$ is clearly one-to-one for a real motion of a continuous fluid, and we can equally consider the inverse mapping $\mathbf{a} = \mathbf{a}(\mathbf{x}, t)$.

The velocity of the particle $\mathbf{a}$ is

$$\mathbf{u}^L(\mathbf{a}, t) = (\partial \mathbf{x}/\partial t)_{\mathbf{a}} = \mathbf{u}(\mathbf{x}, t); \tag{2.96}$$

$\mathbf{u}^L(\boldsymbol{a}, t)$ is the Lagrangian representation, and $\mathbf{u}(\mathbf{x}, t)$ the more usual Eulerian representation. We shall use the superfix L in this way whenever fields are expressed as functions of $(\mathbf{a}, t)$, e.g.

$$\mathbf{B}^L(\mathbf{a}, t) = \mathbf{B}(\mathbf{x}(\mathbf{a}, t), t) \tag{2.97}$$

represents the magnetic field referred to Lagrangian variables. Defining the usual Lagrangian (or material) derivative by

$$\frac{\mathrm{D}}{\mathrm{D}t} = \left(\frac{\partial}{\partial t}\right)_{\mathbf{a}} = \left(\frac{\partial}{\partial t}\right)_{\mathbf{x}} + \mathbf{u} \cdot \nabla, \tag{2.98}$$

it is clear that in particular

$$\frac{\mathrm{D}\mathbf{B}}{\mathrm{D}t} \equiv \frac{\partial \mathbf{B}}{\partial t} + \mathbf{u} \,.\, \nabla \mathbf{B} = \left(\frac{\partial \mathbf{B}^L}{\partial t}\right)_{\mathbf{a}}. \tag{2.99}$$

A material curve C_L is one consisting entirely of fluid particles, and which is therefore convected and distorted with the fluid motion. If p is a parameter on the curve at time $t=0$, so that $\mathbf{a}=\mathbf{a}(p)$, then the parametric representation at time t is given by

$$\mathbf{x} \equiv \mathbf{x}(\mathbf{a}(p), t); \tag{2.100}$$

the curve C_L is closed if $\mathbf{a}(p)$ is a periodic function of p. A material surface S_L may be defined similarly and described in terms of two parameters.

An infinitesimal material line element may be described by the differential

$$\mathrm{d}x_i = E_{ij}(\mathbf{a}, t)\, \mathrm{d}a_j, \qquad E_{ij}(\mathbf{a}, t) = \partial x_i / \partial a_j. \tag{2.101}$$

The symmetric and antisymmetric parts of E_{ij} describe respectively the distortion and rotation of the fluid element initially at $\mathbf{a}$. The material derivative of E_{ij} is

$$\mathrm{D}E_{ij}/\mathrm{D}t = \partial u_i^L / \partial a_j, \tag{2.102}$$

and so it follows that

$$\mathrm{D}\,\mathrm{d}\mathbf{x}/\mathrm{D}t = \mathrm{d}a_j\, \partial \mathbf{u}^L/\partial a_j = (\mathrm{d}\mathbf{x} \,.\, \nabla)\mathbf{u}, \tag{2.103}$$

a result that is equally clear from elementary geometrical considerations.

Change of flux through a moving circuit

Suppose now that

$$\Phi(t) = \int_{S_L} \mathbf{B} \,.\, \mathrm{d}\mathbf{S} = \oint_{C_L} \mathbf{A} \,.\, \mathrm{d}\mathbf{x}, \tag{2.104}$$

where C_L is a material curve spanned by S_L. In order to calculate $\mathrm{d}\Phi/\mathrm{d}t$ we should use Lagrangian variables:

$$\Phi(t) = \oint_{C_L} A_i^L(\mathbf{a}, t)(\partial x_i/\partial a_j)(\mathrm{d}a_j/\mathrm{d}p)\, \mathrm{d}p. \tag{2.105}$$

We can then differentiate under the integral keeping $\mathbf{a}(p)$ constant. This gives, using (2.103) and standard manipulation,

$$\frac{d\Phi}{dt} = \oint_{C_L} \left(\frac{D\mathbf{A}}{Dt} \cdot d\mathbf{x} + \mathbf{A} \cdot (d\mathbf{x} \cdot \nabla)\mathbf{u}\right)$$

$$= \oint_{C_L} \left(\frac{\partial \mathbf{A}}{\partial t} - \mathbf{u} \wedge (\nabla \wedge \mathbf{A}) + \nabla(\mathbf{A} \cdot \mathbf{u})\right) \cdot d\mathbf{x}. \quad (2.106)$$

The term involving $\nabla(\mathbf{A} \cdot \mathbf{u})$ makes zero contribution to the integral since $\mathbf{A} \cdot \mathbf{u}$ is single-valued; and we have therefore

$$\frac{d\Phi}{dt} = \oint_{C_L} \left(\frac{\partial \mathbf{A}}{\partial t} - \mathbf{u} \wedge \mathbf{B}\right) \cdot d\mathbf{x}. \quad (2.107)$$

Faraday's law of induction

In its most fundamental form, Faraday's law states that if $\Phi(t)$ is defined as above for any moving closed curve C_L, then

$$\frac{d\Phi}{dt} = -\oint_{C_L} (\mathbf{E} + \mathbf{u} \wedge \mathbf{B}) \cdot d\mathbf{x}, \quad (2.108)$$

where $\mathbf{E}(\mathbf{x}, t)$ (Vm^{-1}) is the electric field relative to some fixed frame of reference. Comparison of (2.107) and (2.108) then shows that $-\mathbf{E}$ differs from $\partial \mathbf{A}/\partial t$ by at most the gradient of a single-valued scalar $\phi(\mathbf{x}, t)$:

$$\mathbf{E} + \partial \mathbf{A}/\partial t = -\nabla \phi. \quad (2.109)$$

The curl of this gives the familiar Maxwell equation

$$\partial \mathbf{B}/\partial t = -\nabla \wedge \mathbf{E}. \quad (2.110)$$

The corresponding jump condition across discontinuity surfaces is, from (2.108),

$$[\mathbf{n} \wedge (\mathbf{E} + \mathbf{u} \wedge \mathbf{B})] = 0. \quad (2.111)$$

Galilean invariance of the pre-Maxwell equations

The following simple property of the three equations

$$\nabla \cdot \mathbf{B} = 0, \qquad \nabla \wedge \mathbf{B} = \mu_0 \mathbf{J}, \qquad \partial \mathbf{B}/\partial t = -\nabla \wedge \mathbf{E} \quad (2.112)$$

is worth noting explicitly. Under the Galilean transformation

$$\mathbf{x}' = \mathbf{x} - \mathbf{V}t, \qquad t' = t, \quad (2.113)$$

the equations transform to

$$\nabla' \cdot \mathbf{B}' = 0, \qquad \nabla' \wedge \mathbf{B}' = \mu_0 \mathbf{J}', \qquad \partial \mathbf{B}'/\partial t' = -\nabla' \wedge \mathbf{E}' \tag{2.114}$$

where

$$\mathbf{B}' = \mathbf{B}, \qquad \mathbf{J}' = \mathbf{J}, \qquad \mathbf{E}' = \mathbf{E} + \mathbf{V} \wedge \mathbf{B}. \tag{2.115}$$

(These are the non-relativistic limiting forms of the more general Lorentz field transformations of the full Maxwell equations.) It is important to note that **B** and **J** are invariant under Galilean transformation, but that **E** is not. For a fluid moving with velocity $\mathbf{u}(\mathbf{x}, t)$, the field

$$\mathbf{E}' = \mathbf{E} + \mathbf{u} \wedge \mathbf{B} \tag{2.116}$$

is the electric field as measured by an observer moving with the fluid, and the right-hand side of (2.108) can therefore be regarded as (minus) the *effective* electromotive force in the moving circuit.

Ohm's law in a moving conductor

We shall employ throughout the simplest form of Ohm's law which provides the relation between electric current and electric field. In an element of fluid moving with velocity **u**, the relation between the fields $\mathbf{J}'$ and $\mathbf{E}'$ in a frame of reference moving with the element is the same as if the element were at rest (on the assumption that acceleration of the element is insufficient to affect molecular transport processes), and we shall take this relation to be $\mathbf{J}' = \sigma \mathbf{E}'$ where σ is the electric conductivity of the fluid (measured in $\mathrm{A\,V^{-1}\,m^{-1}}$). Relative to the fixed reference frame, this relation becomes

$$\mathbf{J} = \sigma(\mathbf{E} + \mathbf{u} \wedge \mathbf{B}). \tag{2.117}$$

It must be emphasised that, unlike the relations (2.112) which are fundamental, (2.117) is a phenomenological relationship with a limited range of validity. Its justification, and determination of the value of σ in terms of the molecular structure of the fluid, are topics requiring statistical mechanics methods, and are outside the scope of this book.

If we now combine (2.61), (2.109) and (2.117), we obtain immediately

$$\partial \mathbf{A}/\partial t = \mathbf{u} \wedge (\nabla \wedge \mathbf{A}) - \nabla \phi + \lambda \nabla^2 \mathbf{A}, \tag{2.118}$$

where $\lambda = (\mu_0\sigma)^{-1}$ is the *magnetic diffusivity* of the fluid. Clearly, like any other diffusivity, λ has dimensions length^2 time^{-1}; unless explicitly stated otherwise, we shall always assume that λ is uniform and constant. The divergence of (2.118), using $\nabla \cdot \mathbf{A} = 0$, gives

$$\nabla^2\phi = \nabla \cdot (\mathbf{u} \wedge \mathbf{B}). \qquad (2.119)$$

The curl of (2.118) gives the very well-known *induction equation* of magnetohydrodynamics

$$\partial\mathbf{B}/\partial t = \nabla \wedge (\mathbf{u} \wedge \mathbf{B}) + \lambda \nabla^2 \boldsymbol{B}. \qquad (2.120)$$

It is clear that if $\mathbf{u}$ is prescribed, then this equation determines (subject to appropriate boundary conditions) the evolution of $\mathbf{B}(\mathbf{x}, t)$ if $\mathbf{B}(\mathbf{x}, 0)$ is known. We shall consider in detail some of the properties of (2.120) in the following chapter.

2.6. Kinematically possible velocity fields

The velocity field $\mathbf{u}(\mathbf{x}, t)$ is related to the density field in a moving fluid by the equation of conservation of mass

$$\partial\rho/\partial t + \nabla \cdot (\rho\mathbf{u}) = 0. \qquad (2.121)$$

We have also the associated boundary condition

$$\mathbf{u} \cdot \mathbf{n} = 0 \quad \text{on} \quad S_r \qquad (2.122)$$

where S_r is any stationary rigid boundary that may be present. These are both kinematic (as opposed to dynamic) constraints, and we describe the joint field $(\mathbf{u}(\mathbf{x}, t), \rho(\mathbf{x}, t))$ as *kinematically possible* if (2.121) and (2.122) are satisfied. It is of course only a small subset of such fields which are also *dynamically possible* under, say, the Navier–Stokes equations with prescribed body forces; but many useful results can be obtained without reference to the dynamical equations, and these results are generally valid for any kinematically possible flows.

Equation (2.121) may be written in the equivalent Lagrangian form

$$\mathrm{D}\rho/\mathrm{D}t = -\rho\nabla \cdot \mathbf{u}. \qquad (2.123)$$

We shall frequently be concerned with contexts where the fluid may be regarded as incompressible, i.e. for which $\mathrm{D}\rho/\mathrm{D}t = 0$. In this case, a kinematically possible flow $\mathbf{u}(\mathbf{x}, t)$ is simply one which satisfies

$$\nabla \cdot \mathbf{u} = 0, \qquad \mathbf{u} \cdot \mathbf{n} = 0 \quad \text{on } S_r. \tag{2.124}$$

Conservation of mass may equivalently be represented by the Lagrangian equation for mass differentials,

$$\rho(\mathbf{x}, t)\, \mathrm{d}^3\mathbf{x} = \rho(\mathbf{a}, 0)\, \mathrm{d}^3\mathbf{a}. \tag{2.125}$$

Now

$$\varepsilon_{lmn}\, \mathrm{d}^3\mathbf{x} = \varepsilon_{ijk} \frac{\partial x_i}{\partial a_l} \frac{\partial x_j}{\partial a_m} \frac{\partial x_k}{\partial a_n}\, \mathrm{d}^3\mathbf{a}, \qquad (2.126)^5$$

so that (2.125) becomes

$$\rho(\mathbf{x}, t)\varepsilon_{ijk} \frac{\partial x_i}{\partial a_l} \frac{\partial x_j}{\partial a_m} \frac{\partial x_k}{\partial a_n} = \rho(\mathbf{a}, t)\varepsilon_{lmn}. \tag{2.127}$$

For incompressible flow, this of course becomes simply

$$\varepsilon_{ijk} \frac{\partial x_i}{\partial a_l} \frac{\partial x_j}{\partial a_m} \frac{\partial x_k}{\partial a_n} = \varepsilon_{lmn}. \tag{2.128}$$

2.7. Free decay modes

In the absence of fluid motion, a current field $\mathbf{J}(\mathbf{x}, t)$, confined to a finite region V, and its associated magnetic field $\mathbf{B}(\mathbf{x}, t)$, decays under the action of magnetic ('ohmic') diffusion. Consideration of this straightforward effect is a useful preliminary to the topics that will be considered in later chapters. Suppose then that $\mathbf{u} \equiv 0$ in V, so that, from (2.120), $\mathbf{B}$ satisfies the diffusion equation

$$\partial \mathbf{B}/\partial t = \lambda \nabla^2 \mathbf{B} \quad \text{in } V. \tag{2.129}$$

Suppose further that the external region $\hat{V}$ is non-conducting so that

$$\mu_0 \mathbf{J} = \nabla \wedge \mathbf{B} = 0 \quad \text{in } \hat{V}. \tag{2.130}$$

[5] This is equivalent to the statement

$$\mathrm{d}^3\mathbf{x} = J\, \mathrm{d}^3\mathbf{a}, \qquad J = \partial(x_1, x_2, x_3)/\partial(a_1, a_2, a_3).$$

J is the Jacobian of the transformation $\mathbf{x} = \mathbf{x}(\mathbf{a}, t)$.

We have also the boundary conditions

$$[\mathbf{B}]=0 \quad \text{on } S, \qquad \mathbf{B}=O(r^{-3}) \text{ as } r\to\infty, \tag{2.131}$$

where S is the surface of V.

The natural decay modes for this problem are defined by

$$\mathbf{B}(\mathbf{x}, \mathrm{t})=\mathbf{B}^{(\alpha)}(\mathbf{x}) \exp p_\alpha t, \tag{2.132}$$

where $\mathbf{B}^{(\alpha)}(\mathbf{x})$ satisfies

$$\begin{aligned}(\nabla^2-p_\alpha/\lambda)\mathbf{B}^{(\alpha)}&=0 \quad \text{in } V,\\ \nabla\wedge\mathbf{B}^{(\alpha)}&=0 \quad \text{in } \hat{V},\\ [\mathbf{B}^{(\alpha)}]=0 \quad \text{on } S, \qquad \mathbf{B}^{(\alpha)}&=O(r^{-3}) \quad \text{as } r\to\infty.\end{aligned} \tag{2.133}$$

Equations (2.133) constitute an eigenvalue problem, the eigenvalues being p_α and the corresponding eigenfunctions $\mathbf{B}^{(\alpha)}(\mathbf{x})$. These eigenfunctions form a complete set, from the general theory of elliptic partial differential equations, and an initial field $\mathbf{B}(\mathbf{x}, 0)$ corresponding to an arbitrary initial current distribution $\mathbf{J}(\mathbf{x}, 0)$ in V may be expanded as a sum of eigenfunctions:

$$\mathbf{B}(\mathbf{x}, 0)=\sum_\alpha a_\alpha \mathbf{B}^{(\alpha)}(\mathbf{x}). \tag{2.134}$$

For $t>0$, the field is then given by

$$\mathbf{B}(\mathbf{x}, t)=\sum_\alpha a_\alpha \mathbf{B}^{(\alpha)}(\mathbf{x}) \exp p_\alpha t. \tag{2.135}$$

Standard manipulation of (2.133) shows that

$$-p_\alpha=\lambda\int_{V_\infty}(\nabla\wedge\mathbf{B}^{(\alpha)})^2\,\mathrm{d}^3\mathbf{x}\Big/\int_V(\mathbf{B}^{(\alpha)})^2\,\mathrm{d}^3\mathbf{x}, \tag{2.136}$$

where $V_\infty=V\cup\hat{V}$. Hence all the p_α are real and negative, and they may be ordered so that

$$0>p_{\alpha_1}\geqslant p_{\alpha_2}\geqslant p_{\alpha_3}\geqslant\cdots. \tag{2.137}$$

When V is the sphere $r<R$, as in the consideration of the force-free modes of § 2.4, the poloidal and toroidal decomposition is appropriate. Suppose then that

$$\mathbf{B}=\nabla\wedge\nabla\wedge(\mathbf{x}P(\mathbf{x}, t))+\nabla\wedge(\mathbf{x}T(\mathbf{x}, t)), \tag{2.138}$$

with associated current distribution

$$\mathbf{J} = -\nabla \wedge (\mathbf{x}\nabla^2 P) + \nabla \wedge \nabla \wedge (\mathbf{x}\mathrm{T}). \qquad (2.139)$$

The equations (2.129)–(2.131) are then satisfied provided

$$\begin{aligned} &\partial P/\partial t = \lambda \nabla^2 P, \qquad \partial T/\partial t = \lambda \nabla^2 T, \quad \text{in } r < R, \\ &\nabla^2 P = 0, \qquad T = 0, \quad \text{in } r > R, \\ &[P] = [\partial P/\partial r] = [T] = 0 \quad \text{on } r = R. \end{aligned} \qquad (2.140)$$

Toroidal decay modes

The field T may always be expanded in surface harmonics,

$$T(r, \theta, \varphi, t) = \sum T^{(n)}(r, t) S_n(\theta, \varphi), \qquad (2.141)$$

where, from (2.140), $T^{(n)}(r, t)$ satisfies

$$\begin{aligned} &\frac{\partial T^{(n)}}{\partial t} = \lambda \left(\frac{1}{r^2} \frac{\partial}{\partial r} r^2 \frac{\partial T^{(n)}}{\partial r} + \frac{n(n+1)}{r^2} T^{(n)} \right) \quad \text{for } r < R, \\ &T^{(n)} = 0 \quad \text{for } r \geq R. \end{aligned} \qquad (2.142)$$

Putting $T^{(n)}(r, t) = f^{(n)}(r) \exp p_\alpha t$, we obtain a modified form of Bessel's equation for $f^{(n)}(r)$, with solution (regular at $r = 0$)

$$f^{(n)}(r) \propto r^{-1/2} J_{n+\frac{1}{2}}(k_\alpha r), \qquad k_\alpha^2 = -p_\alpha/\lambda. \qquad (2.143)$$

The boundary condition $f^{(n)}(R) = 0$ is then satisfied provided

$$J_{n+\frac{1}{2}}(k_\alpha R) = 0. \qquad (2.144)$$

Let x_{nq} $(q = 1, 2, \ldots)$ denote the zeros of $J_{n+\frac{1}{2}}(x)$ (see table 2.1); then the decay rates $-p_\alpha$ of toroidal modes are given by

$$-p_\alpha = \lambda R^{-2} x_{nq}^2 \qquad (n = 1, 2, \ldots; q = 1, 2, \ldots), \qquad (2.145)$$

where α is now a symbol for the pair (n, q). The general solution for T has the form (in $r < R$)

Table 2.1. *Zeros x_{nq} of $J_{n+\frac{1}{2}}(x)$, correct to 3 decimal places.*

		q					
		1	2	3	4	5	6
	0	π	2π	3π	4π	5π	6π
	1	4·493	7·725	10·904	14·007	17·221	20·371
n	2	5·763	9·095	12·323	15·515	18·689	21·854
	3	6·988	10·417	13·698	16·924	20·122	23·304
	4	8·813	11·705	15·040	18·301	21·525	24·728

$$T(r, \theta, \varphi, t) = \sum_n S_n(\theta, \varphi) \sum_q \exp p_{nq}t\, A_{nq} r^{-1/2} J_{n+\frac{1}{2}}(k_{nq}r). \tag{2.146}$$

Poloidal decay modes

We may similarly expand P in the form

$$P(r, \theta, \varphi, t) = \sum P^{(n)}(r, t) S_n(\theta, \varphi). \tag{2.147}$$

Now, however, since $\nabla^2 P = 0$ for $r > R$, we have

$$P^{(n)}(r, t) = c_n(t) r^{-(n+1)}, \ (r > R). \tag{2.148}$$

Hence continuity of $P^{(n)}$ and $\partial P^{(n)}/\partial r$ on $r = R$ require that

$$\begin{aligned} P^{(n)}(R, t) &= c_n(t) R^{-(n+1)}, \\ (\partial P^{(n)}/\partial r)_{r=R} &= -(n+1) c_n(t) R^{-(n+2)}, \end{aligned} \tag{2.149}$$

or, eliminating $c_n(t)$,

$$\partial P^{(n)}/\partial r + (n+1) R^{-1} P^{(n)} = 0 \quad \text{on } r = R. \tag{2.150}$$

Putting $P^{(n)}(r, t) = g_n(r) \exp p_\alpha t$ for $r < R$, we now obtain

$$g_n(r) \propto r^{-1/2} J_{n+\frac{1}{2}}(k_\alpha r), \qquad k_\alpha^2 = -p_\alpha/\lambda, \tag{2.151}$$

as for the toroidal modes, but now the condition (2.150) reduces to[6]

$$J_{n-\frac{1}{2}}(k_\alpha R) = 0, \tag{2.152}$$

[6] This reduction requires use of the recurrence relation

$$x J_\nu'(x) + \nu J_\nu(x) = x J_{\nu-1}(x).$$

which is to be contrasted with (2.144). The decay rates for the poloidal modes are therefore

$$-p_\alpha = \lambda R^{-2} x^2_{(n-1)q} \qquad (n=1, 2, \ldots; q=1, 2, \ldots), \tag{2.153}$$

and the general solution for P, analogous to (2.146), is

$$P(r, \theta, \varphi, t) = \begin{cases} \sum_n S_n(\theta, \varphi) \sum_q \exp\left(p_{(n-1)q} t\right) B_{nq} r^{-1/2} J_{n+\frac{1}{2}}(k_{(n-1)q} r), & r<R) \\ \sum_n c_n(t) S_n(\theta, \varphi) r^{-(n+1)}, & (r>R) \end{cases} \tag{2.154}$$

where, by (2.149),

$$c_n(t) = \sum_q B_{nq} J_{n+\frac{1}{2}}(k_{(n-1)q} R) R^{n+\frac{1}{2}} \exp\left(p_{(n-1)q} t\right) \tag{2.155}$$

Behaviour of the dipole moment

The slowest decaying mode is the poloidal mode with $n=1$, $q=1$ for which (2.153) gives $p_\alpha = -\lambda R^{-2} x^2_{01}$. This is a mode with dipole structure for $r>R$; if we choose the axis $\theta = 0$ to be in the direction of the dipole moment vector $\boldsymbol{\mu}^{(1)}(t)$, then clearly the angular dependence in the associated contribution to the defining scalar P involves only the particular axisymmetric surface harmonic $S_1(\theta, \varphi) \propto \cos\theta$.

It is interesting to enquire what happens in the case of a magnetic field which is initially totally confined to the conducting region $r<R$ (i.e. $\mathbf{B}(\mathbf{x}, 0) \equiv 0$ for $r>R$). The dipole moment of this field (as well as all the multipole moment tensors) are then evidently zero since the magnetic potential Ψ given by (2.64) must be zero to all orders for $r>R$. It is sufficient to consider the case in which the angular dependence of $\mathbf{B}(\mathbf{x}, 0)$ is the same as that of a dipole; i.e. suppose that only the term with $n=1$ is present in the above analysis for the poloidal field. The dipole moment is clearly related to the coefficient $c_1(t)$. In fact, for $r>R$, using $\nabla^2 P = 0$, we have

$$\mathbf{B}_P = \nabla \wedge \nabla \wedge \mathbf{x} P = \nabla^2 \mathbf{x} P - \nabla\nabla \cdot \mathbf{x} P = -\nabla\Psi, \tag{2.156}$$

where

$$\Psi = -P - (\mathbf{x}\,.\,\nabla)P = -P - r\,\partial P/\partial r, \tag{2.157}$$

and with $P = c_1(t)r^{-2}\cos\theta$ (from (2.148)), this gives $\Psi = c_1(t)r^{-2}\cos\theta$ also. Hence in fact the dipole moment is

$$\boldsymbol{\mu}^{(1)}(t) = c_1(t)\mathbf{i}_z, \tag{2.158}$$

and its variation with time is given by (2.155) with $n = 1$. Under the assumed conditions we must have

$$c_1(0) = \sum_q B_{1q} J_{3/2}(k_{0q}R)R^{3/2} = 0. \tag{2.159}$$

For $t > 0$, $|c_1(t)|$ will depart from zero, rising to a maximum value in a time of order $R^2\lambda^{-1}$, and will then again decay to zero in a time of this same order of magnitude, the term corresponding to $q = 1$ ultimately dominating.

It is important to note from this example that diffusion can result in a temporary increase in the dipole moment as well as leading to its ultimate decay if no regenerative agent is present. It is tempting to think that a linear superposition of exponentially decaying functions must inevitably decrease with time; consideration of the simple function $e^{-t} - e^{-2t}$ will remove this temptation; the function $c_1(t)$ in the above example exhibits similar behaviour.

This possibility of *diffusive increase of the dipole moment* is so important that it is desirable to give it an alternative, and perhaps more transparent, formulation. To this end, we must first obtain an alternative expression for $\boldsymbol{\mu}^{(1)}$, which from (2.73*a*) is given by

$$8\pi\boldsymbol{\mu}^{(1)} = \int_V \mathbf{x} \wedge (\nabla \wedge \mathbf{B})\, d^3\mathbf{x}. \tag{2.160}$$

First we decompose $\mathbf{B}$ into its poloidal dipole ingredient[7] $\mathbf{B}_1$ and the rest, $\mathbf{B}'$ say, i.e. $\mathbf{B} = \mathbf{B}_1 + \mathbf{B}'$, where $\mathbf{B}_1 = O(r^{-3})$, $\mathbf{B}' = O(r^{-4})$ as $r \to \infty$. Since $\nabla \wedge \mathbf{B}' = 0$ for $r > R$, we can rewrite (2.160) in the form

$$8\pi\boldsymbol{\mu}^{(1)} = \int_V \mathbf{x} \wedge (\nabla \wedge \mathbf{B}_1)\, dV + \int_{V_\infty} \mathbf{x} \wedge (\nabla \wedge \mathbf{B}')\, dV, \tag{2.161}$$

[7] By 'dipole ingredient' we shall mean the ingredient having the same angular dependence as a dipole field.

where as usual V_∞ is the whole space. The second integral can be manipulated by the divergence theorem giving

$$\int_{V_\infty} \mathbf{x}\wedge(\nabla\wedge\mathbf{B}')\,\mathrm{d}V=\int_{S_\infty}(\mathbf{x}\wedge(\mathbf{n}\wedge\mathbf{B}')+2(\mathbf{n}\,.\,\mathbf{B}')\mathbf{x})\,\mathrm{d}S, \tag{2.162}$$

and since $\mathbf{B}'=O(r^{-4})$ at infinity this integral vanishes as expected. Similarly the first integral may be transformed using the divergence theorem and we obtain

$$8\pi\boldsymbol{\mu}^{(1)}=\int_S \mathbf{x}\wedge(\mathbf{n}\wedge\mathbf{B}_1)\,\mathrm{d}S+2\int_V \mathbf{B}_1\,\mathrm{d}V, \tag{2.163}$$

where S is the surface $r=R$. Now $\mathbf{n}\wedge\mathbf{B}_1$ is continuous across $r=R$, and on $r=R+$, $\mathbf{B}_1=\nabla(\boldsymbol{\mu}^{(1)}\,.\,\nabla)r^{-1}$; the surface integral in (2.163) may then be readily calculated and it is in fact equal to $(8\pi/3)\boldsymbol{\mu}^{(1)}$; hence (2.163) becomes

$$\boldsymbol{\mu}^{(1)}=\frac{3}{8\pi}\int_V \mathbf{B}_1\,\mathrm{d}V. \tag{2.164}$$

The rate of change of $\boldsymbol{\mu}^{(1)}$ is therefore given by

$$\frac{8\pi}{3}\frac{\mathrm{d}\boldsymbol{\mu}^{(1)}}{\mathrm{d}t}=\int_V\frac{\partial\mathbf{B}_1}{\partial t}\,\mathrm{d}V=\lambda\int_V\nabla^2\mathbf{B}_1\,\mathrm{d}V=\lambda\int_S(\mathbf{n}\,.\,\nabla)\mathbf{B}_1\,\mathrm{d}S. \tag{2.165}$$

Hence change in $\boldsymbol{\mu}^{(1)}$ can be attributed directly to diffusion of $\mathbf{B}_1$ across S due to the normal gradient $(\mathbf{n}\,.\,\nabla)\mathbf{B}_1$ on S.

CHAPTER 3

CONVECTION, DISTORTION AND DIFFUSION OF MAGNETIC FIELD

3.1. Alfvén's theorem and related results

In this section we shall consider certain basic properties of the equations derived in the previous chapter in the idealised limit of perfect conductivity, $\sigma \to \infty$, or equivalently $\lambda \to 0$. First, from (2.108) and (2.117), we have that

$$d\Phi/dt = -\oint_{C_L} \sigma^{-1}\mathbf{J} \,.\, d\mathbf{x}, \tag{3.1}$$

so that in the limit $\sigma \to \infty$, provided $\mathbf{J}$ remains finite on C_L, $\Phi = \text{cst}$. This applies to every closed material curve C_L. In particular, consider a flux-tube consisting of the aggregate of lines of force through a small closed curve. Since any and every curve embracing the flux-tube conserves its flux as it moves with the fluid, it is a linguistic convenience to say that the flux-tube itself moves with the fluid (or is *frozen* in the fluid) and that its flux is conserved. This is Alfvén's theorem (Alfvén, 1942), which is closely analogous to Kelvin's circulation theorem in inviscid fluid dynamics.

A more formal derivation of the 'frozen-field' property can of course be devised. When $\lambda = 0$, the right-hand side of (2.120) can be expanded, giving the equivalent equation

$$D\mathbf{B}/Dt = \mathbf{B} \,.\, \nabla\mathbf{u} - \mathbf{B}\,\nabla \,.\, \mathbf{u}. \tag{3.2}$$

If we combine this with (2.123), we obtain

$$\frac{D}{Dt}\left(\frac{\mathbf{B}}{\rho}\right) = \frac{1}{\rho}\frac{D\mathbf{B}}{Dt} - \frac{\mathbf{B}}{\rho^2}\frac{D\rho}{Dt} = \frac{\mathbf{B}}{\rho} \cdot \nabla\mathbf{u}. \tag{3.3}$$

Hence $\mathbf{B}/\rho$ satisfies the same equation as that satisfied by the line element $d\mathbf{x}(\mathbf{a}, t)$ (2.103) and the solution (c.f. (2.101)) is therefore

$$\frac{B_i(\mathbf{x}, t)}{\rho(\mathbf{x}, t)} = \frac{B_j(\mathbf{a}, 0)}{\rho(\mathbf{a}, 0)} \frac{\partial x_i}{\partial a_j}, \tag{3.4}$$

a result essentially due to Cauchy[1]. Suppose now that the line element $d\mathbf{a}$ at time $t=0$ is directed along a line of force of the field $\mathbf{B}(\mathbf{a}, 0)$, so that $\mathbf{B}(\mathbf{a}, 0) \wedge d\mathbf{a} = 0$. Then at time $t>0$ we have

$$(\mathbf{B}(\mathbf{x}, t) \wedge d\mathbf{x}(\mathbf{a}, t))_i$$

$$= \varepsilon_{ijk} B_j(\mathbf{x}, t)\, dx_k(\mathbf{a}, t)$$

$$= \varepsilon_{ijk} \frac{\rho(\mathbf{x}, t)}{\rho(\mathbf{a}, 0)} B_m(\mathbf{a}, 0) \frac{\partial x_j}{\partial a_m}\, da_n \frac{\partial x_k}{\partial a_n} \quad \text{from (2.101) and (3.4)}$$

$$= \varepsilon_{lmn} \frac{\partial a_l}{\partial x_i} B_m(\mathbf{a}, 0)\, da_n \quad \text{from (2.127).}$$

Hence

$$(\mathbf{B}(\mathbf{x}, t) \wedge d\mathbf{x}(\mathbf{a}, t))_i = (\mathbf{B}(\mathbf{a}, 0) \wedge d\mathbf{a})_l\, \partial a_l / \partial x_i = 0, \qquad (3.5)$$

so that the line element $d\mathbf{x}$ is directed along a line of force of $\mathbf{B}(\mathbf{x}, t)$. It follows that if a material curve C_L coincides with a $\mathbf{B}$-line at time $t=0$, then when $\lambda = 0$ it coincides with a $\mathbf{B}$-line also for all $t>0$, and each $\mathbf{B}$-line may therefore be identified in this way with a material curve.

It is likewise clear from the above discussion that, when $\lambda = 0$, any motion which stretches a line element $d\mathbf{x}$ on a line of force will increase $\mathbf{B}/\rho$ proportionately. In an incompressible flow ($D\rho/Dt = 0$) this means that stretching of $\mathbf{B}$-lines implies proportionate field intensification. This need not be the case in compressible flow; for example, in a uniform spherically symmetric expansion, with velocity field $\mathbf{u} = (\alpha r, 0, 0)$ ($\alpha > 0$) in spherical polars (r, θ, φ), any material line element increases in magnitude linearly with r while the density of a fluid element at position $r(t)$ decreases as r^{-3}; hence the field $\mathbf{B}$ following a fluid element decreases as r^{-2}. Conversely in a spherically symmetric contraction ($\alpha < 0$), the field following a fluid element *increases* as r^{-2}.

Conservation of magnetic helicity

The fact that $\mathbf{B}$-lines are frozen in the fluid implies that the topological structure of the field cannot change with time. One would therefore expect the integrals I_m defined by (2.8) to remain

[1] Cauchy obtained the equivalent result in the context of the vorticity equation for inviscid flow – see § 3.2 below.

constant under any kinematically possible fluid motion, when $\lambda = 0$. The following is a generalisation (Moffatt, 1969) of a result proved by Woltjer (1958).

We first obtain an expression for $D(\rho^{-1}\mathbf{A}\,.\,\mathbf{B})/Dt$. From (2.118) with $\lambda = 0$ we have

$$\frac{DA_i}{Dt} \equiv \frac{\partial A_i}{\partial t} + u_j \frac{\partial}{\partial x_j} A_i = u_j \frac{\partial}{\partial x_i} A_j - \frac{\partial \phi}{\partial x_i}. \tag{3.6}$$

Combining this with (3.3), we have

$$\frac{D}{Dt}\left(\frac{\mathbf{A}\,.\,\mathbf{B}}{\rho}\right) = \mathbf{A}\,.\,\frac{D}{Dt}\left(\frac{\mathbf{B}}{\rho}\right) + \left(\frac{\mathbf{B}}{\rho}\right).\,\frac{D\mathbf{A}}{Dt} = \left(\frac{\mathbf{B}}{\rho}\,.\,\nabla\right)(\mathbf{A}\,.\,\mathbf{u} - \phi). \tag{3.7}$$

Now let the S_m (interior V_m) of (2.8) be a material surface on which (permanently) $\mathbf{n}\,.\,\mathbf{B} = 0$; then, since $D(\rho\, d^3\mathbf{x})/Dt = 0$,

$$\frac{dI_m}{dt} = \int_{V_m} \frac{D}{Dt}\left(\frac{\mathbf{A}\,.\,\mathbf{B}}{\rho}\right)\rho\, d^3\mathbf{x} = \int_{V_m} (\mathbf{B}\,.\,\nabla)(\mathbf{A}\,.\,\mathbf{u} - \phi)\, d^3\mathbf{x}$$

$$= \int_{S_m} (\mathbf{n}\,.\,\mathbf{B})(\mathbf{A}\,.\,\mathbf{u} - \phi)\, dS = 0, \tag{3.8}$$

so that I_m is, as expected, constant.

If $\lambda \neq 0$, this result is of course no longer true; retention of the diffusion terms leads to the equation

$$\frac{dI_m}{dt} = \lambda \int_{V_m} (\mathbf{B}\,.\,\nabla^2\mathbf{A} + \mathbf{A}\,.\,\nabla^2\mathbf{B})\, d^3\mathbf{x}, \tag{3.9}$$

so that I_m in general changes with time under the action of diffusion. This means that the topological structure of the field changes with time and there is no way in which a particular line of force can be 'followed' unambigously from one instant to the next. Attempts have sometimes been made to define an effective 'velocity of slip' $\mathbf{w}(\mathbf{x}, t)$ of field lines relative to fluid due to the action of diffusion; if such a concept were valid then field evolution when $\lambda \neq 0$ would be equivalent to field evolution in a non-diffusive fluid with velocity field $\mathbf{u} + \mathbf{w}$; this would imply conservation of all knots and linkages in field lines which is inconsistent in general with (3.9); it must be concluded that the concept of a 'velocity of slip', although physically appealing, is also dangerously misleading when complicated field structures are considered.

3.2. The analogy with vorticity

The induction equation (2.120),

$$\partial\mathbf{B}/\partial t = \nabla \wedge (\mathbf{u} \wedge \mathbf{B}) + \lambda\nabla^2\mathbf{B} \qquad (\nabla \cdot \mathbf{B} = 0), \tag{3.10}$$

bears a close formal resemblance to the equation for the vorticity $\boldsymbol{\omega} = \nabla \wedge \mathbf{u}$ in the barotropic (pressure $p = p(\rho)$) flow of a fluid with uniform properties under conservative body forces, viz.

$$\partial\boldsymbol{\omega}/\partial t = \nabla \wedge (\mathbf{u} \wedge \boldsymbol{\omega}) + \nu\nabla^2\boldsymbol{\omega} \qquad (\nabla \cdot \boldsymbol{\omega} = 0), \tag{3.11}$$

ν being kinematic viscosity. The analogy, first pointed out by Elsasser (1946) and exploited by Batchelor (1950) in the consideration of the action of turbulence on a weak random magnetic field, is a curious one, in that $\boldsymbol{\omega}$ is related to $\mathbf{u}$ through $\boldsymbol{\omega} = \nabla \wedge \mathbf{u}$, so that (3.11) is a non-linear equation for the evolution of $\boldsymbol{\omega}$, whereas (3.10) is undoubtedly linear in $\mathbf{B}$ when $\mathbf{u}(\mathbf{x}, t)$ is regarded as given. The fact that $\boldsymbol{\omega}$ is restricted (through its additional relationship to $\mathbf{u}$) while $\mathbf{B}$ is not means that the analogy has a sort of one-way character: general results obtained on the basis of (3.10) relating to the $\mathbf{B}$-field usually have a counterpart in the *more particular* context of (3.11). By constrast, results obtained on the basis of (3.11) may not have a counterpart in the *more general* context of (3.10).

We have already noted the parallel between Alfvén's theorem when $\lambda = 0$ and Kelvin's circulation theorem when $\nu = 0$. We have also, when $\nu = 0$, results analogous to the further theorems of § 3.1, viz.

(i) In the notation of § 3.1,

$$\frac{\omega_i(\mathbf{x}, t)}{\rho(\mathbf{x}, t)} = \frac{\omega_j(\mathbf{a}, 0)}{\rho(\mathbf{a}, 0)} \frac{\partial x_i}{\partial a_j}. \tag{3.12}$$

This result, due to Cauchy, is sometimes described as the 'solution' of the vorticity equation: this is perhaps a little misleading, since in the vorticity context $\partial x_i/\partial a_j$ is not known until $\mathbf{x}(\mathbf{a}, t)$ is known, and this can be determined only after $\omega_i(\mathbf{x}, t)$ is determined. Equation (3.12), far from providing a solution of (3.11) (with $\nu = 0$), is rather a reformulation of the equation. Contrast the situation in the magnetic context where $\partial x_i/\partial a_j$ and $B_i(\mathbf{x}, t)$ are truly independent (in so far as Lorentz forces are negligible) and where (3.4) provides a genuine solution of (3.2) or (3.3).

(ii) The integral

$$I_m\{\boldsymbol{\omega}\} = \int_{V_m} \mathbf{u} \cdot \boldsymbol{\omega} \, d^3\mathbf{x} \tag{3.13}$$

is constant if $\nu = 0$ and $\boldsymbol{\omega} \cdot \mathbf{n} = 0$ on the material surface S_m of V_m (Moffatt, 1969). This integral admits interpretation in terms of linkages of vortex tubes (exactly as in § 2.1), and conservation of $I_m\{\boldsymbol{\omega}\}$ is of course attributable to the fact that, when $\nu = 0$ and $p = p(\rho)$, vortex lines are frozen in the fluid. $I_m\{\boldsymbol{\omega}\}$ is the helicity of the velocity field within V_m; we shall use the term 'kinetic helicity' to distinguish it from the magnetic helicity $I_m\{\mathbf{B}\}$ already introduced. Kinetic helicity is of profound importance in dynamo theory, as will become apparent in later chapters.

The relative importance of the two terms on the right of (3.11) is given by the well-known Reynolds number R_e of conventional fluid mechanics: if u_0 is a typical scale for the velocity field $\mathbf{u}$ and l_0 is a typical length-scale over which it varies, then

$$|\nabla \wedge (\mathbf{u} \wedge \boldsymbol{\omega})|/|\nu \nabla^2 \mathbf{u}| = O(R_e), \quad \text{where } R_e = u_0 l_0/\nu. \tag{3.14}$$

Similarly, if l_0 is the scale of variation of $\mathbf{B}$ as well as of $\mathbf{u}$, then the ratio of the two terms on the right of (3.10) is

$$|\nabla \wedge (\mathbf{u} \wedge \mathbf{B})|/|\lambda \nabla^2 \mathbf{B}| = O(R_m), \quad \text{where } R_m = u_0 l_0/\lambda = \mu_0 \sigma u_0 l_0. \tag{3.15}$$

R_m is known as the *magnetic Reynolds number*, and it can be regarded as a dimensionless measure of the fluid conductivity in a given flow situation. If $R_m \gg 1$, then the diffusion term is relatively unimportant, and the frozen-field picture of § 3.1 should be approximately valid. If $R_m \ll 1$, then diffusion dominates, and the ability of the flow to distort the field from whatever distribution it would have under the action of diffusion alone is severely limited.

These conclusions are of course of an extremely preliminary nature and will require modification in particular contexts. Two situations where the estimate (3.15) will be misleading may perhaps be anticipated. First, if the scale L of $\mathbf{B}$ is much greater than the scale l_0 of $\mathbf{u}$, then

$$|\nabla \wedge (\mathbf{u} \wedge \mathbf{B})|/|\lambda \nabla^2 \mathbf{B}| = O(R_m L/l_0), \qquad R_m = u_0 l_0/\lambda. \tag{3.16}$$

Hence, even if $R_m \ll 1$, the induction term $\nabla \wedge (\mathbf{u} \wedge \mathbf{B})$ may nevertheless be of dominant importance if L/l_0 is sufficiently large. Secondly, in any region of rapid variation of $\mathbf{B}$ (e.g. across a thin diffusing current sheet) the relevant scale δ of $\mathbf{B}$ may be *small* compared with l_0; in this case we can have

$$|\nabla \wedge (\mathbf{u} \wedge \mathbf{B})|/|\lambda \nabla^2 \mathbf{B}| = O(R_m \delta/l_0) \quad \text{or} \quad O(R_m(\delta/l_0)^2) \tag{3.17}$$

depending on the precise geometry of the situation. In such layers of rapid change, diffusion can be important even when $R_m \gg 1$. In any event, care is generally needed in the use that is made of estimates of the type (3.15), which should always be subject to retrospective verification.

3.3. The analogy with scalar transport

A further analogy that is sometimes illuminating (Batchelor, 1952) is that between equation (3.10) for the 'transport' of the 'vector contaminant' $\mathbf{B}(\mathbf{x}, t)$ and the equation

$$\mathrm{D}\Theta/\mathrm{D}t \equiv \partial\Theta/\partial t + \mathbf{u} \,.\, \nabla\Theta = \kappa \nabla^2 \Theta, \tag{3.18}$$

which describes the transport of a scalar contaminant $\Theta(\mathbf{x}, t)$ (which may be, for example, temperature or dye concentration) subject to molecular diffusivity κ. The vector $\mathbf{G} = \nabla\Theta$ satisfies the equation

$$\partial \mathbf{G}/\partial t = -\nabla(\mathbf{u} \,.\, \mathbf{G}) + \kappa \nabla^2 \mathbf{G} \quad (\nabla \wedge \mathbf{G} = 0), \tag{3.19}$$

which is the counterpart of (3.10) for an irrotational (rather than a solenoidal) vector field.

When $\kappa = 0$, the Lagrangian solution of (3.18) is simply

$$\Theta(\mathbf{x}, t) = \Theta(\mathbf{a}, 0), \tag{3.20}$$

and surfaces of constant Θ are frozen in the fluid. The counterpart of the magnetic Reynolds number is the Péclet number

$$P_e = u_0 l_0/\kappa, \tag{3.21}$$

and diffusion is dominant or 'negligible' according as $P_e \ll$ or $\gg 1$.

3.4. Maintenance of a flux rope by uniform rate of strain

A simple illustration of the combined effects of convection and diffusion is provided by the action on a magnetic field of the irrotational incompressible velocity field

$$\mathbf{u} = (\alpha x, \beta y, \gamma z), \quad \alpha + \beta + \gamma = 0. \tag{3.22}$$

The rate of strain tensor $\partial u_i/\partial x_j$ is uniform and its principal values are α, β, γ. We suppose further that α is positive and β and γ negative, so that all fluid line elements tend to become aligned parallel to the x-axis. Likewise $\mathbf{B}$-lines tend to become aligned in the same way, so let us suppose that

$$\mathbf{B} = (B(y, z, t), 0, 0). \tag{3.23}$$

Equation (3.10) then has only an x-component which becomes

$$\frac{\partial B}{\partial t} + \beta y \frac{\partial B}{\partial y} + \gamma z \frac{\partial B}{\partial z} = \alpha B + \lambda \nabla^2 B, \tag{3.24}$$

an equation studied in various special cases by Clarke (1964, 1965). It may be easily verified that (3.24) admits the steady solution

$$B(y, z) = B_0 \exp\{-(|\beta| y^2 + |\gamma| z^2)/2\lambda\}, \tag{3.25}$$

representing a flux rope of elliptical structure aligned along the x-axis (fig. 3.1). The total flux in the rope is

$$\Phi = \int_{-\infty}^{\infty} \int_{-\infty}^{\infty} B(y, z)\, \mathrm{d}y\, \mathrm{d}z = \pi B_0 \lambda (\beta\gamma)^{-1/2}. \tag{3.26}$$

Convection of the field towards the axis is exactly balanced by diffusion outwards. It is in fact not difficult to show by Fourier transform methods that (3.25) is the asymptotic steady solution of (3.24) for arbitrary initial conditions; the constant B_0 is related, as in (3.26), to the total initial flux of $\mathbf{B}$ across any plane $x = \text{cst.}$, a quantity that is conserved during the subsequent stretching and diffusion process.

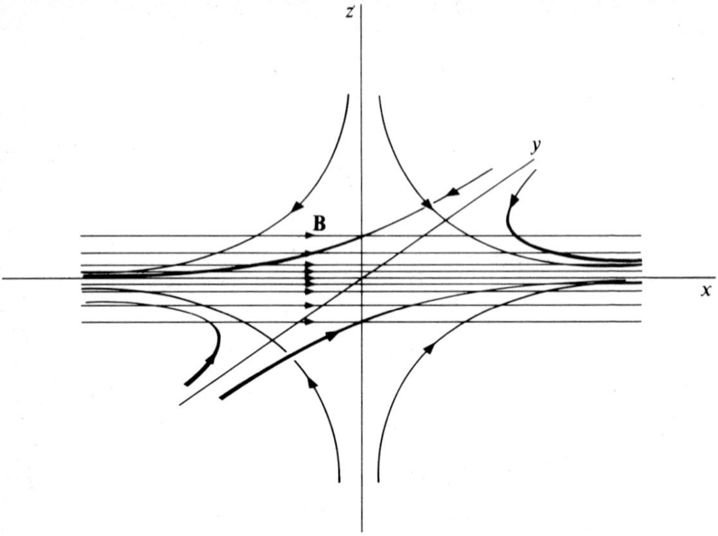

Fig. 3.1 Flux rope maintained by the action of the uniform straining motion (3.22).

If the velocity field is axisymmetric about the x-axis, then $\beta = \gamma = -\frac{1}{2}\alpha$, and in this case (3.25) becomes

$$B(y, z) = B_0 \exp\{-\alpha(y^2+z^2)/4\lambda\}. \tag{3.27}$$

The flux rope has Gaussian structure, with characteristic radius $\delta = O(\lambda/\alpha)^{1/2}$.

3.5. An example of accelerated ohmic diffusion

Suppose now that the uniform strain of § 3.4 is two-dimensional, i.e. that

$$\mathbf{u} = (\alpha x, -\alpha y, 0) \quad (\alpha > 0), \tag{3.28}$$

and that at time $t = 0$

$$\mathbf{B} = (0, 0, B_0 \sin k_0 y). \tag{3.29}$$

For $t > 0$, the **B**-lines (which are parallel to the z-axis) are swept in towards the plane $y = 0$. It is evident that both the wave-number and the amplitude of the field must change with time. We may seek a solution of (3.10) of the form

$$\mathbf{B} = (0, 0, B(t) \sin k(t)y), \tag{3.30}$$

where $k(0)=k_0$, $B(0)=B_0$. Substitution in (3.10) leads to

$$(\mathrm{d}B/\mathrm{d}t)\sin ky+(\mathrm{d}k/\mathrm{d}t)yB\cos ky=\alpha yBk\cos ky-\lambda k^2B\sin ky,$$

and since this must hold for all y,

$$\mathrm{d}k/\mathrm{d}t=\alpha k,\quad \text{and}\quad \mathrm{d}B/\mathrm{d}t=-\lambda k^2B. \tag{3.31}$$

It follows that

$$k(t)=k_0\,\mathrm{e}^{\alpha t},\qquad B(t)=B_0\exp\{-\lambda k_0^2(\mathrm{e}^{2\alpha t}-1)/2\alpha\}. \tag{3.32}$$

The wavelength of the non-uniformity of the **B**-field evidently decreases as $\mathrm{e}^{-\alpha t}$ due to the convection of all variation towards the plane $y=0$; in consequence the natural decay of the field is greatly accelerated. Note that in this case, with **B** oriented along the axis of zero strain rate, there is no tendency to stretch the **B**-lines ($\mathbf{B}\,.\,\nabla\mathbf{u}=0$), but merely a tendency to convect them ($\mathbf{u}\,.\,\nabla\mathbf{B}\neq 0$).

3.6. Equation for vector potential and flux-function under particular symmetries

Suppose now that **u** is a solenoidal velocity field, and let

$$\mathbf{u}=\mathbf{u}_P+\mathbf{u}_T,\qquad \mathbf{B}=\mathbf{B}_P+\mathbf{B}_T \tag{3.33}$$

be the poloidal and toroidal decompositions of **u** and **B**. Suppose further that both **u** and **B** are either two-dimensional (i.e. independent of the Cartesian coordinate z) or axisymmetric (i.e. invariant under rotations about the axis of symmetry Oz). Then it is clear that $\mathbf{u}_T\wedge\mathbf{B}_T=0$ and so

$$\mathbf{u}\wedge\mathbf{B}=(\mathbf{u}_P\wedge\mathbf{B}_T+\mathbf{u}_T\wedge\mathbf{B}_P)+(\mathbf{u}_P\wedge\mathbf{B}_P), \tag{3.34}$$

the first bracketed term on the right being poloidal and the second toroidal. The poloidal ingredient of (3.10) is then

$$\partial\mathbf{B}_P/\partial t=\nabla\wedge(\mathbf{u}_P\wedge\mathbf{B}_P)+\lambda\nabla^2\mathbf{B}_P. \tag{3.35}$$

Writing $\mathbf{B}_P=\nabla\wedge\mathbf{A}_T$, we may 'uncurl' this equation obtaining what is in effect the toroidal ingredient of (2.118),

$$\partial\mathbf{A}_T/\partial t=\mathbf{u}_P\wedge(\nabla\wedge\mathbf{A}_T)+\lambda\nabla^2\mathbf{A}_T, \tag{3.36}$$

there being no toroidal contribution from the term $-\nabla\phi$ of (2.118)[2]. Similarly, the toroidal ingredient of (3.10) is

$$\partial \mathbf{B}_T/\partial t = \nabla \wedge (\mathbf{u}_P \wedge \mathbf{B}_T + \mathbf{u}_T \wedge \mathbf{B}_P) + \lambda \nabla^2 \mathbf{B}_T. \tag{3.37}$$

Equations (3.36) and (3.37) are quite convenient since both $\mathbf{A}_T$ and $\mathbf{B}_T$ have only one component each in both two-dimensional and axisymmetric situations. These situations however now require slightly different treatments.

Two-dimensional case

In this case $\mathbf{A}_T = A(x, y)\mathbf{i}_z$ and $\nabla \wedge \mathbf{A}_T = -\mathbf{i}_z \wedge \nabla A$; hence (3.36) becomes

$$\partial A/\partial t + \mathbf{u}_P \cdot \nabla A = \lambda \nabla^2 A, \tag{3.38}$$

so that A behaves like a scalar quantity (c.f. (3.18)). Similarly, with $\mathbf{B}_T = B(x, y)\mathbf{i}_z$ and $\mathbf{u}_T = u_z(x, y)\mathbf{i}_z$, (3.37) becomes

$$\partial B/\partial t + \mathbf{u}_P \cdot \nabla B = (\mathbf{B}_P \cdot \nabla) u_z + \lambda \nabla^2 B. \tag{3.39}$$

Here B also behaves like a scalar, but with a 'source' term $(\mathbf{B}_P \cdot \nabla) u_z$ on the right-hand side. The interpretation of this term is simply that if u_z varies along a $\mathbf{B}_P$-line, then it will tend to shear the $\mathbf{B}_P$-line in the z-direction, i.e. to generate a toroidal field component.

Axisymmetric case

The differences here are purely associated with the curved geometry, and are in this sense trivial. First, with $\mathbf{A}_T = A(s, z)\mathbf{i}_\varphi$ in cylindrical polar coordinates (z, s, φ) (with $s = r \sin\theta$), we have

$$\nabla^2 \mathbf{A}_T = \mathbf{i}_\varphi (\nabla^2 - s^{-2}) A, \qquad \mathbf{u}_P \wedge (\nabla \wedge \mathbf{A}_T) = -s^{-1}(\mathbf{u}_P \cdot \nabla) s A, \tag{3.40}$$

so that (3.36) becomes

$$\partial A/\partial t + s^{-1}(\mathbf{u}_P \cdot \nabla) s A = \lambda (\nabla^2 - s^{-2}) A. \tag{3.41}$$

[2] This is because ϕ is independent of the azimuth angle φ in an axisymmetric situation. In the two-dimensional case, a uniform electric field E_z in the z-direction could be present, but this requires sources of field 'at infinity', and we disregard this possibility.

Similarly, with $\mathbf{B}_T = B(s, z)\mathbf{i}_\varphi$, $\mathbf{u}_T = u_\varphi(s, z)\mathbf{i}_\varphi$, we have

$$\begin{aligned} \nabla \wedge (\mathbf{u}_P \wedge \mathbf{B}_T) &= -\mathbf{i}_\varphi s(\mathbf{u}_P \,.\, \nabla)(s^{-1}B), \\ \nabla \wedge (\mathbf{u}_T \wedge \mathbf{B}_P) &= \mathbf{i}_\varphi s(\mathbf{B}_P \,.\, \nabla)(s^{-1}u_\varphi), \end{aligned} \tag{3.42}$$

so that (3.37) becomes

$$\partial B/\partial t + s(\mathbf{u}_P \,.\, \nabla)(s^{-1}B) = s(\mathbf{B}_P \,.\, \nabla)(s^{-1}u_\varphi) + \lambda(\nabla^2 - s^{-2})B. \tag{3.43}$$

Again there is a source term in the equation for B, but now it is variation of the angular velocity $\omega(s, z) = s^{-1}u_\varphi(s, z)$ along a $\mathbf{B}_P$-line which gives rise, by field distortion, to the generation of toroidal field. This phenomenon will be studied in detail in § 3.11 below.

Sometimes it is convenient to use the flux-function $\chi(s, z) = sA(s, z)$ (see (2.47)). From (3.41), the equation for χ is

$$\partial\chi/\partial t + (\mathbf{u}_P \,.\, \nabla)\chi = \lambda \mathrm{D}^2\chi, \tag{3.44}$$

where

$$\mathrm{D}^2\chi = s(\nabla^2 - s^{-2})(s^{-1}\chi) = (\nabla^2 - 2s^{-1}\partial/\partial s)\chi. \tag{3.45}$$

The operator D^2, known as the Stokes operator, occurs frequently in problems with axial symmetry. In spherical polars (r, θ, φ) it takes the form

$$\mathrm{D}^2 = \frac{\partial^2}{\partial r^2} + \frac{\sin\theta}{r^2}\frac{\partial}{\partial\theta}\frac{1}{\sin\theta}\frac{\partial}{\partial\theta}. \tag{3.46}$$

Note, from (3.45), that

$$\mathrm{D}^2\chi = \nabla \,.\, \mathbf{f} \quad \text{where } \mathbf{f} = \nabla\chi - 2s^{-1}\chi\mathbf{i}_s. \tag{3.47}$$

3.7. Field distortion by differential rotation

By *differential rotation*, we shall mean an incompressible velocity field axisymmetric about, say, Oz, and with circular streamlines about this axis. Such a motion has the form (in cylindrical polars)

$$\mathbf{u} = \omega(s, z)\mathbf{i}_z \wedge \mathbf{x}. \tag{3.48}$$

If $\nabla\omega = 0$, then we have rigid body rotation which clearly rotates a magnetic field without distortion. If $\nabla\omega \neq 0$, lines of force are in general distorted in a way that depends both on the appropriate value of R_m (§ 3.3) and on the orientation of the field relative to the vector $\mathbf{i}_z$. The two main possibilities are illustrated in fig. 3.2. In (*a*),

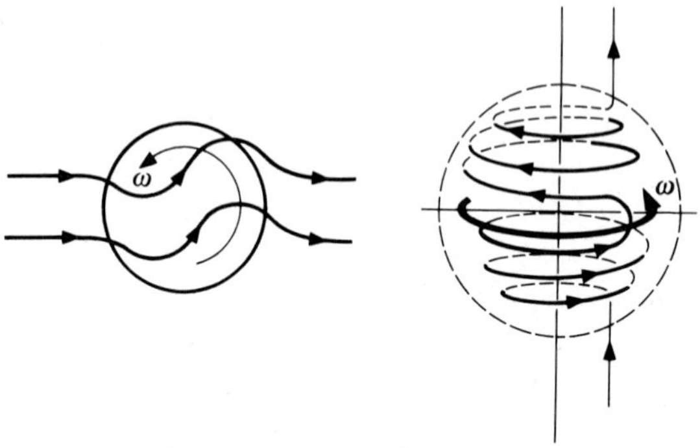

(*a*) (*b*)

Fig. 3.2 Qualitative action of differential rotation on an initially uniform magnetic field; (*a*) rotation vector $\omega\mathbf{k}$ perpendicular to field; (*b*) rotation vector $\omega\mathbf{k}$ parallel to field.

ω is a function of s alone, and the **B**-field lies in the $x-y$ plane perpendicular to $\mathbf{i}_z$; the effect of the motion, neglecting diffusion, is to wind the field into a tight double spiral in the $x-y$ plane. In (*b*), $\omega=\omega(r)$, where $r^2=s^2+z^2$, and **B** is initially axisymmetric and poloidal; the effect of the rotation, neglecting diffusion, is to generate a toroidal field, the typical **B**-line becoming helical in the region of differential rotation.

Both types of distortion are important in the solar context, and possibly also in the geomagnetic context, and have been widely studied. We discuss first in the following two sections the type (*a*) distortion (first studied in detail by Parker, 1963), and the important related phenomenon of flux expulsion from regions of closed streamlines.

3.8. Effect of plane differential rotation on an initially uniform field

Suppose then that $\omega=\omega(s)$, so that the velocity field given by (3.48) is independent of z, and suppose that at time $t=0$ the field $\mathbf{B}(\mathbf{x}, 0)$ is uniform and equal to $\mathbf{B}_0$. We take the axis Ox in the direction of $\mathbf{B}_0$. For $t>0$, $\mathbf{B}=-\mathbf{i}_z \wedge \nabla A$, where, from (3.38), A satisfies

$$\partial A/\partial t+\omega(s)(\mathbf{x}\wedge\nabla A)_z=\lambda\nabla^2 A. \tag{3.49}$$

It is natural to use plane polar coordinates defined here by

$$x=s\cos\varphi, \qquad y=s\sin\varphi, \tag{3.50}$$

in terms of which (3.49) becomes

$$\partial A/\partial t+\omega(s)\,\partial A/\partial\varphi=\lambda\nabla^2 A. \tag{3.51}$$

The initial condition $\mathbf{B}(\mathbf{x},0)=\mathbf{B}_0$ is equivalent to

$$A(s,\varphi,0)=B_0 s\sin\varphi, \tag{3.52}$$

and the relevant solution of (3.51) clearly has the form

$$A(s,\varphi,t)=\operatorname{Im} B_0 f(s,t)\,\mathrm{e}^{\mathrm{i}\varphi}, \tag{3.53}$$

where

$$\frac{\partial f}{\partial t}+\mathrm{i}\omega(s)f=\lambda\left(\frac{1}{s}\frac{\partial}{\partial s}s\frac{\partial f}{\partial s}-\frac{1}{s^2}\right)f, \tag{3.54}$$

and

$$f(s,0)=s. \tag{3.55}$$

The initial phase

When $t=0$, the field $\mathbf{B}$ is uniform and there is no diffusion; it is therefore reasonable to anticipate that diffusion will be negligible during the earliest stages of distortion. With $\lambda=0$ the solution of (3.54) satisfying the initial condition (3.55) is $f(s,t)=s\,{}^{-\mathrm{i}\omega(s)t}$, so that from (3.53)

$$A(s,\varphi,t)=B_0 s\sin(\varphi-\omega(s)t). \tag{3.56}$$

This solution is of course just the Lagrangian solution $A(\mathbf{x},t)=A(\mathbf{a},0)$, since for the motion considered, the particle whose coordinates are (s,φ) at time t originated from position $(s,\varphi-\omega(s)t)$ at time zero. The components of $\mathbf{B}=-\mathbf{i}_z\wedge\nabla A$ are now given by

$$\left.\begin{aligned} B_s&=s^{-1}\,\partial A/\partial\varphi=B_0\cos(\varphi-\omega(s)t),\\ B_\varphi&=-\partial A/\partial s=-B_0\sin(\varphi-\omega(s)t)\\ &\qquad +B_0 s\omega'(s)t\cos(\varphi-\omega(s)t).\end{aligned}\right\} \tag{3.57}$$

If $\omega'(s)=0$, i.e. if the motion is a rigid body rotation, then as expected the field is merely rotated with the fluid. If $\omega'(s)\neq 0$, the φ-component of $\mathbf{B}$ increases linearly with time as a result of the stretching process.

We may estimate from (3.56) just for how long the effects of diffusion are negligible; for from (3.56)

$$\lambda\nabla^2 A = -\lambda B_0 s^{-2}(s^3\omega')' t \cos(\varphi-\omega(s)t) - \lambda B_0 s\omega'^2 t^2 \sin(\varphi-\omega(s)t), \quad (3.58)$$

while

$$\omega\, \partial A/\partial\varphi = B_0\omega s \cos(\varphi-\omega(s)t). \quad (3.59)$$

For the purpose of making estimates, suppose that $\omega(s)$ is a reasonably smooth function, and let

$$\omega_0 = \max|\omega(s)|, \qquad \omega_0/s_0 = \max|\omega'(s)|. \quad (3.60)$$

It is clear that $\lambda\nabla^2 A$ is negligible compared with $\omega\,\partial A/\partial\varphi$ provided the coefficients of both the cosine and the sine terms on the right of (3.58) are small compared with the coefficient $B_0\omega s = O(B_0\omega_0 s_0)$ of the cosine term in (3.59); this leads to the conditions

$$\omega_0 t \ll R_m \quad \text{and} \quad \omega_0 t \ll R_m^{1/2}, \quad (3.61)$$

where $R_m = \omega_0 s_0^2/\lambda$ is the appropriate magnetic Reynolds number. If $R_m \ll 1$, then the more stringent condition is $\omega_0 t \ll R_m$, so that diffusion is negligible during only a small fraction of the first rotation period. If $R_m \gg 1$ however, the condition $\omega_0 t \ll R_m^{1/2}$ is the more stringent, but, even so, diffusion is negligible during a large number of rotations. Note that in this case the field is greatly intensified before diffusion intervenes; from (3.57), when $R_m \gg 1$ and $\omega_0 t = O(R_m^{1/2})$, B_φ is dominated by the part linear in t which gives

$$|\mathbf{B}|_{\max} = O(R_m^{1/2})B_0. \quad (3.62)$$

This gives an estimate of the maximum value attained by $|\mathbf{B}|$ before the process is influenced by diffusion.

The estimates (3.61b) and (3.62) differ from estimates obtained by E. N. Parker[3] (1963) who observed that when $R_m \gg 1$, the radial distance between zeros of the field B_φ is (from (3.57)) $\Delta s = O(s_0/\omega_0 t)$, so that the time characteristic of field diffusion is

$$t_d = O((\Delta s)^2/\lambda) = O(s_0^2/\lambda \omega_0^2 t^2). \tag{3.63}$$

Parker argued that diffusion should be negligible for all $t \ll t_d$, a condition that becomes

$$\omega_0 t \ll R_m^{1/3}, \tag{3.64}$$

in contrast to (3.61*b*); the corresponding maximum value of $|\mathbf{B}|$ becomes

$$|\mathbf{B}|_{\max} = O(R_m^{1/3}) B_0, \tag{3.65}$$

in contrast to (3.62). One can equally argue however that diffusion should be negligible for so long as $t_d \ll t_{in}$ where t_{in} is a time characteristic of the induction process; defining this by

$$t_{in} = |\mathbf{B}|/|\nabla \wedge (\mathbf{u} \wedge \mathbf{B})| \tag{3.66}$$

explicit evaluation from (3.57) gives

$$t_{in} = O(\omega_0^{-1}). \tag{3.67}$$

The condition $t_d \ll t_{in}$ restores the estimates (3.61*b*) and (3.62). The difference between $O(R_m^{1/2})$ and $O(R_m^{1/3})$ is not very important for modest values of R_m, but becomes significant if $R_m > 10^6$, say.

The ultimate steady state

It is to be expected that when $t \to \infty$ the solution of (3.54) will settle down to a steady form $f_1(s)$ satisfying

$$\frac{i\omega(s)}{\lambda} f_1 = \left(\frac{1}{s} \frac{d}{ds} s \frac{d}{ds} - \frac{1}{s^2} \right) f_1, \tag{3.68}$$

[3] Weiss (1966) also obtained estimates similar to those obtained by Parker, but for a velocity field consisting of a periodic array of eddies; the estimates (3.61) and (3.62) should in fact apply to this type of situation also. The numerical results presented by Weiss for values of R_m up to 10^3 are consistent with (3.65) rather than (3.62); but it is difficult to be sure that the asymptotic regime (for $R_m \to \infty$) has been attained. The matter perhaps merits further study.

and that, provided $\omega(s) \to 0$ as $s \to \infty$, the outer boundary condition should be that the field at infinity is undisturbed, i.e.

$$f_1(s) \sim s \quad \text{as} \quad s \to \infty. \tag{3.69}$$

The situation is adequately illustrated by the particular choice[4]

$$\frac{\omega(s)}{\lambda} = \begin{cases} k_0^2 & (s < s_0), \\ 0 & (s > s_0). \end{cases} \tag{3.70}$$

where k_0 is constant. The solution of the problem (3.68)–(3.70) is then straightforward:

$$f_1(s) = \begin{cases} s + Cs^{-1} & (s > s_0), \\ DJ_1(ps) & (s < s_0), \end{cases} \tag{3.71}$$

where $p = (1-\mathrm{i})k_0/\sqrt{2}$. The constants C and D are determined from the conditions that B_s and B_φ (and hence f_1 and f_1') should be continuous across $s = s_0$; these conditions yield

$$D = \frac{2}{pJ_0(ps_0)}, \qquad C = \frac{s_0(2J_1(ps_0) - ps_0J_0(ps_0))}{pJ_0(ps_0)}, \tag{3.72}$$

and this completes the formal determination of $f_1(s)$ from which A and hence B_s and B_φ may be determined. The **B**-lines are drawn in fig. 3.3 for $R_m = 1$, 10 and 25; note the increasing degree of distortion as R_m increases.

The nature of the solution is of particular interest when $R_m \gg 1$; in this situation $|ps_0| \gg 1$, and the asymptotic formulae

$$J_0(z) \sim (2/\pi z)^{1/2} \sin(z + \pi/4), \quad J_1(z) \sim -(2/\pi z)^{1/2} \cos(z + \pi/4) \tag{3.73}$$

[4] Note that for this discontinuous choice, there can be no initial phase of the type discussed above; diffusion must operate as soon as the motion commences to eliminate the incipient singularity in the magnetic field on $s = s_0$. Note also that it is only the variation with s of the ratio ω/λ that affects the ultimate field distribution; in particular if $\omega = 0$ for $s > s_0$, then λ may be an arbitrary (strictly positive) function of s for $s > s_0$ without affecting the situation.

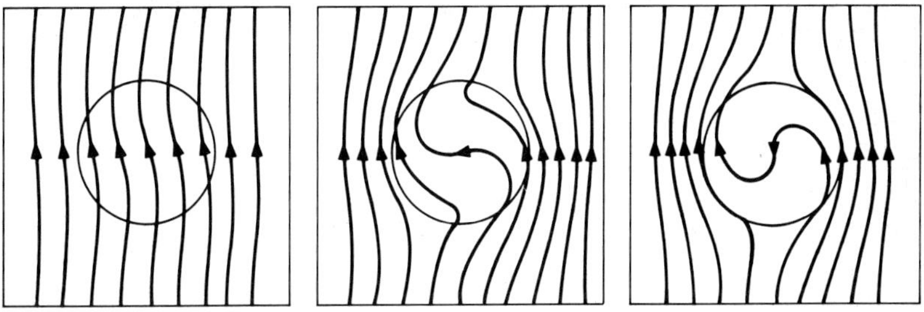

(a) $R_m = 1$ (b) $R_m = 10$ (c) $R_m = 25$

Fig. 3.3 Ultimate steady state field distributions for three values of R_m; when R_m is small the field distortion is small, while when R_m is large the field tends to be excluded from the rotating region. The sense of the rotation is anticlockwise. [Curves computed by R. H. Harding.]

may be used both in (3.72) with $z = ps_0$, and in (3.71)[5] with $z = ps$. After some simplification the resulting formula for A from (3.53) takes the form

$$A \sim \begin{cases} B_0\left(s - \dfrac{s_0^2}{s}\right)\sin\varphi + \dfrac{2B_0 s_0^2}{k_0 s}\sin\left(\varphi + \dfrac{\pi}{4}\right) & (s > s_0), \\ \left(\dfrac{2B_0}{k_0}\right)\exp\left(-\dfrac{k_0(s_0 - s)}{\sqrt{2}}\right)\sin\left(\varphi + \dfrac{k_0(s_0 - s)}{\sqrt{2}} + \dfrac{\pi}{4}\right) & (s < s_0). \end{cases} \tag{3.74}$$

In the limit $R_m = \infty (k_0 = \infty)$, this solution degenerates to

$$A \sim \begin{cases} B_0\left(s - \dfrac{s_0^2}{s}\right)\sin\varphi & (s > s_0), \\ 0 & (s < s_0). \end{cases} \tag{3.75}$$

The lines of force $A =$ cst. are then identical with the streamlines of an irrotational flow past a cylinder. In this limit of effectively infinite conductivity the field is totally excluded from the rotating region $s < s_0$; the tangential component of field suffers a discontinuity across the surface $s = s_0$ which consequently supports a current sheet.

This form of field exclusion is related to the skin effect in conventional electromagnetism. Relative to axes rotating with

[5] There is a small neighbourhood of $s = 0$ where, strictly, the asymptotic formulae (3.73) may not be used, but it is evident from the nature of the result (3.74) that this is of no consequence.

angular velocity ω_0, the problem is that of a field rotating with angular velocity $-\omega_0$ outside a cylindrical conductor. (As observed in the footnote on p. 58, the conductivity is irrelevant for $s > s_0$ in the steady state so that we may treat the medium as insulating in this region.) A rotating field may be decomposed into two perpendicular components oscillating out of phase, and at high frequencies these oscillating fields are excluded from the conductor. The same argument of course applies to the rotation of a conductor of any shape in a magnetic field, when the medium outside the conductor is insulating; at high rotation rate, the field is always excluded from the conductor when it has no component parallel to the rotation vector.

The additional terms in (3.74) describe the small perturbation of the limiting form (3.75) that results when the effects of finite conductivity in the rotating region are included. The field does evidently penetrate a small distance δ into this region, where

$$\delta = O(k_0^{-1}) = O(R_m^{-1/2})s_0. \tag{3.76}$$

The current distribution (confined to the region $s < s_0$) is now distributed through a layer of thickness $O(\delta)$ in which the field falls to an effectively zero value. The behaviour is already evident in the field line pattern for $R_m = 25$ in fig. 3.3(*c*).

The intermediate phase

The full time-dependent problem described by (3.54) and (3.55) has been solved by R. L. Parker (1966) for the case of a rigid body rotation $\omega = \omega_0$ in $s < s_0$ and zero conductivity ($\lambda = \infty$) in $s > s_0$. In this case, there are no currents for $s > s_0$ so that $\nabla^2 A = 0$ in this region (for all t), and hence (cf. 3.71*a*)

$$f(s, t) = s + C(t)s^{-1}. \tag{3.77}$$

This function satisfies

$$f + s\,\partial f/\partial s = 2s, \tag{3.78}$$

and so continuity of f and $\partial f/\partial s$ across $s = s_0$ provides the boundary condition

$$f + s_0\,\partial f/\partial s = 2s_0 \quad \text{on } s = s_0 \tag{3.79}$$

for the solution of (3.54). Setting $f = f_1(s) + g(s, t)$, the transient function $g(s, t)$ may be found as a sum of solutions separable in s

and t. The result (obtained by Parker by use of the Laplace transform) is

$$g(s,t) = \sum_{n=1}^{\infty} \frac{4s_0 \exp\left(-i\omega_0 t - (\omega_0 t\sigma_n^2/R_m)\right) J_1(\sigma_n s/s_0)}{\sigma_n^2 (1 + (\sigma_n^2/R_m)) J_1(\sigma_n)}, \tag{3.80}$$

where σ_n is the nth zero of $J_0(\sigma)$.

The lines of force $A = \text{cst.}$ as computed by Parker for $R_m = 100$ and for various values of $\omega_0 t$ during the first revolution of the cylinder are reproduced in fig. 3.4. Note the appearance of closed

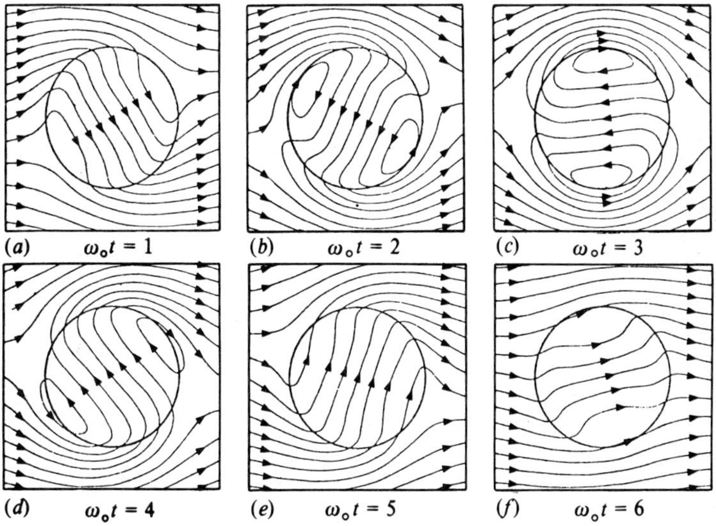

Fig. 3.4 Development of lines of force $A = \text{cst.}$ due to rotation of cylinder with angular velocity ω_0; the sense of rotation is clockwise. The sequence (a)–(f) shows one almost complete rotation of the cylinder, with magnetic Reynolds number $R_m = \omega_0 a^2/\lambda = 100$. (From Parker, 1966.)

loops[6] when $\omega_0 t \approx 2$ and the subsequent disappearance when $\omega_0 t \approx 5$; this process is clearly responsible for the destruction of flux within the rotating region. The process is repeated in subsequent revolu-

[6] This manifestation of diffusion effects at a time rather earlier than the discussion preceding (3.61) would suggest is presumably attributable (at least in part) to the assumed discontinuity of ω at $s = s_0$.

tions, flux being repeatedly expelled until the ultimate steady state is reached. Parker has in fact shown that, when $R_m = 100$, closed loops appear and disappear during each of the first fifteen revolutions of the cylinder but not subsequently. He has also shown that the number of revolutions during which the closed loop cycle occurs increases as $R_m^{3/2}$ for large R_m; in other words it takes a surprisingly long time for the field to settle down in detail to its ultimate form.

3.9. Flux expulsion for general flows with closed streamlines

A variety of solutions of (3.38) have been computed by Weiss (1966) for steady velocity fields representing either a single eddy or a regular array of eddies. The computed lines of force develop in much the same way as described for the particular flow of the previous section, closed loops forming and decaying in such a way as to gradually expel all magnetic flux from any region in which the streamlines are closed. The following argument (Proctor, 1975), analogous to that given by Batchelor (1956) for vorticity, shows why the field must be zero in the final steady state in any region of closed streamlines in the limit of large R_m (i.e. $\lambda \to 0$).

We consider a steady incompressible velocity field derivable from a stream function $\psi(x, y)$:

$$\mathbf{u} = (\partial\psi/\partial y, -\partial\psi/\partial x, 0). \tag{3.81}$$

In the limit $\lambda \to 0$ and under steady conditions, (3.38) becomes $\mathbf{u} \cdot \nabla A = 0$, and so A is constant on streamlines, or equivalently

$$A = A(\psi). \tag{3.82}$$

If λ were exactly zero, then any function $A(x, y)$ of the form (3.82) would remain steady. However, the effect of non-zero λ is to eliminate any variation in A across streamlines. To see this, we integrate the exact steady equation

$$\mathbf{u} \cdot \nabla A \equiv \nabla \cdot (\mathbf{u}A) = \lambda \nabla^2 A \tag{3.83}$$

over the area inside any closed streamline C. Since $\mathbf{n} \cdot \mathbf{u} = 0$ on C, where $\mathbf{n}$ is normal to C, the left-hand side integrates to zero, while the right-hand side becomes (with s representing arc length)

$$\oint_C \lambda \mathbf{n} \cdot \nabla A \, ds = \lambda A'(\psi) \oint_C (\partial\psi/\partial n) \, ds = \lambda K_C A'(\psi), \tag{3.84}$$

where K_C is the circulation round C. It follows that $A'(\psi) = 0$; hence $A = \text{cst.}$, and so $\mathbf{B} \equiv 0$ throughout the region of closed streamlines.

We have seen in § 3.8 that the flux does in fact penetrate a distance $\delta = O(l_0 R_m^{-1/2})$ into the region of closed streamlines, where l_0 is the scale of this region. Within this thin layer, the diffusion term in (3.83) is $O(\lambda A/\delta^2)$ and this is of the same order of magnitude as the convective term $\mathbf{u} \cdot \nabla A = O(u_0 A/l_0)$.

The phenomenon of flux expulsion has interesting consequences when a horizontal band of eddies acts on a vertical magnetic field. Fig. 3.5, reproduced from Weiss (1966), shows the steady state field

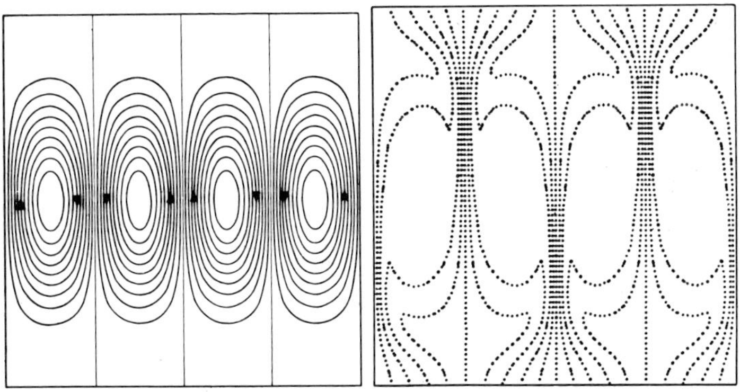

(a) (b)

Fig. 3.5 Concentration of flux into ropes by a convective layer ($R_m = 10^3$); (a) streamlines $\psi = \text{cst.}$ where ψ is given by (3.85); (b) lines of force of the resulting steady magnetic field. (From Weiss, 1966.)

structure when

$$\psi(x, y) = -(u_0/4\pi l_0)(1-(4y^2/l_0^2))^4 \sin(4\pi x/l_0), \qquad (3.85)$$

and when $R_m = u_0 l_0/\lambda = 10^3$. The field is concentrated into sheets of flux along the vertical planes between neighbouring eddies. These sheets have thickness $O(R_m^{-1/2})$, and the field at the centre of a sheet is of order of magnitude $R_m^{1/2} B_0$ where B_0 is the uniform vertical field far from the eddies; this result follows since the total vertical magnetic flux must be independent of height. This behaviour is comparable with that described by the flux rope solution (3.25), particularly if $\beta = 0$ in that solution, when the 'rope' becomes a 'sheet'.

3.10. Expulsion of poloidal fields by meridional circulation

We consider now the axisymmetric analogue of the result obtained in the preceding section. Let $\mathbf{u}$ be a steady poloidal axisymmetric velocity field with Stokes stream function $\psi(s, z)$ and let $\mathbf{B}$ be a poloidal axisymmetric field with flux-function $\chi(s, z, t)$. Then from (3.44), we have

$$\mathrm{D}\chi/\mathrm{D}t \equiv \partial\chi/\partial t + \mathbf{u}\,.\,\nabla\chi = \lambda\,\mathrm{D}^2\chi, \tag{3.86}$$

with the immediate consequence that when $\lambda = 0$, in Lagrangian notation, $\chi(\mathbf{x}, t) = \chi(\mathbf{a}, 0)$. In a region of closed streamlines in meridian planes, steady conditions are therefore possible in the limit $R_m = \infty$ only if

$$\chi = \chi(\psi(s, z)). \tag{3.87}$$

Again, as in the plane case, the effect of weak diffusion is to eliminate any variation of χ as a function of ψ. This may be seen as follows.

Using $\nabla\,.\,\mathbf{u} = 0$ and the representation (3.47) for $\mathrm{D}^2\chi$, the exact steady equation for χ may be written

$$\nabla\,.\,(\mathbf{u}\chi) = \lambda\nabla\,.\,(\nabla\chi - 2s^{-1}\chi\mathbf{i}_s). \tag{3.88}$$

Let C be any closed streamline in the $s-z$ (meridian) plane, and let S and $\mathcal{T}$ be the surface and interior of the torus described by rotation of C about Oz. Then $\mathbf{u}\,.\,\mathbf{n} = 0$ on S, and integration of (3.88) throughout $\mathcal{T}$ leads to

$$\int_S \mathbf{n}\,.\,\nabla\chi\,\mathrm{d}S = \int_S 2s^{-1}\chi\mathbf{i}_s\,.\,\mathbf{n}\,\mathrm{d}S. \tag{3.89}$$

With $\chi = \chi(\psi)$, and noting that $\mathbf{i}_s\,.\,\mathbf{n} = \mathbf{i}_z\,.\,\mathbf{t}$ on S, where $\mathbf{t}$ is a unit vector tangent to C, (3.89) gives

$$\chi'(\psi)\int \mathbf{n}\,.\,\nabla\psi\,\mathrm{d}S = 4\pi\chi\oint_C \mathbf{i}_z\,.\,\mathrm{d}\mathbf{x} = 4\pi\chi\oint_C \mathrm{d}z = 0, \tag{3.90}$$

so that $\chi'(\psi)=0$ and hence $\mathbf{B}\equiv 0$ in the region of closed streamlines.

Poloidal magnetic flux is therefore expelled by persistent meridional circulation from regions of closed meridional streamlines in much the same manner as for the plane two-dimensional configuration of §3.9.

3.11. Generation of toroidal field by differential rotation

Consider now an axisymmetric situation in which the velocity field is purely toroidal, i.e.

$$\mathbf{u}=\mathbf{u}_T=s\omega(s, z)\mathbf{i}_\varphi, \tag{3.91}$$

and a steady poloidal field $\mathbf{B}_P(s, z)$ is maintained by some unspecified mechanism. From (3.37), the toroidal field $\mathbf{B}_T$ then evolves according to the equation

$$\partial\mathbf{B}_T/\partial t=\nabla\wedge(\mathbf{u}_T\wedge\mathbf{B}_P-\lambda\nabla\wedge\mathbf{B}_T), \tag{3.92}$$

or equivalently, with $\mathbf{B}_T=B\mathbf{i}_\varphi$, from (3.43),

$$\partial B/\partial t=s(\mathbf{B}_P\,.\,\nabla)\omega+\lambda(\nabla^2-s^{-2})B. \tag{3.93}$$

Note first that if ω is constant on $\mathbf{B}_P$-lines so that $\mathbf{B}_P\,.\,\nabla\omega=0$ and if $B=0$ at time $t=0$, then $B=0$ for $t>0$ also. This is the *law of isorotation*, one of the earliest results of magnetohydrodynamics (Ferraro, 1937). In the light of Alfvén's theorem, the result is of course self-evident: if ω is constant on $\mathbf{B}_P$-lines, then each $\mathbf{B}_P$-line is rotated without distortion about the axis Oz, and there is no tendency to generate toroidal field.

If $s(\mathbf{B}_P\,.\,\nabla)\omega\neq 0$, then undoubtedly a field $B(s, z, t)$ does develop from a zero initial condition according to (3.93). It is not immediately clear whether a net flux of $\mathbf{B}_T$ across the whole meridian plane can develop by this mechanism. Let S_m denote the meridian plane $(0\leqslant s<\infty,\ -\infty<z<\infty)$ and let C_m denote its boundary consisting of the z-axis and a semi-circle at infinity. Then integration of (3.92) over S_m gives

$$\frac{\mathrm{d}}{\mathrm{d}t}\int_{S_m}\mathbf{B}_T\,.\,\mathrm{d}\mathbf{S}=\oint_{C_m}(\mathbf{u}_T\wedge\mathbf{B}_P-\lambda\nabla\wedge\mathbf{B}_T)\,.\,\mathrm{d}\mathbf{x}. \tag{3.94}$$

We shall assume that $\omega(s, z)$ is finite on the axis $s=0$ and, for the sake of simplicity, identically zero outside some sphere of finite radius R; then $\mathbf{u}_T \equiv 0$ on C_m; moreover, as will become apparent from the detailed solutions that follow, $\nabla \wedge \mathbf{B}_T = O(r^{-3})$ as $r=(s^2+z^2)^{1/2} \to \infty$, so that (3.94) becomes

$$\frac{\mathrm{d}}{\mathrm{d}t}\int_{S_m} \mathbf{B}_T \,.\, \mathrm{d}\mathbf{S} = -\int_{-\infty}^{\infty} \lambda\left(\frac{1}{s}\frac{\mathrm{d}}{\mathrm{d}s} sB\right)_{s=0} \mathrm{d}z$$

$$= -\int_{-\infty}^{\infty} \mu_0\lambda\,(\mathbf{J}_P \,.\, \mathbf{i}_z)_{s=0}\, \mathrm{d}z. \qquad (3.95)$$

Hence a net toroidal flux *can* develop, but only as a result of diffusion, and only if the term $s(\mathbf{B}_P \,.\, \nabla)\omega$ in (3.93) is not antisymmetric about any plane $z=$ cst. (in which case B would be similarly antisymmetric, and its total flux would vanish trivially).

Again, as in the discussion at the end of chapter 2, the crucially important role of diffusion is evident. If $\lambda=0$, then, although toroidal field develops, the integrated toroidal flux across meridian planes remains equal to zero. Only if $\lambda \neq 0$ is it possible for *net* toroidal flux to develop. Note incidentally that (3.92) and (3.95) are equally valid if λ is an arbitrary function of s and z.

Let us now examine in detail the behaviour of solutions of (3.93), with initial condition $B=0$ at $t=0$. As in the discussion of § 3.8, there is an initial phase when diffusion effects may be neglected and an ultimate steady state in which diffusion effects are all-important.

(i) The initial phase

Putting $\lambda=0$, the solution of (3.93) is simply

$$B(s, z, t) = s(\mathbf{B}_P \,.\, \nabla)\omega t. \qquad (3.96)$$

Physically, it is as if the $\mathbf{B}_P$-lines are gripped by the fluid and 'cranked' round the z-axis in regions where ω is greatest. In the important special case when $\mathbf{B}_P = B_0\mathbf{i}_z$, B_0 being constant, (3.96) becomes

$$B(s, z, t) = sB_0(\partial\omega/\partial z)t, \qquad (3.97)$$

and it is evident that if ω is symmetric about the plane $z=0$ then B is antisymmetric, and vice versa. In general if the $\mathbf{B}_P$-lines are symmetrical about the plane $z=0$, then B exhibits the opposite symmetry from the product $B_z\omega$.

The neglected diffusion term of (3.93) has order of magnitude $\lambda B/l_0^2$, where l_0 is the length-scale of the region over which ω varies appreciably. This becomes comparable with the retained term $s(\mathbf{B}_P\,.\,\nabla)\omega$ when $t=O(\lambda/l_0^2)$; if $\omega=O(\omega_0)$ and $|\nabla\omega|=O(\omega_0/l_0)$, then at this stage (contrast (3.62))

$$B=O(R_m)B_0 \quad \text{where } R_m=\omega_0 l_0^2/\lambda. \tag{3.98}$$

(ii) The ultimate steady state

When $t \gg \lambda/l_0^2$, B may be expected to reach a steady state given by

$$(\nabla^2 - s^{-2})B = -\lambda^{-1}s(\mathbf{B}_P\,.\,\nabla)\omega. \tag{3.99}$$

To solve this, note first that if $\mathbf{B}_T=\nabla\wedge(\mathbf{x}T)$ then (equation (2.44)) $B=-\partial T/\partial\theta$, and it follows easily that

$$(\nabla^2 - s^{-2})B = -\partial(\nabla^2 T)/\partial\theta. \tag{3.100}$$

Suppose that T has the expansion

$$T=\sum_n f_n(r)P_n(\cos\theta). \tag{3.101}$$

Then

$$\nabla^2 T=\sum_n g_n(r)P_n(\cos\theta), \qquad g_n(r)=r^{-2}[(r^2 f_n')'-n(n+1)f_n]. \tag{3.102}$$

If the right-hand side of (3.99) has the expansion

$$-\lambda^{-1}s(\mathbf{B}_P\,.\,\nabla)\omega=-\sum_1^\infty g_n(r)\,\mathrm{d}P_n(\cos\theta)/\mathrm{d}\theta, \tag{3.103}$$

then (3.101) gives the appropriate form of $T(r,\theta)$, and the corresponding $B(r,\theta)$ is given by

$$B(r,\theta)=-\sum_1^\infty f_n(r)\,\mathrm{d}P_n(\cos\theta)/\mathrm{d}\theta. \tag{3.104}$$

If $(\mathbf{B}_P\,.\,\nabla)\omega$ is regular at $r=0$, then clearly from (3.103) $g_n(0)=0$ for each n. We shall moreover suppose that, for each n,

$g_n(r) = O(r^{-3})$ at most as $r \to \infty$[7]. The solution $f_n(r)$ of (3.102*b*) for which $f_n(0)$ is finite and $f_n(\infty)$ is then

$$f_n(r) = \frac{-1}{2n+1}\left\{\frac{1}{r^{n+1}}\int_0^r x^{n+2} g_n(x)\,\mathrm{d}x + r^n \int_r^\infty \frac{g_n(x)}{x^{n-1}}\,\mathrm{d}x\right\}. \tag{3.105}$$

A case of particular interest is that in which $\nabla \wedge \mathbf{B}_P = 0$, so that $\mathbf{B}_P = -\nabla\Psi$, $\nabla^2\Psi = 0$. Then Ψ, being axisymmetric and finite at $r = 0$, has the expansion

$$\Psi = \sum_{m=1}^{\infty} \Psi_m, \qquad \Psi_m = -A_m r^m P_m(\cos\theta). \tag{3.106}$$

Let us suppose further, to simplify matters, that ω is a function of $r = (s^2 + z^2)^{1/2}$ only. Then

$$\begin{aligned} s(\mathbf{B}_P \,.\, \nabla)\omega &= -r\sin\theta \frac{\partial\Psi}{\partial r}\omega'(r) \\ &= \sum_{m=1}^{\infty} mA_m r^m \omega'(r)\sin\theta\, P_m(\cos\theta) \\ &= -\sum_{m=1}^{\infty} \frac{mA_m}{2m+1} r^m \omega'(r)\frac{\mathrm{d}}{\mathrm{d}\theta}(P_{m+1}(\cos\theta) - P_{m-1}(\cos\theta)). \end{aligned}$$

(3.107)[8]

Hence comparing with (3.103), we have for $n = 1, 2, \ldots,$

$$\lambda g_n(r) = -\left\{\frac{(n-1)A_{n-1}r^{n-1}}{2n-1} - \frac{(n+1)A_{n+1}r^{n+1}}{2n+3}\right\}\omega'(r), \tag{3.108}$$

wherein we may take $A_0 = 0$. From (3.105), the corresponding

[7] This assumption is of course much less restrictive than the assumption $g_n(r) \equiv 0$ for $r > R$, but it includes this possibility; the effect of the differential rotation is *localized* provided the $g_n(r)$ fall off sufficiently rapidly with r.

[8] Here we use the recurrence relation

$$(1 + 2m)P_m(\mu) = P'_{m+1}(\mu) - P'_{m-1}(\mu).$$

expression for $f_n(r)$ (after integrating by parts and simplifying) is

$$f_n(r) = \frac{-(n-1)A_{n-1}}{(2n-1)r^{n+1}} \int_0^r x^{2n}\omega(x)\,\mathrm{d}x + \frac{(n+1)A_{n+1}}{(2n+1)r^{n+1}} \int_0^r x^{2n+2}\omega(x)\,\mathrm{d}x$$
$$+\frac{2(n+1)A_{n+1}}{(2n+3)(2n+1)r^n} \int_r^\infty x\omega(x)\,\mathrm{d}x. \qquad (3.109)$$

In the particular case when $\mathbf{B}_P = B_0\mathbf{i}_z$, we have only the term $\Psi = \Psi_1$, with $A_1 = B_0$. Then from (3.104) and (3.109),

$$B(r,\theta) = -\tfrac{1}{3}\lambda^{-1}B_0 \sin\theta\cos\theta\, r^{-3} \int_0^r x^4\omega(x)\,\mathrm{d}x. \qquad (3.110)$$

Note that the condition $\omega = O(x^{-6})$ as $x \to \infty$ is sufficient in this case to ensure that $B = O(r^{-3})$ at infinity.

Similarly, if $\Psi = -A_2 r^2 P_2(\cos\theta)$, then B is given by

$$B(r,\theta) = \lambda^{-1}A_2\left\{\frac{2}{3}r^{-2}\int_0^r x^4\omega(x)\,\mathrm{d}x + \frac{4}{15}r\int_r^\infty x\omega(x)\,\mathrm{d}x\right\}\sin\theta$$
$$+\frac{1}{5}\lambda^{-1}A_2\left\{r^{-4}\int_0^r x^6\omega(x)\,\mathrm{d}x\right\}(1-15\cos^2\theta)\sin\theta. \qquad (3.111)$$

The most important thing to notice about this rather complicated expression is its asymptotic behaviour as $r \to \infty$: if $\omega(x) = O(x^{-6})$ as $x \to \infty$, then

$$B(r,\theta) \sim \frac{2A_2\sin\theta}{3\lambda r^2}\int_0^\infty x^4\omega(x)\,\mathrm{d}x \quad \text{as} \quad r \to \infty. \qquad (3.112)$$

This is to be contrasted with the behaviour (3.110) (which implies $B \propto r^{-3}$ as $r \to \infty$) in the former case. The expression (3.111) exhibits an *infinite* toroidal flux over the meridian plane S_m. As is clear from the introductory discussion in this section, this flux arises through the action of diffusion, which of course has an infinite time to operate before the steady field (3.111) can be established throughout all space.

The slower decrease of B with r given by (3.112) as compared with that given by (3.110) is attributable to the symmetry of the

B-field about the plane $z=0$ $(\theta=\pi/2)$; the simple $\sin\theta$ dependence of (3.112) makes the associated B less vulnerable to the influence of diffusion than the more complicated $\sin\theta\cos\theta$ structure of (3.110) and the field therefore diffuses further from the region of differential rotation. This has the important consequence that, in general, if $\mathbf{B}_P$ is non-uniform, it is the gradient of $\mathbf{B}_P$ in the neighbourhood of the centre of rotation (rather than its local average value) that determines the toroidal field generated at a very great distance.

Finally note that in all cases, in the steady state,

$$\max|B| = O(R_m)|\mathbf{B}_P|. \tag{3.113}$$

This means that, unlike the situation considered in § 3.8, the toroidal field here increases to values of order R_m and then levels off through the diffusion process without further change in order of magnitude, the whole process taking a time of order l_0^2/λ. There is no suggestion of any flux expulsion mechanism here; flux expulsion does not occur if the 'applied field' $\mathbf{B}_P$ is symmetric about the axis of rotation.

The analysis given above can be modified to cope with the situation when the poloidal field $\mathbf{B}_P$ has non-axisymmetric as well as axisymmetric ingredients (see Herzenberg & Lowes, 1957, for the case of a rigid spherical rotator imbedded in a solid conductor). The analysis is complicated by the appearance of spherical Bessel functions in the inversion of the operator $\nabla^2 - s^{-2}$, but the result is not unexpected: the non-axisymmetric ingredients are expelled from the rotating region when $R_m \gg 1$, and the axisymmetric ingredient is distorted, without expulsion, in the manner described above.

3.12. Topological pumping of magnetic flux

A fundamental variant of the flux expulsion mechanism discussed in §§ 3.8–3.10 has been discovered by Drobyshevski & Yuferev (1974). This study was motivated by the observation that in steady thermal convection between horizontal planes, the lower plane being heated uniformly, the convection cell pattern generally exhibits what may be described as a topological asymmetry about the centre-plane: fluid generally rises at the centre of the convection cells and falls on the periphery, so that regions of rising fluid are separated from each other whereas regions of falling fluid are all connected. The reason for this type of behaviour must be sought in the non-linear dynamical stability characteristics of the problem: it certainly cannot be explained in terms of linear stability analysis since if $\mathbf{u}(\mathbf{x}, t)$ is any velocity field satisfying linearised stability equations about a state of rest, then $-\mathbf{u}(\mathbf{x}, t)$ is another solution. However, it would be inappropriate to digress in this manner here; in the spirit of the present kinematic approach, let us simply assume that a steady velocity field $\mathbf{u}(\mathbf{x})$ exhibiting the above kind of topological asymmetry is given, and we consider the consequences for an initial horizontal magnetic field $\mathbf{B}$ which is subject to diffusion and to convection by $\mathbf{u}(\mathbf{x})$.

Suppose that the fluid is contained between the two planes $z = 0, z_0$. Near the upper plane $z = z_0$ the rising fluid must diverge, so that the flow is everywhere directed towards the periphery of the convection cells. A horizontal $\mathbf{B}$-line near $z = z_0$ will then tend to be distorted by this motion so as to lie everywhere near cell peripheries, where it can then be convected downwards. A horizontal $\mathbf{B}$-line near $z = 0$, by contrast, cannot be so distorted as to lie everywhere in a region of rising fluid, since these regions are disconnected. Hence a $\mathbf{B}$-line *cannot* be convected upwards (although loops of field can be lifted by each rising blob of fluid). It follows that, as far as the horizontal average $\mathbf{B}_0(z, t) = \langle \mathbf{B}(x, y, z, t) \rangle$ is concerned, there is a valve effect which permits downward transport but prohibits upward transport. This effect will be opposed by diffusion; but one would expect on the basis of this physical argument that an equilibrium distribution $\mathbf{B}_0(z)$ will develop, asymmetric about $z = \frac{1}{2}z_0$, and with greater flux in the lower half, the degree of asymmetry being related to the relevant magnetic Reynolds number.

A regular cell pattern over the horizontal plane can be characterised by cell boundaries that are either triangular, square or hexagonal. In normal Bénard convection, the hexagonal pattern is preferred (again for reasons associated with the non-linear dynamics of the system). A velocity field with square cell boundaries however allows simpler analysis, and the qualitative behaviour is undoubtedly the same whether hexagons or squares are chosen.

Drobyshevski & Yuferev chose as their velocity field

$$\mathbf{u}=u_0[-\sin x'(1+\tfrac{1}{2}\cos y')\cos z', -(1+\tfrac{1}{2}\cos x')\sin y'\cos z', (\cos x'+\cos y'+\cos x'\cos y')\sin z'], \qquad (3.114)$$

where $\mathbf{x}'=\pi\mathbf{x}/z_0$. This velocity field satisfies $\nabla\,.\,\mathbf{u}=0$, $\mathbf{n}\,.\,\mathbf{u}=0$ on $z'=0, \pi$, has square cell boundaries at $x'=(2n+1)\pi$, $y'=(2m+1)\pi$ $(n, m=0, \pm1, \pm2, \ldots)$, and has the basic topological asymmetry referred to above. The discrete regions of rising fluid are the interiors of the closed curves $\cos x'+\cos y'+\cos x'\cos y'=0$; these closed curves are centred at the cell centres, and are contained in the annular regions defined by

$$\tfrac{2}{3}z_0 \leqslant [(x-2nz_0)^2+(y-2mz_0)^2]^{1/2} \leqslant z_0/\sqrt{2}, \qquad (3.115)$$

i.e. they approximate to circles of radii $0{\cdot}68z_0$.

Under steady conditions, the magnetic field $\mathbf{B}(\mathbf{x})$ must satisfy (3.10) with $\partial\mathbf{B}/\partial t=0$. Putting $\mathbf{u}'=\mathbf{u}/u_0$, and dropping the dashes on $\mathbf{u}'$ and $\mathbf{x}'$, this becomes,

$$\nabla^2\mathbf{B}=-R_m\nabla\wedge(\mathbf{u}\wedge\mathbf{B}), \qquad R_m=z_0u_0/\pi\lambda. \qquad (3.116)$$

In order to solve this equation, we need boundary conditions on $\mathbf{B}$; following Drobyshevski & Yuferev, we assume that the solid regions $z<0$ and $z>\pi$ are perfect electrical conductors in which $\mathbf{E}$ and $\mathbf{B}$ vanish; then from (2.5),

$$B_z=0 \quad \text{on} \quad z=0, \pi; \qquad (3.117)$$

also, from (2.111) and (2.117), we have that $\mathbf{n}\wedge\mathbf{J}=0$ on the boundaries, or equivalently, since $\mu_0\mathbf{J}=\nabla\wedge\mathbf{B}$,

$$\partial B_x/\partial z=\partial B_y/\partial z=0 \quad \text{on} \quad z=0, \pi. \qquad (3.118)$$

(There will be a surface current on the boundaries under these conditions.) Finally we suppose that in the absence of fluid motion (or equivalently if $R_m=0$) the field $\mathbf{B}$ is uniform and in the x-direction, $(B_0, 0, 0)$; the flux $\Phi_0=z_0B_0$ is then trapped between the perfectly conducting planes, and the problem is to determine the distribution of mean flux according to (3.116) when $R_m\neq0$.

If $R_m \ll 1$, the problem can be solved in power series

$$\mathbf{B}(\mathbf{x}) = \sum_{n=0}^{\infty} R_m^n \mathbf{B}^{(n)}(\mathbf{x}), \tag{3.119}$$

where $\mathbf{B}^{(0)} = (B_0, 0, 0)$ and, from (3.116),

$$\nabla^2 \mathbf{B}^{(n+1)} = -\nabla \wedge (\mathbf{u} \wedge \mathbf{B}^{(n)}) \quad (n = 0, 1, 2, \ldots). \tag{3.120}$$

Since the right-hand side is always a space-periodic function, inversion of the operator ∇^2 simply requires repeated use of the identity

$$\nabla^{-2} \cos(lx+p)\cos(my+q)\cos(nz+r)$$
$$= (l^2+m^2+n^2)^{-1} \cos(lx+p)\cos(my+q)\cos(nz+r). \tag{3.121}$$

The horizontal averages $\mathbf{B}_0^{(n)}(z) = \langle \mathbf{B}^{(n)}(x, y, z)\rangle$ can then be constructed. The procedure is somewhat tedious, and it is necessary to go as far as the term $\mathbf{B}_0^{(3)}(z)$ before the pumping effect appears. The result at that level of approximation is $\mathbf{B}_0(z) = B_0(z)\mathbf{i}_x$, where

$$B_0(z) = B_0\left(1 + \frac{7R_m^2}{48}\cos 2z + \frac{R_m^3}{240}(28\cos z - 3\cos 3z) + O(R_m^4)\right). \tag{3.122}$$

Here, the term of order R_m^2 is *symmetric* about the centre plane $z = \frac{1}{2}$; at this level there is therefore symmetrical flux expulsion of the type obtained for two-dimensional flows by Weiss (1966) (but here of course the effect is weak when $R_m \ll 1$). The term of order R_m^3 is however antisymmetric about $z = \frac{1}{2}$, and shows as expected that there is a net downward transport of flux; in fact the difference in flux between the lower half and the upper half is, returning to dimensional variables,

$$\Delta\Phi = \left(\int_0^{z_0/2} - \int_{z_0/2}^{z_0}\right) B_0(z)\,dz = B_0 z_0(0{\cdot}24 R_m^3 + O(R_m^5)). \tag{3.123}$$

It may be noted that if the expansion (3.122) is continued, only terms involving odd powers of R_m can contribute to the asymmetric flux pumping effect since the even power terms are invariant under the sign change $u_0 \to -u_0$.

The function $B_0(z)$ was computed directly from (3.116) (by a computational method involving truncation of Fourier series) by Drobyshevski & Yuferev for 5 values of R_m, the maximum value being $16/\pi \approx 5{\cdot}09$. The computed curves are shown in fig. 3.6, and

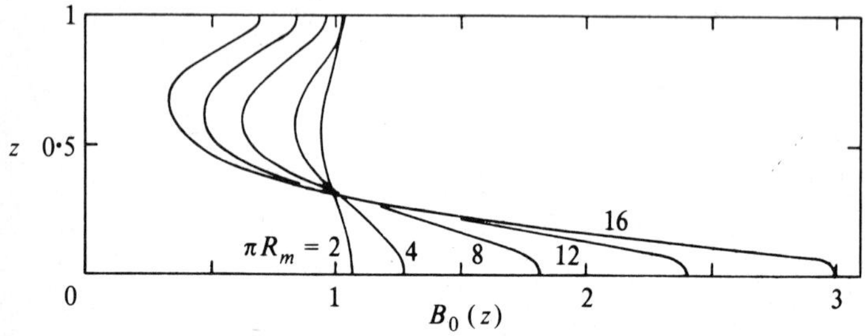

Fig. 3.6 Mean field distribution for different values of R_m as computed from (3.116) (from Drobyshevski & Yuferev, 1974). Note (i) the almost symmetric, but weak, field expulsion effect when $\pi R_m = 2$; (ii) the strong concentration of field near the lower boundary when $\pi R_m = 16$.

the concentration of flux in the lower half of the gap is clearly evident; the increase in this concentration as R_m increases is particularly striking.

For $R_m \gg 1$, one would expect the mean flux to be concentrated in a thin layer of thickness $\delta = O(R_m^{-1/2})z_0$ on the lower boundary. A boundary layer analysis should then be valid. The three-dimensionality of the velocity field however makes this a difficult problem which has not yet been solved. Nor have numerical computations yet been carried beyond the value $R_m = 5{\cdot}09$ mentioned above.

Further aspects of the flux pumping phenomenon have however been investigated by Proctor (1975), who has pointed out that asymmetric pumping can occur even when the topological distinction between upward and downward moving fluid is absent. Proctor has analysed the effect of *two-dimensional* motions in detail, and has shown that lack of *geometrical* symmetry about the midplane is sufficient to lead to a net transport of flux either up or down; e.g. if $\langle w^3 \rangle \neq 0$, where w is the vertical velocity at the midplane, then there will be a net transport, which Proctor describes as *geometrical* (as opposed to topological) pumping. When $R_m \gg 1$ however, he shows that this type of two-dimensional geometrical pumping is so weak as to be negligible, whereas it is precisely in this limit that the Drobyshevski & Yuferev mechanism may be expected to be most effective.

CHAPTER 4

THE MAGNETIC FIELD OF THE EARTH

4.1. Planetary magnetic fields in general

The dramatic achievements of launched satellite programmes over the last few years now make it possible to see the Earth and its magnetic field in the proper context of planetary magnetic fields in general. Certain obviously relevant properties of the five planets nearest to the Sun (the only ones for which magnetic field measurements are at present available)[1] are summarised in table 4.1. Of these five, Jupiter has the strongest mean surface field (as measured by the quantity μ/R^3) followed by the Earth, Mercury and Mars in that order. The field of Venus, if it exists at all, is extremely weak and below the threshold of present detectors.

Table 4.1. *Properties of the planets Mercury, Venus, Earth, Mars and Jupiter. The magnetic data for Mercury are derived from Ness et al.* (1975), *for Mars from Dolginov et al.* (1973), *and for Jupiter from Warwick* (1963) *and Smith et al.* (1974). *The rotation of Venus is retrograde relative to its sense of rotation round the Sun. The rotation of the other four planets are all prograde. The quantity μ/R^3 in the final column may be regarded as a measure of mean surface field strength.*

Planet	Radius R km	Mean density kg m^{-3}	Rotation period days	Angular velocity Ω s^{-1}	Dipole moment μ gauss km^3	μ/R^3 gauss
Mercury	2440	5400	59	$1{\cdot}23\times10^{-6}$	$4{\cdot}8\times10^{7}$	$3{\cdot}3\times10^{-3}$
Venus	6050	5200	243	$2{\cdot}99\times10^{-7}$	$<4\times10^{7}$	$<1{\cdot}8\times10^{-4}$
Earth	6380	5500	1·0	$7{\cdot}27\times10^{-5}$	$8{\cdot}05\times10^{10}$	$3{\cdot}11\times10^{-1}$
Mars	3390	3900	1·026	$7{\cdot}09\times10^{-5}$	$2{\cdot}47\times10^{7}$	$6{\cdot}36\times10^{-4}$
Jupiter	71400	1300	0·41	$1{\cdot}77\times10^{-4}$	$1{\cdot}31\times10^{15}$	3·61

It is now widely agreed that the mechanism of generation and maintenance of the Earth's field is to be sought in the inductive

[1] Brown (1975) has interpreted hectometric radio emission from Saturn in terms of a magnetic field about 12% that of Jupiter.

motion, strongly influenced by Coriolis forces, in a liquid core. Seismological studies, coupled with knowledge of the density distribution within the Earth and the relative abundances of chemical compounds of which it is composed, lead to the conclusion that the gross structure of the Earth is as indicated in fig. 4.1(a) (see, for

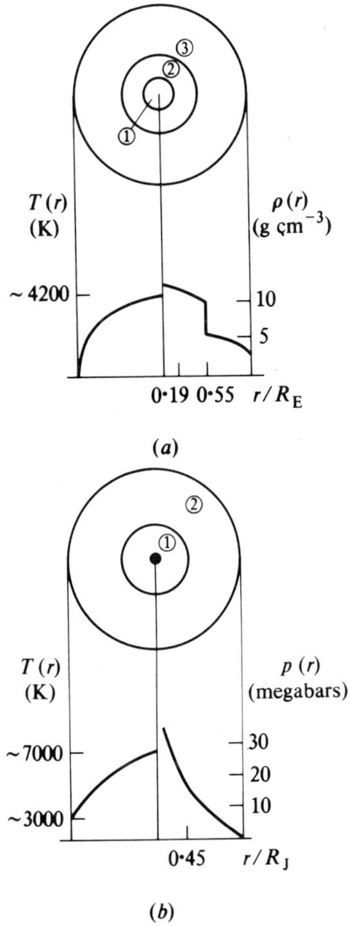

Fig. 4.1 Interior structures of Earth and Jupiter. (a) Earth, $R_E = 6380$ km: 1. solid inner core; iron/nickel alloy; 2. liquid outer core, iron and some lighter elements; 3. solid mantle, ferro-magnesium silicates (Jacobs, 1975). (b) Jupiter, $R_J = 71460$ km: 1. liquid core, helium/hydrogen alloy; 2. solid hydrogen; there may be a solid inner core for $r/R_J \leqslant 0{\cdot}1$, consisting of helium and heavier elements. (After the model of Smoluchowski, 1971.)

example, Jacobs, 1975). If we take $R_E = 6380$ km as the mean radius of the Earth (there are of course slight departures from exact sphericity), then there are distinct phase transitions at $r = R_C$ (the Lehmann–Gutenberg discontinuity) and at $r = R_I$, where

$$R_C \approx 0{\cdot}55\, R_E, \qquad R_I \approx 0{\cdot}19\, R_E. \tag{4.1}$$

The radius $r = R_C$ marks the core boundary; the outer core, $R_I < r < R_C$, consists of molten metal, probably iron with a weak admixture of lighter elements, sulphur, carbon or silicon. The argument for sulphur (Murthy & Hall, 1970) is based on the relative depletion of sulphur in the outer layers of the Earth and the presumption that processes of chemical and gravitational differentiation in the early stages of the Earth's history could have led to relatively high sulphur concentration in the central regions.

The inner core $r < R_I$ is solid, most probably an alloy of iron and nickel, presumably formed by slow crystallisation from the outer core. The mantle $r > R_C$ is also solid (although subject to visco-plastic deformation on time-scales of the order of millions of years); ferro-magnesium silicates of composition (MgO FeO)SiO_2, where Fe and Mg are freely interchangeable, are the most likely constituents (Hide, 1956).

From the standpoint of dynamo theory in the terrestrial context, we are therefore faced with the problem of fluid flow in a rotating spherical shell $R_I < r < R_C$, and the electric currents and magnetic fields that such flow may generate. The fluid may reasonably be regarded as incompressible in the dynamo context and, in a purely kinematic approach, any kinematically possible velocity fields, satisfying merely $\nabla \cdot \mathbf{u} = 0$, and $\mathbf{n} \cdot \mathbf{u} = 0$ on $r = R_I$, R_C, may be considered. At a subsequent stage it is of course essential to consider the nature of the forces (or of the sources of energy) that may be available to drive the motions. We defer to § 4.4 further consideration of the physical state of the Earth's interior.

The inference that planetary dynamo action requires both a conducting fluid core and a 'sufficient' degree of rotation (i.e. a sufficient influence of Coriolis forces) is to some extent supported (in a most preliminary way) by the information contained in table 4.1. Venus, with approximately the same radius and mean density as the Earth, has presumably a comparable structure; it rotates very slowly, however, as compared with the Earth, and it is reasonable to suppose that this is why it exhibits no magnetic field. Mars on the other hand rotates at approximately the same angular velocity as the Earth; however its substantially lower mean density indicates a much smaller abundance of iron-rich compounds, and so a much smaller molten metal core (if it has one at all). Mercury rotates slowly, and its quite significant dipole moment (relative to its size) is therefore something of a surprise; intermediate between the Earth and Venus in both rotation rate and mean surface fields strength, Mercury may well provide a key test case for dynamo models.

Finally, Jupiter is of unique interest in the dynamo context. Its low mean density indicates an internal constitution totally different from that of the four inner planets. Its hypothetical structure, again based on total mass and general arguments concerning relative abundance of elements in the proto-planetary medium, is indicated in fig. 4.1(*b*). The vast bulk of the planet consists of liquid hydrogen with possibly a small admixture of helium; in the core region $r \lesssim 46000$ km (excluding a very small central region where heavier elements may be concentrated), high pressure (of the order of 3×10^6 atmospheres and greater) causes dissociation of the hydrogen molecules into atoms, i.e. hydrogen is then in its liquid metallic phase with an electrical conductivity comparable with that of other liquid metals. The planet rotates at more than twice the angular velocity of the Earth, and Coriolis forces are undoubtedly important in its internal dynamics. It is reasonable to anticipate that Jupiter's magnetic field, like the Earth's, is attributable to dynamo action in its liquid conducting core region.

4.2. Spherical harmonic analysis of the Earth's field

The magnetic field at the surface of the Earth is due in part to currents in the interior and in part to currents in the outer conducting layers of the Earth's atmosphere. Measurements of all three field components on the surface provide a means of separating out these contributions, and it has been demonstrated by this process that by far the dominant contribution is of internal origin. For detailed numbers, the reader is referred to the classical treatise on geomagnetism by Chapman & Bartels (1940); see also Hide & Roberts (1961).

The magnetic potential due to the internal currents has the spherical polar expansion (cf. (2.64))

$$\Psi(r, \theta, \varphi) = -R_E \sum_{n=1}^{\infty} (R_E/r)^{n+1} S_n(\theta, \varphi) = \sum_{n=1}^{\infty} \Psi^{(n)}, \tag{4.2}$$

where

$$S_n(\theta, \varphi) = \sum_{m=0}^{n} (g_n^m(t) \cos m\varphi + h_n^m(t) \sin m\varphi) P_n^m(\cos\theta), \tag{4.3}$$

and where $P_n^m(\cos\theta)$ is the associated Legendre polynomial with the normalisation

$$\int_{-1}^{1} P_n^m(\cos\theta) P_{n'}^{m'}(\cos\theta)\, d(\cos\theta) = \frac{1}{2n+1} \delta_{nn'} \delta_{mm'}. \tag{4.4}$$

The coefficients $g_n^m(t)$, $h_n^m(t)$ are the 'geomagnetic elements' in conventional notation. The axis of reference $\theta = 0$ is taken to be the geographical axis (i.e. the axis of rotation) of the Earth; the origin of longitude ($\varphi = 0$) is at Greenwich.

The potential Ψ is conveniently split into dipole and non-dipole ingredients:

$$\Psi = \Psi_d + \Psi_{nd}, \qquad \Psi_d = \Psi^{(1)}, \qquad \Psi_{nd} = \sum_{n=2}^{\infty} \Psi^{(n)}. \tag{4.5}$$

The dipole ingredient corresponds to a fictitious dipole $\boldsymbol{\mu}$ at the Earth's centre where, with Cartesian coordinates (x, y, z) related to (r, θ, φ) by (2.20),

$$\mu_x = R_E^3 g_1^1, \qquad \mu_y = R_E^3 h_1^1, \qquad \mu_z = R_E^3 g_1^0. \tag{4.6}$$

This dipole makes an angle $\psi(t)$ with the axis Oz where

$$\tan\psi = ((g_1^1)^2 + (h_1^1)^2)^{1/2} / g_1^0. \tag{4.7}$$

Using the figures in the first two rows of table 4.2, which shows the values of g_n^m, h_n^m and their time derivatives for $n, m \leqslant 8$ as given by Barraclough *et al.* (1975), (4.7) gives $\psi \approx 11°$.

The series (4.2) converges quite rapidly for $r = R_E$. If the source currents were strictly confined to the core region $r < R_C$, then the series would converge for $r > R_C$. In fact, the mantle is by no means

Table 4.2. *Geomagnetic elements* $g_n^m(t)$, $h_n^m(t)$ *and their rates of change at epoch* 1975·0 *for* $n, m \leq 8$. *The unit of field measurement is* $1\gamma = 10^{-5}$ gauss $= 10^{-9}$ Wb m^{-2}. (*From Barraclough et al.*, 1975.)

m	n	g_n^m (γ)	h_n^m (γ)	$\dot{g}_n^m$ (γ yr^{-1})	$\dot{h}_n^m$ (γ yr^{-1})
0	1	−30103·6		26·8	
1	1	−2016·5	5682·6	10·0	−10·1
0	2	−1906·7		−25·0	
1	2	3009·9	−2064·7	0·3	−2·8
2	2	1633·0	−58·1	5·5	−18·9
0	3	1278·2		−3·8	
1	3	−2142·0	−329·8	−10·5	7·2
2	3	1254·7	265·9	−4·7	2·8
3	3	831·0	−227·0	−4·7	−6·4
0	4	946·9		−0·9	
1	4	792·5	193·4	−2·2	5·4
2	4	443·8	−265·8	−4·0	0·7
3	4	−403·9	53·0	−2·1	2·6
4	4	212·5	−285·2	−4·6	−0·7
0	5	−220·6		0·2	
1	5	351·4	24·5	−1·0	0·9
2	5	262·3	148·4	1·3	2·6
3	5	−63·8	−161·3	−2·1	−2·7
4	5	−157·5	−83·4	−0·6	1·3
5	5	−40·2	92·3	1·3	1·1
0	6	44·1		0·6	
1	6	69·9	−11·2	0·9	−0·3
2	6	27·7	100·4	2·3	−0·2
3	6	−194·3	77·6	3·5	0·2
4	6	−0·9	−40·3	0·0	−1·6
5	6	3·8	−7·9	0·8	0·4
6	6	−108·7	15·6	−0·4	2·0
0	7	71·5		−0·4	
1	7	−53·3	−76·6	−0·2	−1·2
2	7	2·3	−24·7	−0·5	−0·2
3	7	13·4	−4·5	0·3	0·0
4	7	−6·4	7·0	0·8	0·3
5	7	3·2	24·5	0·6	−0·6
6	7	17·0	−21·8	0·5	0·0
7	7	−5·9	−12·9	−0·8	1·2
0	8	11·0		0·4	
1	8	5·1	4·9	0·3	−0·2

Table 4.2.—*contd.*

m	n	g_n^m (γ)	h_n^m (γ)	$\dot{g}_n^m$ (γ yr^{-1})	$\dot{h}_n^m$ (γ yr^{-1})
2	8	−2·6	−13·9	0·0	−0·3
3	8	−12·6	5·0	0·4	−0·3
4	8	−13·8	−18·0	−0·2	−0·3
5	8	−0·1	5·7	−0·4	0·5
6	8	−2·4	14·5	0·6	−0·5
7	8	12·3	−11·1	−0·3	−0·6
8	8	4·9	−16·7	0·0	0·5

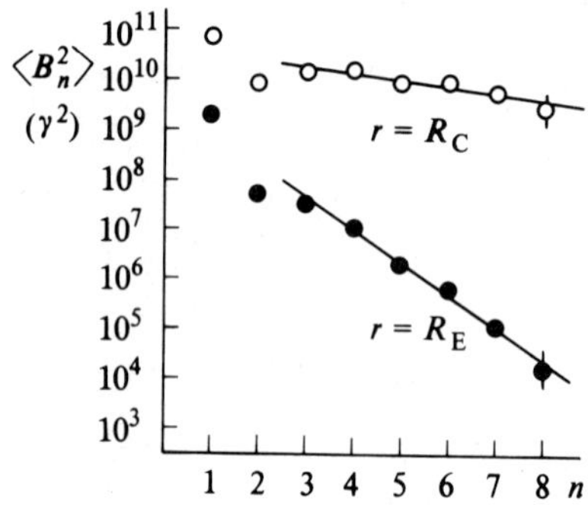

(*a*)

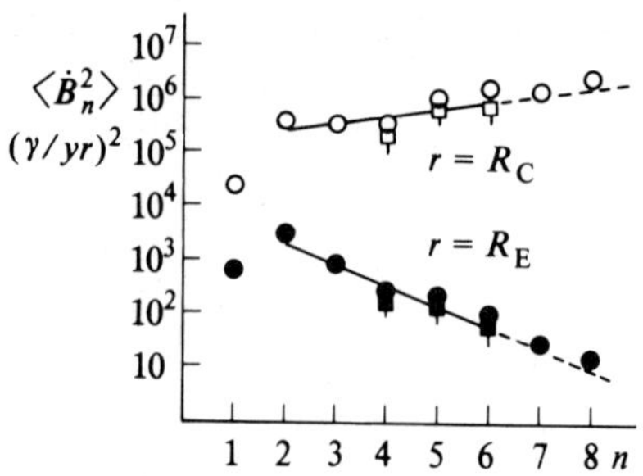

(*b*)

Fig. 4. 2 Contributions (*a*) to mean-square field and (*b*) to mean-square rate of change of field on the surfaces $r = R_E$ and $r = R_C$. $\mathbf{B}_n = -\nabla\Psi^{(n)}$ and $\langle \ldots \rangle$ indicates an average over a spherical surface. The unit of field strength is $1\gamma = 10^{-5}$ gauss. The squared points in (b) indicate values of $\langle \dot{B}_n^2 \rangle$ corrected for random errors in the secular variation coefficients $\dot{g}_n^m$, $\dot{h}_n^m$. (From Lowes, 1974; data from International Geomagnetic Reference Field 1965–70.)

a perfect insulator, although its conductivity is certainly much less than that of the core, and some current may leak out from the core to the mantle. It is not therefore safe to use the expansion (4.2) in the neighbourhood of $r = R_C$. Nevertheless it is interesting to plot the mean square contributions $\langle B_n^2 \rangle$, where $\mathbf{B}_n = -\nabla\Psi^{(n)}$ and the angular brackets represent averaging over a sphere $r = \text{cst.}$, for $n = 1, 2, \ldots, 8$ (Lowes, 1974) – see fig. 4.2. On $r = R_E$, the convergence is convincing, while on $r = R_C$, as expected, convergence is certainly not obvious; indeed it is evident that the harmonics up to $n = 8$ make roughly equal contributions to the mean square field on $r = R_C$. Thus although the field at the Earth's surface is certainly predominantly dipole in structure, the evidence is that this situation does not persist down to the neighbourhood of the core boundary. This difference between the field structure on $r = R_E$ and $r = R_C$ is even more marked in terms of the mean square of the time derivatives $\dot{B}_n^2$; in this case, on $r = R_C$, the contributions actually show a distinct increase with increasing n (fig. 4.2(b)); the values for $n = 7, 8$ in this figure are however uncertain due to random errors in the secular variation coefficients.

The non-dipole field is generally depicted by magnetic maps, i.e. by a set of contours of, say, constant vertical magnetic field (usually obtained on the basis of annual averages). Comparison of such maps at say 10-year intervals provides qualitative evidence of the evolution of the non-dipole field with time. Since $\langle \mathbf{B}_{nd}^2 \rangle^{1/2} \approx 0{\cdot}02$ gauss and $\langle \dot{\mathbf{B}}_{nd}^2 \rangle^{1/2} \approx 0{\cdot}0005$ gauss/year, the time-scale for this *secular variation* at any fixed location is of order 40 years. The magnetic contours evolve on this time-scale, with centres of activity growing and decaying like isobars on a weather chart. A distinct ingredient of this evolution is a *westward drift* of the non-dipole field at a rate that has been estimated by Bullard *et al.* (1950) as $0{\cdot}18 \pm 0{\cdot}015$ degrees of longitude per year (i.e. about 7° in 40 years).

4.3. Variation of the dipole field over long time-scales

Measurements of the coefficients $g_n^m(t)$, $h_n^m(t)$ have been available only since Gauss. During the period 1835–1945, the dipole field decreased in rms intensity from $0{\cdot}464$ gauss to $0{\cdot}437$ gauss, indicating a time-scale of variation of order 2000 years; (this is probably an underestimate – see below). For evidence of field evolution in eras prior to Gauss, we have two important sources of information. First and most important, the science of rock magnetism (palaeomagnetism) provides estimates of **B** at geological epochs when rocks cooled below the Curie point; since some rocks are known on the basis of radiometric dating to be as old as $3{\cdot}5\times10^9$ years, this provides information about the Earth's field from the earliest stages in its history, the estimated age of the Earth being of order 5×10^9 years. Studies of the magnetisation of sea-floor sediments provides similar information. Secondly, archaeomagnetism (i.e. the study of the remanent magnetism acquired by clay pots, kilns and other objects baked by man) gives information about the field intensity and orientation over the last 4000 years or so. The information from both sources is reliable only if the samples analysed can be accurately dated. Dating of ancient rocks is accurate only to within about 3%, so that, roughly speaking, the further back we go in geological time, the more uncertain the picture becomes.

Nevertheless, certain broad conclusions are now widely accepted from such studies (see, for example, Bullard, 1968; Jacobs, 1976). First, from the archaeomagnetic studies, it has been shown that the intensity fluctuates on a time-scale of order 10^4 years, and that it has been decreasing over the last few thousand years from a level about 50% greater than the present level. The direction of the dipole moment also changes (dipole wobble), though at a rate small compared with the westward drift of the non-dipole field; the indications are that the time average of the dipole moment $\boldsymbol{\mu}(t)$ over periods of order 10^3–10^4 years is accurately in the north–south direction. The smallness of the angle $\psi\approx11°$ between the present magnetic axis and the geographic axis is an indication of the relevance of rotational constraints on the inductive fluid motions in the core; the fact that the long-time average of ψ is apparently zero, or near zero, provides more emphatic evidence that these rotational constraints have a strong bearing on the magnetic field generation problem.

More dramatically, the palaeomagnetic studies provide clear evidence of reversals in the polarity of the Earth's field that have occurred repeatedly during the Earth's history. Fig. 4.3 (derived from Cox, 1969) shows the record of variations in polarity of the

Earth's field over the recent geological past, i.e. the last 4 million years. The reversals apparently occur randomly in time, the typical period between reversals being of order 2×10^5 years. This period is however highly variable; going back further in geological time, there was a conspicuous long interval of 5×10^7 years throughout the Permian era (between 280 and 230 million years ago) when no reversals occurred (Irving, 1964). There was also a distinct change in the statistics of the occurrence of reversals about 45 million years ago (Jacobs, 1976).

The details of individual reversals, which take place on time-scales of order 10^4 years, have been described by Cox (1972). During this process, the field intensity first decreases to about one quarter its usual value over a period of several thousand years; the field direction then undergoes several swings up to 30° in direction before following an irregular route to its reversed direction; finally the intensity recovers in the reversed direction to its original level.

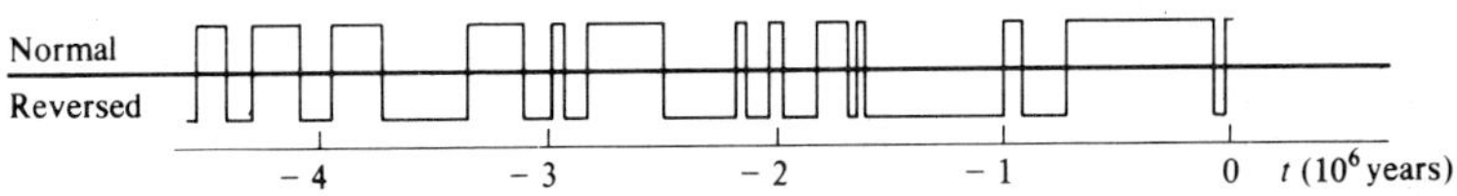

Fig. 4.3 Record of reversals of the Earth's dipole polarity over the last 4×10^6 years (derived from Cox, 1969). The figure indicates only the direction of the dipole vector, not its intensity.

4.4. Parameters and physical state of the lower mantle and core

Apart from chemical constitution, the fundamental thermodynamic variables in the interior of the Earth are density and temperature from which, in principle, other properties such as electrical conductivity may be deduced (although this requires bold extrapolation of curves based on available laboratory data). The density $\rho(r)$ as inferred from seismic data increases with depth monotonically from about $3{\cdot}4\ \mathrm{g\,cm^{-3}}$ in the upper mantle to about $5{\cdot}5\ \mathrm{g\,cm^{-3}}$ in the lower mantle at the core–mantle interface (see fig. 4.1(*a*)). It then jumps to about 10 g cm^{-3} across the core-mantle interface and increases to approximately 12 g cm^{-3} at the interface between inner and outer core. There may be a further jump in density at this interface, but an upper bound of 1·9 g cm^{-3} can be put on this jump (Bolt & Qamar, 1970). The central density in the inner core is believed to be of order 13 or 14 g cm^{-3}.

The temperature $T(r)$ likewise rises monotonically with increasing depth to a value of the order of 4200–4300 K in the inner core; this estimate is based on knowledge of the melting-point temperature of iron under high pressure and the reasonable assumption that the temperature at the surface of the inner core is precisely the melting-point temperature corresponding to the local hydrostatic pressure (Higgins & Kennedy, 1971). Since the outer core is liquid, clearly the temperature there must everwhere be above the local melting-point temperature. If the inner and outer cores are in chemical and thermal equilibrium at the interface this implies a severe restriction on possible temperature distributions in the outer core. In particular Higgins & Kennedy (1971) have argued, on the basis of extrapolation of melting-point curves for iron to high pressures, that the melting-point gradient is considerably less than the adiabatic temperature gradient corresponding to neutral convective stability conditions; they further argue that in these circumstances the actual temperature will be very nearly equal to the melting-point temperature $T_m(r)$, the resulting temperature gradient being then insufficient to generate convection currents. In a further examination of the problem (Kennedy & Higgins, 1973), the same authors confirmed their earlier conclusion as regards the outer two-thirds of the outer core, but conceded that uncertainties in the various parameters involved could possibly allow a superadiabatic temperature gradient in the inner one-third; turbulent convection would then tend to restore an adiabatic mean temperature distribution in this regime.

The Higgins & Kennedy argument does however rest on the assumption of chemical and thermal equilibrium at the interface between the inner solid and outer liquid core, and it is not quite clear that this assumption is justified. The presence of sulphur in the outer core would imply a lower melting point on the liquid side of the interface than on the solid side, to some extent relaxing the constraint on temperature mentioned above. Against this must be set the argument that the concentration of sulphur (the lighter element) presumably increases with r, providing a further stabilising contribution to the density variation.

The fact that the outer core might be thermally stably stratified was recognised by Braginskii (1964*c*), who argued that convection might nevertheless be driven by the upward flotation of lighter elements (Braginskii argued in terms of silicon) during the process of crystallisation of iron from the outer core onto the inner core. A similar process of density differentiation had been earlier suggested by Urey (1952): precipitation of iron from the mantle into the outer core and subsequent sedimentation of the iron towards the inner core (through the lighter core fluid mixture) could also lead to the generation of convection currents. In both processes, gravitational energy is released and a proportion of this is potentially available for conversion to magnetic energy by the dynamo mechanism. At the same time, both processes lead to a decrease in the moment of inertia C of the Earth about its axis of spin, and so to an increase in its angular velocity (in so far as the influence of external torques can be neglected) which should in principle be detectable from study of minute changes in the 'length of day'. Unfortunately many competing mechanisms can cause small variations in the Earth's angular velocity (Munk & MacDonald, 1960), and the effect of gravitational sedimentation in the outer core has not yet been unambiguously separated from other effects; indeed it seems likely that it will be swamped by the more important *retarding* effect associated with tidal friction.

Even if the outer core were stably stratified, this does not of course mean that all radial motions are impossible; internal gravity waves (modified by Coriolis forces and possibly also by Lorentz forces) can propagate in a stably stratified medium, and such waves will in general be generated by perturbation forces either in the fluid interior or on its boundary. There is some evidence that the core–mantle boundary is not smooth, but is bumpy on a tangential scale of the order of hundreds of kilometres and on a radial scale of one or two kilometres (see § 4.6 below); if this is the case, then it is easy to visualise that disturbances to an otherwise smooth core flow could well originate at the core–mantle interface.

The electrical conductivity σ of the outer core was estimated by Bullard & Gellman (1954) to be $3\times10^5\,\Omega^{-1}\,m^{-1}$, an estimate acknowledged to be uncertain by a factor 3 either way; with $\mu_0=4\pi\times10^{-7}\,H\,m^{-1}$, this gives a value for the magnetic diffusivity $\lambda=(\mu_0\sigma)^{-1}=2\cdot6\,m^2\,s^{-1}$. The value of the conductivity in the lower mantle is estimated (e.g. Braginskii & Nikolaychik, 1973) to be $\sigma_m\approx2\times10^3\,\Omega^{-1}\,m^{-1}$, so that $\lambda_m=(\mu_0\sigma_m)^{-1}\approx2\times10^3\,m^2\,s^{-1}$, and

$$\sigma/\sigma_m=\lambda_m/\lambda\approx1\cdot5\times10^2. \tag{4.8}$$

Other estimates (e.g. Roberts, 1972*b*) place this ratio nearer to 10^3. In any event, the ratio is large enough for it to be reasonable in a first approximation to treat the mantle as an insulator ($\sigma_m\approx0$). Some phenomena however (e.g. the electromagnetic coupling between core and mantle – see e.g. Loper, 1975) depend crucially on the leakage of electric current from core to mantle, a process that is controlled by the weak mantle conductivity, which must naturally in such contexts be retained in the analysis.

The kinematic viscosity ν of the outer core has been described (Roberts & Soward, 1972) as the worst determined quantity in the whole of geophysics: it was placed in the range $10^{-7}<\nu\ll10^5\,m^2\,s^{-1}$ by Hide (1956), the upper limit being determined by the fact that seismic P-waves are known to traverse the core without appreciable damping. A figure near to the lower limit has been frequently adopted as being the most reasonable estimate; e.g. Loper (1975) takes $\nu=4\times10^{-6}\,m^2\,s^{-1}$, quoting again uncertainty by a factor of 3 either way. With this estimate, and the above estimate for λ, we have for the dimensionless ratio ν/λ (the *magnetic Prandtl number*) the estimate

$$\nu/\lambda\approx1\cdot5\times10^{-6}, \tag{4.9}$$

which may be compared with the value 1.5×10^{-7} for mercury under normal laboratory conditions. The small value of ν/λ suggests that the dominant contribution to energy dissipation in the outer core will be ohmic rather than viscous, and that (except possibly in thin shear layers at the solid boundaries of the outer core or in its interior) viscous effects will be negligible in the governing dynamical equations.

4.5. The need for a dynamo theory for the Earth

If induction effects in the core were non-existent or negligible, the free decay time for any current distribution predominantly confined to the core would (§ 2.7) be of order $t_d = R_C^2/\lambda\pi^2$, which, with $R_C = 3500$ km, $\lambda = 2{\cdot}6\ m^2\,s^{-1}$ gives $t_d \approx 4{\cdot}7 \times 10^{11}\ s \approx 1{\cdot}5 \times 10^4$ years. As mentioned in § 4.3, studies of rock magnetism indicate that the field of the Earth has existed at roughly its present intensity (except possibly during rapid reversals) on a geological time-scale of order 10^9 years. It therefore clearly cannot be regarded as a relic of a field trapped during accretion of the Earth from interplanetary matter; such a field could not have survived throughout the long history of the Earth in the absence of any regenerative mechanism.

On the other hand, the fact that the field exhibits time variation on scales extremely short compared with geological time-scales of order 10^6 years and greater (e.g. the secular variation of the non-dipole field with characteristic time of order 40 years) indicates rather strongly that such variations are not attributable to general evolutionary properties of the Earth on these geological time-scales, but rather to relatively rapid processes most probably associated directly with core fluid motions.

The rate of westward drift ($0{\cdot}18°$ per year) suggests velocities of order $u_c \approx 4 \times 10^{-4}\ \mathrm{m\,s^{-1}}$ near the core–mantle interface, of the core relative to the mantle (on the simplistic picture that magnetic perturbations are convected by the core fluid). A characteristic length-scale for magnetic perturbations associated with the secular variation is $l_c \approx 10^3$ km. A magnetic Reynolds number can be constructed on the basis of these figures:

$$R_m = u_c l_c/\lambda \approx 150. \qquad (4.10)$$

This is by no means infinite, but is perhaps large enough to justify the frozen-field assumption for magnetic perturbations, at least in a first approximation.

4.6. The core–mantle interface

A remarkable discovery was published by Hide & Malin (1970) to the effect that the pattern of the non-dipole magnetic potential Ψ_{nd}

over the surface of the Earth exhibits a strong correlation with the pattern of the potential Φ_g of gravitational fluctuations. Defining this correlation by

$$R(\varphi_0) = \langle \Psi_{nd}(R_E, \theta, \varphi - \varphi_0)\Phi_g(R_E, \theta, \varphi)\rangle / \langle \Psi_{nd}^2 \rangle^{1/2} \langle \Phi_g^2 \rangle^{1/2}, \quad (4.11)$$

where the angular brackets indicate averaging over the sphere $r = R_E$, Hide & Malin found, by analysis of data in which only surface harmonics up to $n = 4$ were considered significant, that for the year 1965, $R(\varphi_0)$ was maximal for $\varphi_0 = 160° = \hat{\varphi}_0$ say, and that then $R(\hat{\varphi}_0) = 0{\cdot}84$; here $\hat{\varphi}_0$ represents the shift to the east in the geomagnetic potential Ψ_{nd} required to maximise the correlation. Analysis of data for earlier decades indicated that $\hat{\varphi}_0$ is not constant in time, but rather increases at a rate of about $0{\cdot}27°$ per year, presumably a manifestation of the westward drift of the magnetic potential relative to the (fixed) gravitational potential. A degree of correlation as high as 0·84 is statistically unlikely unless the fluctuations in the two fields can be traced to some common influence; and it was suggested by Hide & Malin that this common influence might be attributed to undulations and irregularities ('bumps') on the core–mantle interface. Such bumps could well be a consequence of low-speed thermal convection in the lower mantle[2]. The density jump across this interface certainly implies that bumps will lead to gravity perturbations vertically above them; moreover any core flow over the bumps in the presence of a magnetic field will generate magnetic perturbations which may well be shifted in phase relative to the bumps by convective or wave propagation effects. The statistical significance of the Hide–Malin correlation has been questioned by Khan (1971) and Lowes (1971) and reasserted by Hide & Malin (1971); ultimate certainty in the matter will perhaps have to await independent evidence on the structure of the core–mantle interface. At this stage, we can merely say that it is at least highly plausible that bumps of the order of one or two kilometres in height (below the level of resolution of seismic waves) may be present on the interface, and that if so these will undoubtedly influence both fields in the manner described above; we shall consider a detailed model in § 10.8.

[2] Whether the associated gravitational stresses could be supported by the material of the lower mantle is debatable.

The possible presence of bumps on the interface is of crucial importance in the problem of calculating the torque exerted by the core on the mantle (Hide, 1969). It seems intuitively likely that bumps may greatly increase the torque over the value calculated on the basis of a smooth interface (Loper, 1975); however, a recent calculation by Anufriyev & Braginskii (1975) provides some indication that the interaction is in fact severely limited by the presence of any ambient horizontal magnetic field, at any rate when this is strong.

4.7. Precession of the Earth's angular velocity vector

The angular velocity of the Earth's mantle $\mathbf{\Omega}(t)$ is not quite steady, but is subject to slow changes and weak perturbations in both direction and magnitude (Munk & Macdonald, 1960). Chief among these in its relevance to core dynamics is the slow precession of $\mathbf{\Omega}$ about the normal to the plane of the Earth's rotation about the Sun caused by the net torque exerted by the Sun and the Moon on the Earth's equatorial bulge. The period of precession is known from astronomical observations to be 25 800 years, the vector $\mathbf{\Omega}$ describing a cone of semi-angle 23·5° over this period. The precessional angular velocity $\mathbf{\Omega}_p$ has magnitude $7{\cdot}71\times10^{-12}\ \mathrm{s}^{-1}$. Precession has been advocated by Malkus (1963, 1968) as providing the most plausible source of energy for core motions, a view that has however been contested by Rochester *et al.* (1975).

It is easy to see why precession must cause some departures from rigid body rotation in the liquid core region. Firstly, since the mean density of the core is substantially greater than the mean density of the mantle, the dynamic ellipticity $\varepsilon_c=(C_c-A_c)/C_c$ of the core (where A_c and C_c are its equatorial and polar moments of inertia respectively) is less than the corresponding quantity ε_m for the mantle; in fact $\varepsilon_c\approx\frac{3}{4}\varepsilon_m$. If the core and the mantle were dynamically uncoupled, they would then precess at different angular velocities proportional to ε_c and ε_m respectively, i.e. the precessional angular velocity of the core would be $\frac{3}{4}\mathbf{\Omega}_p$, so that after about 10^5 years the angular velocities of core and mantle, though equal in magnitude, would be quite different in direction. This would imply large relative velocities between core and mantle of order $\Omega R_C\approx200\ \mathrm{m\,s}^{-1}$, for which there is no evidence whatsoever. The inference is that core and mantle are *not* dynamically uncoupled – indeed there is no good reason to expect that they should be – but are quite strongly coupled through both viscous and electromagnetic transfer of angular momentum across the interface. This coupling must act in such a way as to equalise the precessional angular velocity of core and mantle, although in the absence of perfect coupling the mean angular velocity of the core may be expected to lag behind the angular velocity of the mantle in its precessional orbit. A difference in angular velocity requires a boundary layer at the core–mantle interface, a phenomenon studied in the purely viscous context by Stewartson & Roberts (1963), Toomre (1966) and Busse (1968) and in the hydromagnetic context by Rochester (1962), Roberts (1972*b*), and Loper (1975).

It would be inappropriate to enter at this stage into any of the detailed calculations undertaken by the above authors – this would require a disproportionate digression from the main theme of this monograph. The influence of precession is merely mentioned here as one element of the complicated dynamical background that will have to be fully understood before the dynamo problem in the terrestrial context can be regarded as solved. There is still wide disagreement about what the dominant source of energy for core motions may be, although convection due to buoyancy forces (of thermal or sedimentary origin as discussed in § 4.4) and precessional torques appear to be the principal candidates for serious consideration. The most recent detailed analysis of the problem of coupling between mantle and core (Loper, 1975) indicates that the rate of supply of energy through the mechanism of precessional coupling is a factor of order 10^{-3} smaller than the estimated rate of ohmic dissipation in the core; Loper also makes the point that a fraction of this energy supplied (necessarily 100% in a steady state model!) is dissipated in the boundary layers through which angular momentum is transferred from mantle to core. The concept of a precessionally driven dynamo rests however on the existence of a turbulent flow in the core arising from instabilities of these boundary layers and the secondary flows to which they give rise (Malkus, 1968), and it is by no means clear that Loper's arguments can be applied to this situation. There are as yet no detailed dynamo models that rely on precession as the main source of energy, and the question of whether such a dynamo is possible even in principle (setting aside the question of the energy budget in the terrestrial context) must await further developments in numerical and experimental modelling.

CHAPTER 5

THE SOLAR MAGNETIC FIELD

5.1. Introduction

As for the case of the Earth, it is now generally accepted among astrophysicists that the origin of the Sun's magnetic field, which is highly variable both in space and in time, must be sought in inductive motions; these are localised in its outer convective zone, which extends from the visible surface of the Sun ($r = R_\odot = 6{\cdot}96 \times 10^5$ km) down to about $r = 0{\cdot}8\ R_\odot$. The natural decay time for the fundamental dipole mode is of order 4×10^9 years (Wrubel, 1952), which is of the same order as the age of the solar system. If the solar magnetic field were steady on historic time-scales there would be no need to seek for a renewal mechanism; the field could simply be a 'fossil' relic of a field frozen into the solar gas during the initial process of condensation from the galactic medium. Even local time-dependent phenomena, such as the evolution of sunspots (see § 5.2 below), could be regarded as transient and localised events occurring in the presence of such a fossil field, and this was in fact the widely accepted view until the mid-1950s. The development of the solar magnetograph (Babcock & Babcock, 1955) permitting direct measurement of the weak general poloidal field of the Sun, and the discovery of reversals of this field (see § 5.4 below), first in the period 1957–8 and again in 1969–71 (in both cases either at or just after periods of maximum sunspot activity), have now led to the view that the 22-year sunspot cycle is in fact a particular manifestation of a roughly periodic evolution of the Sun's general field. Such a periodic behaviour cannot possibly be interpreted in terms of a fossil theory. Curiously, whereas dynamo theory was originally conceived to explain the persistence of cosmic magnetic fields over very long time-scales, in the case of the solar field it is now invoked to explain the extremely rapid variations (on the cosmic time-scale) that at present attract no other equally plausible explanation.

In this chapter we shall review some of the observed features of solar activity, in order to provide some background for the theoretical studies to be described, particularly those of chapters 7, 9 and 11. This review is necessarily highly selective; a comprehensive (though now somewhat dated) treatment of solar physics and of related observational problems and techniques may be found in Kuiper (1953); recent developments may be found in the Proceedings of the IAU Symposium 'Basic Mechanisms of Solar Activity' held at Prague in 1975 (Bumba & Kleczek, 1976).

5.2. Observed velocity fields

The Sun rotates about an axis inclined at an angle $7° 15'$ to the normal of the Earth's orbital plane, with a period of approximately 27 days. The rotation rate is however non-uniform, being greater (by about 4 days) at the equatorial plane than at the poles. The angular velocity $\omega(\theta)$ on the surface $r = R_\odot$ as a function of heliographic latitude θ may be represented by the formula

$$\omega(\theta) = (2{\cdot}78 + 0{\cdot}35 \cos^2\theta - 0{\cdot}44 \cos^4\theta) \times 10^{-6}\,\text{rad s}^{-1}, \qquad (5.1)$$

(Howard & Harvey, 1970). This result is based on measurements of the Doppler shift in spectral lines observed near the solar limbs at different latitudes. Small fluctuations $\omega'(\theta, t)$ superposed on (5.1) have also been detected.

The visible surface of the Sun is not uniformly bright, but exhibits a regular pattern of granulation, the scale of this pattern (i.e. the mean radius of the granules) being approximately 1000 km. The details of the pattern change on a time-scale of the order of a few minutes. Velocity fluctuations (again detected by means of Doppler shift measurements) of the order of 1 km s^{-1} are associated with the granulation, and are the surface manifestation of convective turbulence which penetrates to a depth of the order of $1{\cdot}5 \times 10^5$ km (Weiss, 1976; Gough & Weiss, 1976) below the solar surface. Velocity fluctuations exist on larger scales (e.g. velocities of order $0{\cdot}1$ km s^{-1} are associated with 'supergranular' and 'giant cell' structures on scales of order 10^4–10^5 km) and likewise velocity fluctuations may be detected on all scales down to the limit of resolution (of order 10^2 km). If we adopt the scales

$$l_g \sim 10^3\,\text{km}, \qquad u_g \sim 1\,\text{km s}^{-1} \qquad (5.2)$$

as being characteristic of the most energetic ingredients of the turbulent convection in the outermost layers of the Sun, then, with estimates of the kinematic viscosity and magnetic diffusivity given by

$$\nu \sim 10^{-8}\, km^2\, s^{-1}, \qquad \lambda \sim 10^{-1}\, km^2\, s^{-1}, \tag{5.3}$$

we may construct a Reynolds number R_e and magnetic Reynolds number R_m:

$$R_e = \frac{u_g l_g}{\nu} \sim 10^{11}, \qquad R_m = \frac{u_g l_g}{\lambda} \sim 10^4. \tag{5.4}$$

If magnetic effects were dynamically negligible, then the traditional Kolmogorov picture of turbulence (see chapter 11) would imply the existence of a continuous spectrum of velocity fluctuations on all scales down to the inner Kolmogorov scale $l_v \sim R_e^{-3/4} l_g \sim 1\ cm$, a factor 10^7 below the limit of resolution! Lack of an adequate theory for the effects of small-scale turbulence has generally led to a simple crude representation of the mean effects of the turbulence in terms of a turbulent (or 'eddy') viscosity ν_e and magnetic diffusivity λ_e; if these are supposed to incorporate all the effects of turbulence on scales smaller than the granular scale l_g, then they are given in order of magnitude by

$$\nu_e \sim \lambda_e \sim u_g l_g \sim 10^3\, km^2\, s^{-1}. \tag{5.5}$$

The precise mechanism by which turbulence can lead to a 'cascade' of both kinetic and magnetic energies towards smaller and smaller length-scales and ultimately to the very small scales on which viscous and ohmic dissipation take place presents a difficult problem, some aspects of which will be considered in chapter 11.

5.3. Sunspots and the solar cycle

Sunspots are dark spots (typically of order 10^4 km in diameter) that appear and disappear on the surface of the Sun mainly within ±35° of the equatorial plane $\theta = 90°$. The number of spots visible at any time varies from day to day and from year to year, the most striking feature being the current periodicity (with approximate period 11 years) in the annual mean sunspot number (see fig. 5.1)[1]. This figure also shows signs of a weak longer term periodicity with period of order 80 years.

Sunspots commonly occur in pairs roughly aligned along a line of latitude $\theta = \text{cst.}$, and they rotate with an angular velocity a little greater than that given by (5.1) (see e.g. Durney, 1976), the leading spot of a pair being slightly nearer the equatorial plane (in general) than the following spot. The typical distance between spots in a pair (or more complicated spot group) is of order 10^5 km.

Formation of sunspot pairs is essentially a purely magneto-hydrodynamic phenomenon, which may be understood in physical terms (Parker, 1955*a*) as follows. Suppose that a weak poloidal magnetic field of, say, dipole symmetry is present, and is maintained by some mechanism as yet unspecified. Any differential rotation that is present in the convective zone of the Sun (due to redistribution of angular momentum by thermal turbulence) will then tend to generate a toroidal field which is greater by a large factor ($O(R_m)$ if the mechanism of § 3.11 is of dominant importance) than the poloidal field. Consider then a tube of strong toroidal flux ($B_\varphi \sim 10^4$ gauss) immersed at some depth in the convective zone. If this tube is in dynamic equilibrium with its surroundings then the total pressure $p + (2\mu_0)^{-1}B_\varphi^2$ inside the tube must equal the fluid pressure p_0 outside; hence the fluid pressure is less inside the tube than outside, and so, if there is a simple monotonic relationship between pressure and density, the density inside must also be less than the density outside. The tube is then buoyant relative to its surroundings (like a toroidal bubble) and may be subject to instabilities if the decrease of density with height outside the flux tube is not too great. This phenomenon (described as 'magnetic buoyancy' by Parker, 1955*a*) will be considered in some detail in § 10.7. For the moment it is enough to say that instability can manifest itself in the form of large kinks in the toroidal tube of force, which may rise and break

[1] The current periodicity may not be a permanent feature of solar behaviour; early astronomical records, discussed in detail by Eddy (1976), indicate in particular that the 70-year period 1645–1715 was anomalous in that very few sunspots were recorded and no periodicity was apparent. There is also evidence (Eddy, Gilman & Trotter, 1976) that the angular velocity gradient $\omega'(\theta)$ in low latitudes was greater during this period than at present by a factor about 3!

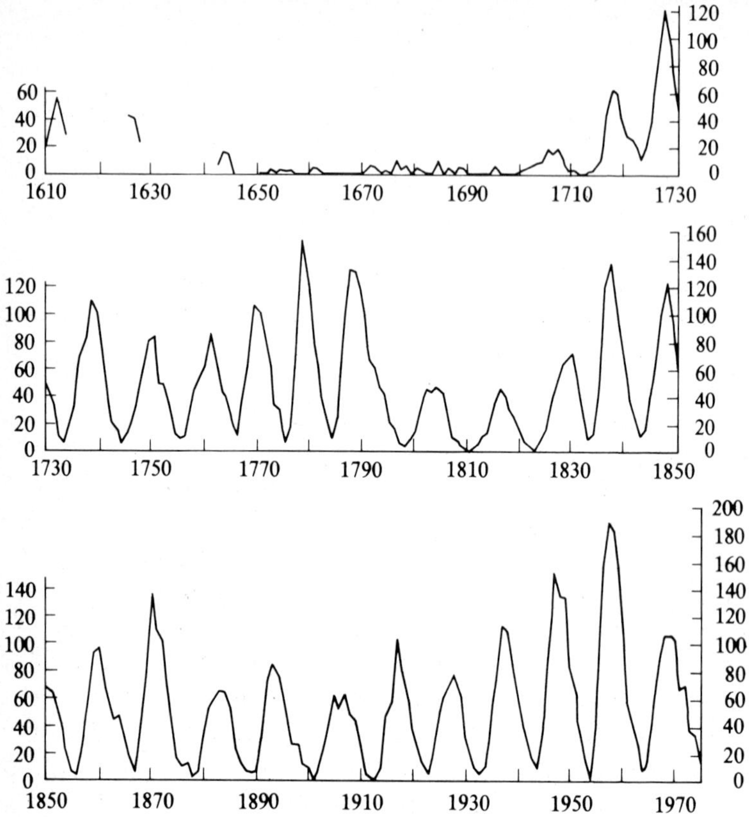

Fig. 5.1 Annual mean sunspot number, 1610–1975. During the last 250 years there has been a characteristic periodicity, with period approximately 11 years. The latter half of the seventeenth century was anomalous in that very few sunspots were recorded. Observations prior to 1650 were unsystematic and there are large gaps in the data. (Courtesy of John A. Eddy.)

through the solar surface (fig. 5.2). Such a rise is associated with a vertical stretching of the lines of force in the tube, as exemplified in the idealised solution of the induction equation discussed in § 3.4. Also, since fluid in the neighbourhood of the rising kink may be expected to expand on rising (the total vertical extent of the phenomenon being large compared with the scale-height) conservation of angular momentum will make this fluid rotate in a left-handed sense (in the northern hemisphere) as it rises, thus

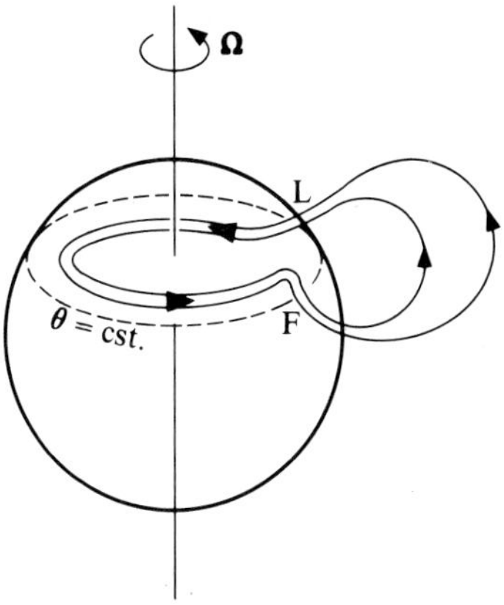

Fig. 5.2 Schematic representation of the formation of a sunspot pair by eruption of a subsurface toroidal tube of force. The vertical field is intensified at L and F where the leading and following spots are formed. Expansion of the erupting gas leads to a deficit of the local vertical component of angular momentum relative to the rotating Sun, and so to a twist of the sunspot pair LF relative to the surface line of latitude $\theta = \text{cst}$. (After Parker, 1955*a*.)

providing a natural explanation for the preferred tilt of a spot pair relative to the line of latitude as mentioned above.

The strong localised vertical fields thus created may be expected to suppress thermal turbulence and therefore to decrease dramatically the transport of heat to the solar surface – hence the darkening of a sunspot relative to its surroundings.

Magnetic fields were first detected in sunspots by Hale (1908); fields measured are typically of the order of 10^3 gauss in a sunspot, and may be as large as 4000 gauss. The polarity of the sunspot fields (i.e. whether the field direction is radially outwards or inwards) is almost invariably consistent with the physical picture described above, i.e. in any sunspot pair, the field polarity is positive in one sunspot and negative in the other sunspot of the pair; moreover all pairs in one hemisphere (with few exceptions) have the same polarity sense, indicating eruption from a subsurface toroidal field that is coherent over the hemisphere, and pairs of sunspots in opposite hemispheres generally have opposite senses, indicating that the toroidal field is generally antisymmetric about the equatorial plane.

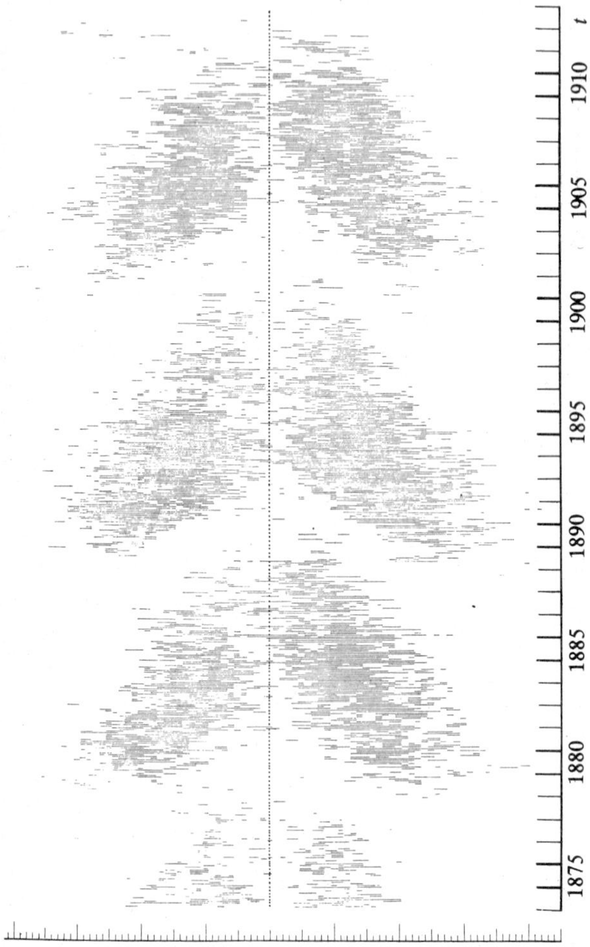

Fig. 5.3 The butterfly diagram showing the incidence of sunspots as a function of colatitude $\theta' = 90° - \theta$ and time t for the years 1874–1913; a vertical segment is included for each degree interval and for each 27-day solar rotation period if and only if one or more sunspots is observed in the interval during the rotation. The 11-year cycle and the migration of the 'active regions' towards the equatorial plane are clearly revealed. (Maunder, 1913.)

As mentioned earlier, sunspots appear only within about $\pm 35°$ of the equatorial plane $\theta = 90°$. In fact the distribution of spots in latitude varies periodically in time with the same period (~ 11 years) as that of the sunspot cycle. This behaviour is generally represented by means of the celebrated 'butterfly diagram' introduced by Maunder (1904, 1913). Maunder's (1913) diagram for the years 1874–1913 is reproduced in fig. 5.3; this diagram is constructed by inserting, for each rotation of the Sun, a set of vertical line segments corresponding to the bands of latitude in which sunspots are observed during the period of rotation. The 11-year periodicity is again clearly apparent on such a diagram; also a migration of the sunspot pattern towards the equatorial plane is evident, reflecting a migration of the underlying toroidal field, if the above qualitative description of sunspot formation is correct.

The sense of polarity of sunspot pairs in either hemisphere is observed to change from one 11-year cycle to the next. It may be inferred that the underlying toroidal magnetic field is periodic in time with period approximately $2 \times 11 = 22$ years. If this toroidal field is generated from a poloidal field by steady differential rotation, then the poloidal field must also be time-periodic with the same period. The observational evidence for this behaviour is discussed in the following section.

5.4. The general poloidal magnetic field of the sun

The solar magnetograph (Babcock & Babcock, 1953) has been used since 1953 to provide daily charts of the general magnetic field of the Sun. Since 1966, the magnetograph signals of the Mt. Wilson Observatory have been recorded on magnetic tape for subsequent analysis (Howard, 1974). Fig. 5.4 shows records of the line-of-sight component of magnetic field averaged over different bands of latitude, and smoothed over the 27-day solar rotation period. Although averaged to this extent, the mean field still exhibits variations in time that apparently have a random ingredient, with excursions however being limited to the range ± 2 gauss. If the signal is further averaged over a period of one year, certain definite trends become apparent. It is clear for example that the field within $30°$ of the South Pole, when thus averaged, was positive in 1968, but

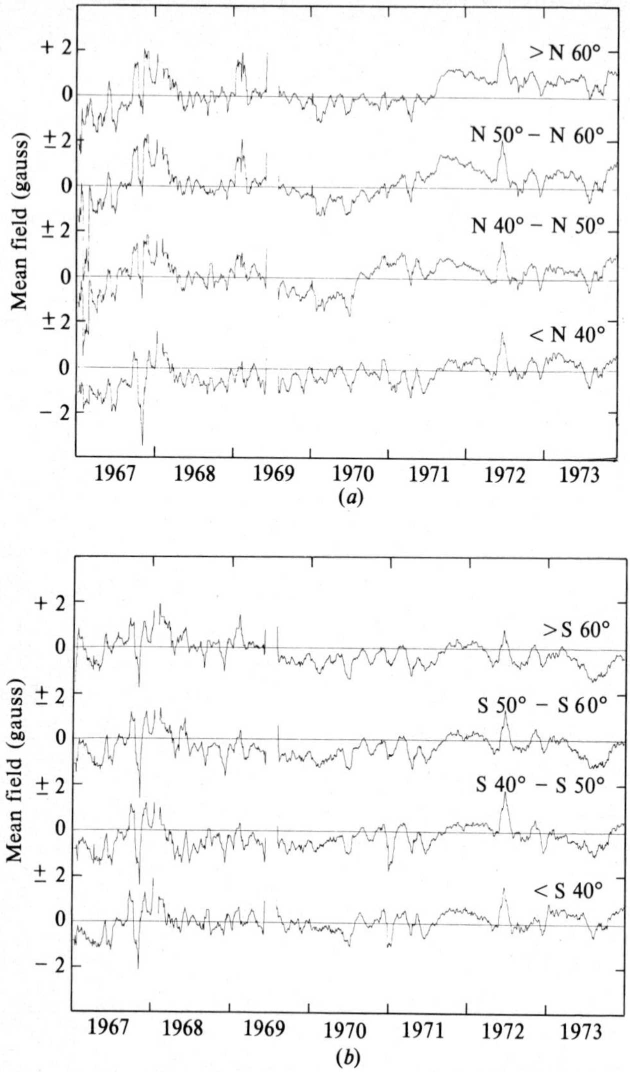

Fig. 5.4 Records (from Mt. Wilson Observatory) of line-of-sight component of the Sun's surface magnetic field, averaged over bands of latitude, and smoothed over the 27-day rotation period. (*a*) Northern hemisphere. (*b*) Southern hemisphere. (From Howard, 1974.)

negative in each of the years 1969–73. Similarly the north-polar field was negative in the mean from January 1970 to July 1971, but positive in the mean from August 1971 till the end of 1973. In this sense, it may be said that the south-polar field reversed direction around the beginning of 1969 and the north-polar field reversed direction around July 1971; this type of reversal activity occurred shortly after a period of maximum sunspot activity.

Similar reversals were recorded (Babcock, 1959) during the previous period of maximum sunspot activity 1957–9: the south-polar field reversed (from negative to positive) early in 1957, and the north-polar field reversed (from positive to negative) around November 1958.

These observations suggest that, although there are clearly strong random effects at work in the evolution of the solar magnetic field, there is nevertheless a significant coupling between the poloidal and toroidal ingredients of the field, and that these both exhibit (in a suitably averaged sense) time-periodic behaviour with period of order 22 years. At any rate, from the observations there is certainly sufficient motivation in studying closely any dynamo models which involve coupling between toroidal and poloidal field ingredients and which are capable of predicting such time-periodic behaviour.

The coupling between mean poloidal field evolution and sunspot activity is rather strikingly revealed in fig. 5.5 (from Stix, 1976) which shows curves of constant radial field $B_r(\theta, t) = \text{cst.}$ superposed on the butterfly diagram for the period 1954–75. The radial field was obtained from spherical harmonic coefficients determined by Altschuler *et al.* (1974) from the magnetograph recordings of the Mt. Wilson Observatory. During the half-cycle 1964–75, the polarity of the leading spots in the northern hemisphere was negative indicating that B_φ was positive in the northern hemisphere (and similarly negative in the southern hemisphere). During this same period, B_r was negative in the northern hemisphere (in the sunspot zone) and positive in the southern hemisphere. The fact that B_r and B_φ are apparently *out of phase* has important implications for possible dynamo mechanisms, as pointed out by Stix (1976). These implications will be discussed later in the context of particular dynamo models (see § 9.12).

Let us now consider briefly the detailed spatial structure of the radial component of magnetic field as observed over the solar disc. Increasing refinement in spectroscopic detection techniques can

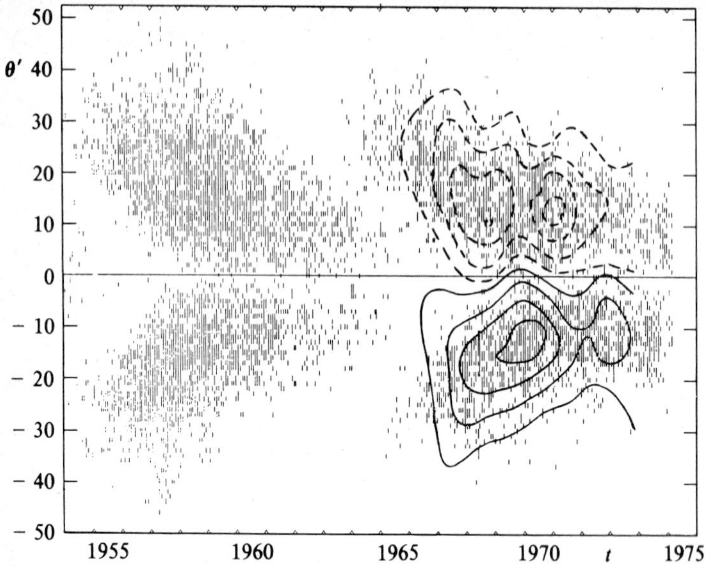

Fig. 5.5 Butterfly diagram for 1954 to 1975 (Mt. Wilson Observatory) and contours of constant radial field component. The levels of the curves are approximately ±0·17, ±0·50, ±0·83 and ±1·16 gauss, positive for the solid and negative for the dashed curves. (From Stix, 1976.)

now reveal the fine structure of this field down to scales of the order of 1000 km and less. The remarkable fact is that, although the spatially averaged radial field is of the order of 1 or 2 gauss (see fig. 5.4), it is by no means uniformly spread over the solar surface, but appears to be concentrated in 'flux elements' with diameters of order 200 km and less, in which field strengths are typically of the order of 1000–2000 gauss (Stenflo, 1976). Such flux elements naturally occupy a very small fraction (~0·1%) of the solar surface; the number of elements required to give the total observed flux is of order 10^5. The process of flux concentration can be understood in terms of the 'flux expulsion' mechanism discussed in § 3.9; there are however difficulties in accepting this picture, in that flux expulsion as described in § 3.9 occurs effectively when the flow pattern is steady over a time-scale large compared with the turn-over time of the constituent eddies, whereas observation of the granulation pattern suggests unsteadiness on a time-scale t_g somewhat *less* than the turn-over time given by $l_g/u_g \sim 10^3$ s. The problem of the fine-scale structure of the solar field presents a challenging problem that is by no means as yet fully resolved.

Our main concern however in subsequent chapters will be with the evolution of the *mean* magnetic field (the mean being defined either spatially over scales large compared with l_g or temporally over scales large compared with t_g). The difficulties associated with the fine-scale structure are to some extent concealed in this type of treatment; nevertheless the treatment is justified in that the first requirement in the solar (as in the terrestrial) context is to provide a theoretical framework for the treatment of the *gross* properties of the observed field; the treatment of the fine-structure must at this stage be regarded to some extent as a secondary problem, although in fact the two aspects are inextricably linked, and understanding of the dynamo mechanism will be complete only when the detailed fine-structure processes and their cumulative effects are fully understood.

5.5. Magnetic stars

The magnetic field of the Sun is detectable only by virtue of its exceptional proximity to the Earth as compared with other stars. Magnetic fields of distant stars are detectable only if strong enough to provide significant Zeeman splitting of their spectral lines; and this phenomena can be observed only for stars whose spectral lines are sharply defined and not smeared out by other effects such as rapid rotation. The dramatic discovery (Babcock, 1947) of a magnetic field of the order of 500 gauss in the star 78 Virginis heralded a new era in the subject of cosmical electrodynamics. This star is one of the class of peculiar A-type (Ap) stars, which exhibit, via their spectroscopic properties, unusual chemical composition relative to the Sun. It is now known (see e.g. Preston, 1967) that 78 Virginis is typical of Ap stars whose spectral lines are not too broadened by rotation to make Zeeman splitting detection impracticable: virtually all such stars exhibit magnetic fields in the range 100–30000 gauss; these fields are variable in time, and a proportion of these 'magnetic stars' show periodic behaviour with period typically of the order of several days.

The fields of these stars have an order of magnitude that is consistent with the simple and appealing conjecture that the stellar field is formed by compression of the general galactic field (which is of order 10^{-5}–10^{-6} gauss) during the process of gravitational condensation of the star from the interstellar medium. For a sphere of gas of radius R, with mass M and trapped flux F, this gives simple relations for the mean density ρ and mean surface field strength B:

$$Br^2 \sim F, \qquad \rho r^3 \sim M, \qquad B \sim (F/M^{2/3})\rho^{2/3}, \tag{5.6}$$

and hence

$$B_s/B_g = (\rho_s/\rho_g)^{2/3}, \tag{5.7}$$

where the suffixes s and g refer to the stellar and galactic fields respectively. A compression ratio of order 10^{15} would thus be sufficient to explain the order of magnitude of the observed fields ($B_s/B_g \sim 10^{10}$). In fact ρ_s/ρ_g is considerably greater than 10^{15}, and the problem is rather to explain the inferred loss of magnetic flux during the process of gravitational condensation and subsequent early stage of stellar evolution (see e.g. Mestel, 1967).

As in the case of the Sun, the natural (ohmic) decay time for the field of a magnetic star is of the same order as (or even greater than) the life-time of the star, and a possible view is that the field is simply a 'fossil' relic of the field created during the initial condensation of the star. If the dipole moment of the star is inclined to the rotation axis, the observed periodicities can be explained in a natural way in terms of a rotating dipole (the 'oblique rotator' model). Some of the observations (e.g. the irregular field variations observed in at least some *Ap* stars) cannot however be explained in terms of the oblique rotator model, and it seems likely that dynamo processes, similar to those occurring in the Sun, may play an important part in determining field evolution in most cases. As pointed out by Preston (1967), a dynamo theory and an oblique rotator theory are not necessarily mutually exclusive possibilities; it is more likely that both types of process occur and interact in controlling the evolution of the observed fields.

The simple relationship (5.7) suggests that, as massive stars evolve into yet more compressed states, in so far as magnetic flux remains trapped during the later stages of evolution, so the associated surface fields should be intensified. Very strong magnetic fields (of order 10^7–10^8 gauss) have in fact been detected in white dwarfs (see e.g. Landstreet & Angel, 1974; Angel, 1975). Fields of the order of 10^{12} gauss and greater are to be expected under the extreme condensation conditions of neutron stars (see e.g. Woltjer, 1975). In these situations however it seems unlikely that hydromagnetic effects remain of any importance; the fields observed can most reasonably be interpreted as fossil relics of fields compressed from earlier eras of stellar evolution.

CHAPTER 6

LAMINAR DYNAMO THEORY

6.1. Formal statement of the kinematic dynamo problem

As in previous chapters, V will denote a bounded region in $\mathbf{R}^3$, S its surface, and $\hat{V}$ the exterior region extending to infinity. Conducting fluid of uniform magnetic diffusivity $\lambda\,(0<\lambda<\infty)$ is confined to V, and the medium in $\hat{V}$ will be supposed insulating, so that the electric current distribution $\mathbf{J}(\mathbf{x}, t)$ is also confined to V. Let $\mathbf{u}(\mathbf{x}, t)$ be the fluid velocity, satisfying

$$\mathbf{u} \,.\, \mathbf{n} = 0 \quad \text{on } S. \tag{6.1}$$

Steady velocity fields $\mathbf{u}(\mathbf{x})$ are of particular interest; we shall also be concerned in subsequent chapters with turbulent velocity fields having statistical properties that are time independent. For the most part we shall limit attention to incompressible flows for which $\nabla \,.\, \mathbf{u} = 0$; we shall also assume, unless explicitly stated otherwise, that $\mathbf{u}$ is differentiable in V, and that the total kinetic energy $E(t)$ of the motion is permanently bounded:

$$E(t) = \tfrac{1}{2}\int_V \rho \mathbf{u}^2 \, \mathrm{d}V \leqslant E_0. \tag{6.2}$$

Concerning the magnetic field $\mathbf{B}(\mathbf{x}, t)$, we assume that this is produced entirely by the current distribution $\mathbf{J}$, which is without artificial singularities; then $\mathbf{B}$ is also without singularities, and it satisfies the outer condition

$$\mathbf{B} = O(r^{-3}) \ \text{ as } \ r = |\mathbf{x}| \to \infty. \tag{6.3}$$

The condition $\nabla \,.\, \mathbf{B} = 0$ is of course always satisfied.

The field $\mathbf{B}$ then evolves according to the equations

$$\begin{aligned} \partial \mathbf{B}/\partial t &= \nabla \wedge (\mathbf{u} \wedge \mathbf{B}) + \lambda \nabla^2 \mathbf{B} \quad && \text{in } V, \\ \nabla \wedge \mathbf{B} &= 0 && \text{in } \hat{V}, \\ [\mathbf{B}] &= 0 && \text{across } S, \end{aligned} \tag{6.4}$$

and subject to an initial condition, compatible with (6.4), of the form

$$\mathbf{B}(\mathbf{x}, 0) = \mathbf{B}_0(\mathbf{x}). \tag{6.5}$$

The total magnetic energy $M(t)$ is given by

$$M(t) = (2\mu_0)^{-1} \int_{V_\infty} \mathbf{B}^2 \, \mathrm{d}V, \tag{6.6}$$

where V_∞ as usual represents the whole space. Under the assumed conditions, $M(t)$ is certainly finite, and we suppose that $M(0) = M_0 > 0$. We know that if $\mathbf{u} \equiv 0$, then $M(t) \to 0$ as $t \to \infty$, the time-scale for this natural process of ohmic decay (§ 2.7) being $t_d = L^2/\lambda$ where L is a scale characterising V. A natural definition of dynamo action is then the following: for given V and λ, the velocity field $\mathbf{u}(\mathbf{x}, t)$ *acts as a dynamo* if $M(t) \nrightarrow 0$ as $t \to \infty$, i.e. if it successfully counteracts the erosive action of ohmic dissipation. Under these circumstances, $M(t)$ may tend to a constant (non-zero) value, or may fluctuate about such a value either regularly or irregularly, or it may increase without limit[1].

A given velocity field $\mathbf{u}(\mathbf{x}, t)$ may act as a dynamo for some, but not all, initial field structures $\mathbf{B}_0(\mathbf{x})$ and for some, but not all, values of the parameter λ. The field $\mathbf{u}(\mathbf{x}, t)$ may be described as *capable of dynamo action* if there exists an initial field structure $\mathbf{B}_0(\mathbf{x})$ and a finite value of λ for which, under the evolution defined by (6.3)–(6.5), $M(t) \nrightarrow 0$ as $t \to \infty$. Under this definition, a velocity field is either capable of dynamo action or it is not, and a primary aim of dynamo theory must be to develop criteria by which a given velocity field may be 'tested' in this respect.

6.2. Rate of strain criterion

Since magnetic field intensification is associated with stretching of magnetic lines of force, it is physically clear that a necessary condition for successful dynamo action must be that in some sense (for given λ) the rate of strain associated with $\mathbf{u}(\mathbf{x}, t)$ must be sufficiently intense. The precise condition ((6.14) below) was

[1] Such 'unphysical' behaviour would imply a growing importance of the Lorentz force $\mathbf{J} \wedge \mathbf{B}$ whose effect on $\mathbf{u}(\mathbf{x}, t)$ would ultimately have to be taken into account.

obtained by Backus (1958). Suppose for simplicity that V is the sphere $r<R$, and that $\mathbf{u}$ is steady and solenoidal and zero on $r=R$. The rate of change of magnetic energy is given by

$$\frac{\mathrm{d}M}{\mathrm{d}t}=-\frac{1}{\mu_0}\int_{V_\infty}\mathbf{B}\,.\,(\nabla\wedge\mathbf{E})\,\mathrm{d}V=-\frac{1}{\mu_0}\int_{V_\infty}\mathbf{E}\,.\,(\nabla\wedge\mathbf{B})\,\mathrm{d}V,\quad(6.7)$$

since there is zero flux of the Poynting vector $\mathbf{E}\wedge\mathbf{B}$ out of the sphere at infinity. Hence since $\nabla\wedge\mathbf{B}=0$ for $r>R$,

$$\mathrm{d}M/\mathrm{d}t=\mathscr{P}-\mathscr{J},\quad(6.8)$$

where

$$\mu_0\mathscr{P}=\int_V(\mathbf{u}\wedge\mathbf{B})\,.\,(\nabla\wedge\mathbf{B})\,\mathrm{d}V=\int_V\mathbf{B}\,.\,\nabla\wedge(\mathbf{u}\wedge\mathbf{B})\,\mathrm{d}V$$
$$=\int_V\mathbf{B}\,.\,(\mathbf{B}\,.\,\nabla)\mathbf{u}\,\mathrm{d}V,\quad(6.9)$$

and

$$\mu_0\mathscr{J}=\lambda\int_V(\nabla\wedge\mathbf{B})^2\,\mathrm{d}V=\mu_0\int_V\sigma^{-1}\mathbf{J}^2\,\mathrm{d}V.\quad(6.10)$$

Here, $\mathscr{P}$ represents the rate of production of magnetic energy by the velocity field $\mathbf{u}$, and $\mathscr{J}$ represents the rate of ohmic dissipation.

Bounds may be put on both these integrals as follows. First, let $e_{ij}=\frac{1}{2}(u_{i,j}+u_{j,i})$ be the rate of strain tensor[2] with $e_{ii}=\nabla\,.\,\mathbf{u}=0$. Then from (6.9),

$$\mu_0\mathscr{P}=\int_V B_iB_je_{ij}\,\mathrm{d}V\leqslant e_m\int_V\mathbf{B}^2\,\mathrm{d}V\leqslant 2e_m\mu_0M,\quad(6.11)$$

where $e_m=\max_{\mathbf{x}\in V}(|e_{11}|,|e_{22}|,|e_{33}|)$. Secondly, by standard methods of the calculus of variations, it may easily be seen that the quotient $\mathscr{J}/M$ is minimised when $\mathbf{B}$ has the the simplest free decay mode structure discussed in § 2.7, and consequently (from (2.136), and using $x_{01}=\pi$) that

$$\mathscr{J}\geqslant 2\pi^2(\lambda/R^2)M.\quad(6.12)$$

Hence from (6.8), (6.11) and (6.12),

$$\mathrm{d}M/\mathrm{d}t\leqslant 2(e_m-\pi^2\lambda/R^2)M,\quad(6.13)$$

[2] A suffix after a comma will denote space differentiation, e.g. $u_{i,j}=\partial u_i/\partial x_j$.

and so M certainly decays to zero (and the motion fails as a dynamo) if $e_m < \lambda\pi^2/R^2$. Conversely a necessary (though by no means sufficient) condition for dynamo action is that

$$e_m R^2/\lambda \geqslant \pi^2. \tag{6.14}$$

Equality is possible in (6.14) only if e_{ij} is everywhere uniform, $\mathbf{B}$ is everywhere aligned with the direction of maximum rate of strain and the structure of $\mathbf{B}$ is that of a free decay mode. In general these conditions cannot be simultaneously satisfied, and it is likely that the condition (6.14) grossly underestimates the order of magnitude of the rate-of-strain intensity needed for successful dynamo action.

An alternative, though weaker, form of the criterion (6.14) results from the fact that if $u_m = \max_{\mathbf{x},\mathbf{x}' \in V} |\mathbf{u}(\mathbf{x}) - \mathbf{u}(\mathbf{x}')|$, then $e_m \leqslant u_m/R$, so that (6.14) implies

$$R_m = u_m R/\lambda \geqslant \pi^2. \tag{6.15}$$

The magnetic Reynolds number based on maximum relative velocity must therefore be sufficiently large if dynamo action is to be possible.

6.3. Rate of change of dipole moment

We have seen in § 2.7 that the dipole moment of the current distribution in V may be expressed in the form

$$\boldsymbol{\mu}^{(1)}(t) = \frac{3}{8\pi}\int_V \mathbf{B}(\mathbf{x}, t)\,\mathrm{d}V. \tag{6.16}$$

Its rate of change is therefore given by

$$\frac{8\pi}{3}\frac{\mathrm{d}\boldsymbol{\mu}^{(1)}}{\mathrm{d}t} = -\int_V \nabla\wedge\mathbf{E}\,\mathrm{d}V = -\int_S \mathbf{n}\wedge\mathbf{E}\,\mathrm{d}S. \tag{6.17}$$

With $\mathbf{E} = -\mathbf{u}\wedge\mathbf{B} + \lambda\nabla\wedge\mathbf{B}$, and $\mathbf{u}\,.\,\mathbf{n} = 0$ on S, this becomes

$$\frac{8\pi}{3}\frac{\mathrm{d}\boldsymbol{\mu}^{(1)}}{\mathrm{d}t} = \int_S \mathbf{u}(\mathbf{n}\,.\,\mathbf{B})\,\mathrm{d}S - \lambda\int_S \mathbf{n}\wedge(\nabla\wedge\mathbf{B})\,\mathrm{d}S. \tag{6.18}$$

In the perfectly conducting limit ($\lambda = 0$), the dipole moment $|\boldsymbol{\mu}^{(1)}|$ can therefore be increased by any motion $\mathbf{u}$ having the property that $(\mathbf{u}\,.\,\boldsymbol{\mu}^{(1)})(\mathbf{n}\,.\,\mathbf{B}) \geqslant 0$ at all points of S; i.e. the motion on S must be such as to sweep the flux lines towards the polar regions defined by the direction of $\boldsymbol{\mu}^{(1)}$ (fig. 6.1). As pointed out by Bondi & Gold

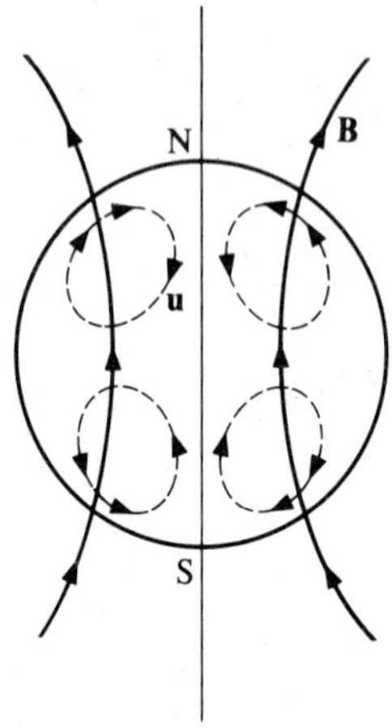

Fig. 6.1 The motion illustrated tends to increase the dipole moment due to the sweeping of surface field towards the North and South poles (Bondi & Gold, 1950).

(1950), possible increase of $|\boldsymbol{\mu}^{(1)}|$ by this mechanism is limited, and $|\boldsymbol{\mu}^{(1)}|$ in fact reaches a maximum finite value when the flux lines crossing S are entirely concentrated at the poles[3].

A small but non-zero diffusivity λ may totally transform the situation since, as shown in § 2.7, diffusion even on its own may temporarily increase $|\boldsymbol{\mu}^{(1)}|$, if $(\mathbf{n} \,.\, \nabla)\mathbf{B}$ is appropriately distributed over S. A *sustained* increase in $|\boldsymbol{\mu}^{(1)}|$ may be envisaged if the role of the velocity $\mathbf{u}$ is to maintain a distribution of $(\mathbf{n} \,.\, \nabla)\mathbf{B}$ over S that implies diffusive increase of $|\boldsymbol{\mu}^{(1)}|$. Thus, paradoxically, diffusion must play the primary role in increasing $|\boldsymbol{\mu}^{(1)}|$ (if this increase is to be sustained), and induction, though impotent in this respect in isola-

[3] Bondi & Gold argued that this type of limitation in the growth of $|\boldsymbol{\mu}^{(1)}|$ when $\lambda = 0$ need not apply if V has toroidal (rather than spherical) topology, and they quoted the homopolar disc dynamo as an example of a dynamo of toroidal topology which will function when $\lambda = 0$. As pointed out in § 1.1, however, even this simple system requires $\sigma < \infty$ (i.e. $\lambda > 0$) in the disc if the flux $\Phi(t)$ across it is to change with time; and the same *flux limitation* will certainly apply for general toroidal systems.

tion, plays a crucial subsidiary role in making (sustained) diffusive increase of $|\boldsymbol{\mu}^{(1)}|$ a possibility.

6.4. The impossibility of axisymmetric dynamo action

When both **u** and **B** (and the associated vectors **A**, **E**, **J**) have a common axis of symmetry Oz, we have seen (§ 3.6) that the toroidal component of Ohm's law becomes

$$\frac{\partial \mathbf{A}_T}{\partial t} = \mathbf{u}_P \wedge \mathbf{B}_P - \lambda \nabla \wedge \mathbf{B}_P, \tag{6.19}$$

the corresponding equation for the flux function being

$$\frac{\partial \chi}{\partial t} + \mathbf{u}_P \,.\, \nabla \chi = \lambda\, \mathrm{D}^2 \chi. \tag{6.20}$$

The absence of any 'source' term in this scalar equation is an indication that a steady state is not possible, a result established by Cowling (1934).

The following proof is a slight modification of that given by Braginskii (1964*a*). We shall allow λ to be an axisymmetric function of position satisfying

$$\mathbf{u}_P \,.\, \nabla \lambda = 0, \tag{6.21}$$

i.e. λ is constant on streamlines of the $\mathbf{u}_P$-field. (Note that (6.19) and (6.20) remain valid when λ is non-uniform.) It is natural to impose the condition (6.21), since variations in conductivity (and so in λ) tend to be convected with the fluid, so that $\lambda(\mathbf{x})$ can be steady only if (6.21) is satisfied. The condition (6.21) covers the situation when λ is constant in V and is an arbitrary axisymmetric function of position in $\hat{V}$, and in particular includes the limiting case when the medium in $\hat{V}$ is insulating.

We now multiply (6.20) by $\lambda^{-1}\chi$ and integrate throughout all space; since $\lambda^{-1}\chi \mathbf{u}_P \,.\, \nabla \chi = \frac{1}{2}\nabla \,.\, (\mathbf{u}_P \lambda^{-1} \chi^2)$, using (6.21) and $\nabla \,.\, \mathbf{u}_P = 0$, we obtain

$$\frac{\mathrm{d}}{\mathrm{d}t}\int_{V_\infty} \frac{1}{2\lambda}\chi^3\,\mathrm{d}V + \int_{S_\infty} \frac{1}{2\lambda}\chi^2 \mathbf{u}_p \,.\, \mathbf{n}\,\mathrm{d}S = \int_{V_\infty} \chi\,\mathrm{D}^2\chi\,\mathrm{d}V. \tag{6.22}$$

For a field that is at most dipole at infinity, $\chi = O(r^{-1})$ as $r \to \infty$, and the surface integral vanishes provided $\lambda^{-1}|\mathbf{u}_P| \to 0$ as $r \to \infty$, a condition that is of course trivially satisfied when $\mathbf{u}_P \equiv 0$ outside a finite volume V. Moreover, using (3.47) we have

$$\int_{V_\infty} \chi\, \mathrm{D}^2\chi\, \mathrm{d}V = \int_{V_\infty} \chi\nabla\,.\,\mathbf{f}\, \mathrm{d}V = -\int_{V_\infty} \mathbf{f}\,.\,\nabla\chi\, \mathrm{d}V$$
$$= -\int_{V_\infty} (\nabla\chi)^2\, \mathrm{d}V, \qquad (6.23)$$

where we have used $\chi = O(r^{-1})$, $|\mathbf{f}| = O(r^{-2})$ to discard integrals over the surface at infinity; the final step also requires use of the identity $\chi^2\nabla\,.\,(\mathbf{i}_s/s) = 0$. Hence (6.22) becomes

$$\frac{\mathrm{d}}{\mathrm{d}t}\int_{V_\infty} \frac{1}{2\lambda}\chi^2\, \mathrm{d}V = -\int_{V_\infty} (\nabla\chi)^2\, \mathrm{d}V = -\int_{V_\infty} s^2\mathbf{B}_P^2\, \mathrm{d}V. \qquad (6.24)$$

It is clear from this equation that a steady state with $\chi \not\equiv 0$ is not possible and that ultimately

$$\int s^2\mathbf{B}_P^2\, \mathrm{d}V \to 0. \qquad (6.25)$$

Excluding the unphysical possibility that singularities in $\mathbf{B}_P$ develop as $t \to \infty$, we are driven to the conclusion that $\mathbf{B}_P \to 0$ everywhere.

Meridional circulation $\mathbf{u}_P$ cannot therefore prevent the decay of an axisymmetric poloidal field[4]. As is clear from the structure of (6.20), the role of $\mathbf{u}_P$ is to redistribute the poloidal flux but it cannot regenerate it. If the meridional circulation is strong, it seems likely that the process of redistribution will in general lead to greatly accelerated decay for the following reason. We have seen in § 3.10 that when $R_m \gg 1$ (R_m being based on the scale a and intensity u_0 of the meridional circulation), poloidal flux tends to be excluded from regions of closed streamlines of the $\mathbf{u}_P$-field. In a time of order $R_m^{1/2}t_0$, where $t_0 = a/u_0$ is the 'turnover time' of meridional eddies, the scale of the field $\mathbf{B}_P$ is reduced by distortion from $O(a)$ to $O(\delta)$ where $\delta = R_m^{-1/2}a$; the subsequent characteristic decay time is of order $\delta^2/\lambda = O(R_m^{-1})a^2/\lambda = O(t_0)$. The total time-scale for the

[4] It must be emphasised that the result as proved here relates to a situation in which both **B** and **u** are axisymmetric with the same axis of symmetry. Steady maintenance of a non-axisymmetric **B**-field by an axisymmetric **u**-field is not excluded; indeed an example of such a dynamo will be considered in § 6.10.

process of winding and decay is then of order $R_m^{1/2}t_0 = R_m^{-1/2}a^2/\lambda$, and this is much less than the natural decay time a^2/λ (when $R_m \gg 1$).

A modest amount of meridional circulation ($R_m = O(1)$ or less) may on the other hand lead to a modest delaying action in the decay process (Backus, 1957).

Ultimate decay of the toroidal field

Under axisymmetric conditions, the toroidal component of the induction equation is

$$\frac{\partial B}{\partial t} + (\mathbf{u}_P \,.\, \nabla)B = s(\mathbf{B}_P \,.\, \nabla)\omega - \nabla \wedge \lambda(\nabla \wedge B\mathbf{i}_\varphi), \tag{6.26}$$

the notation being as in § 3.6. We have seen that while $\mathbf{B}_P \neq 0$, the source term $s(\mathbf{B}_P \,.\, \nabla)\omega$ on the right of (6.26) can lead to the generation of a strong toroidal component $B\mathbf{i}_\varphi$. Ultimately however, $\mathbf{B}_P$ tends uniformly to zero, and from (6.26) we then deduce that

$$\frac{\mathrm{d}}{\mathrm{d}t}\int \tfrac{1}{2}B^2 \,\mathrm{d}V = -\int \lambda(\nabla \wedge B\mathbf{i}_\varphi)^2 \,\mathrm{d}V, \tag{6.27}$$

and, by arguments similar to those used above, that

$$\mathbf{B}_T = B\mathbf{i}_\varphi \to 0 \quad \text{everywhere.} \tag{6.28}$$

6.5. Cowling's neutral point argument

The proof of the impossibility of axisymmetric dynamo action given in the previous section rests on the use of global properties of the field, as determined by the integrals appearing in (6.22). It is illuminating to supplement this type of proof with a purely local argument, as devised by Cowling (1934). The flux-function χ is zero at infinity and is zero (by symmetry) on the axis Oz. If $\mathbf{B}_P$ (and so χ) is not identically zero, there must exist at least one point, N say, in the (s, z) plane where χ is maximal or minimal; at N, $\mathbf{B}_P$ vanishes and the $\mathbf{B}_P$-lines are closed in the neighbourhood of N, i.e. N is an O-type neutral point of the field $\mathbf{B}_P$.

Let C_ε be a closed $\mathbf{B}_P$-line near N of small length ε, and let S_ε be the surface in the (s, z) plane spanning C_ε (fig. 6.2). Suppose that the

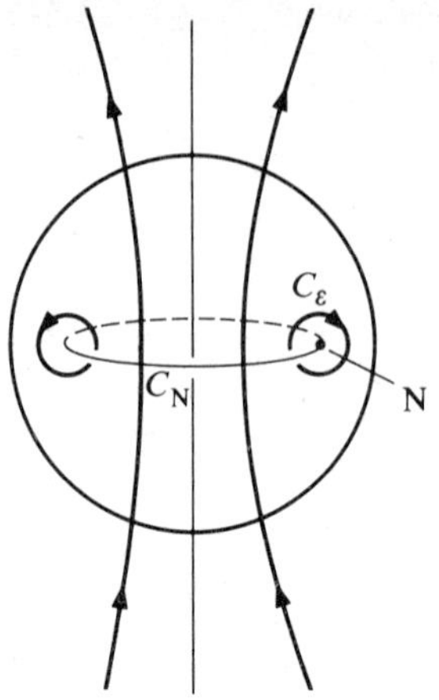

Fig. 6.2 In the neighbourhood of the neutral point N of the $\mathbf{B}_P$-field, induction must fail to maintain the field against ohmic decay.

field $\mathbf{B}_P$ is steady. Then from (6.19), with $\partial \mathbf{A}_T/\partial t = 0$,

$$\int_{S_\varepsilon} (\mathbf{u}_P \wedge \mathbf{B}_P) \,.\, d\mathbf{S} = \oint_{C_\varepsilon} \lambda \mathbf{B}_P \,.\, d\mathbf{x}. \tag{6.29}$$

Let B_ε be the average value of $|\mathbf{B}_P|$ on C_ε; the right-hand side of (6.29) is then $\lambda_N \varepsilon B_\varepsilon$ to leading order, where λ_N is the value of λ at N. Moreover, since $B_\varepsilon \to 0$ as $\varepsilon \to 0$, the mean value of $|\mathbf{B}_P|$ over S_ε is evidently less than B_ε when ε is sufficiently small; hence

$$\int_{S_\varepsilon} (\mathbf{u}_P \wedge \mathbf{B}_P) \,.\, d\mathbf{S} \leqslant U B_\varepsilon S_\varepsilon, \tag{6.30}$$

where U is the maximum value of $|\mathbf{u}|$ in V. Hence

$$\lambda_N \varepsilon \leqslant U S_\varepsilon. \tag{6.31}$$

But, as $\varepsilon \to 0$, $S_\varepsilon = O(\varepsilon^2)$ and this is clearly incompatible with (6.31) for any finite values of U/λ_N. The implication is that the induction effect, represented by $\mathbf{u}_P \wedge \mathbf{B}_P$ in (6.19), cannot compensate for the diffusive action of the term $-\lambda \nabla \wedge \mathbf{B}_P$ in the neighbourhood of the neutral point. Of course if there is more than one such O-type neutral point, then the same behaviour occurs in the neighbourhood of each.

The circle C_N obtained by rotating the point N about the axis of symmetry is a closed **B**-line (on which $\mathbf{B}_P = 0$, $\mathbf{B}_T \neq 0$) and the failure of dynamo action in the neighbourhood of N may equally be interpreted in terms of the inability of the induced electromotive force $\mathbf{u} \wedge \mathbf{B}$ to drive current along such a closed **B**-line. In general, if we integrate Ohm's law (2.117) along a closed **B**-line on the assumption that **E** is a steady electrostatic field (therefore making no contribution to the integral) we obtain trivially

$$\oint_{\mathbf{B}\text{-line}} \lambda (\nabla \wedge \mathbf{B}) \,.\, \mathrm{d}\mathbf{x} = 0. \tag{6.32}$$

Hence if $\mathbf{B} \,.\, (\nabla \wedge \mathbf{B})$ is non-zero over any portion of a closed **B**-line, then positive values must be compensated by negative values so that (6.32) can be satisfied.

More generally, let S_B (interior V_B) be any closed 'magnetic surface' on which $\mathbf{n} \,.\, \mathbf{B} = 0$. Then, from $\lambda \nabla \wedge \boldsymbol{B} = -\nabla\phi + \mathbf{u} \wedge \mathbf{B}$, we have

$$\int_{V_B} \lambda \mathbf{B} \,.\, (\nabla \wedge \mathbf{B}) \,\mathrm{d}V = -\int_{V_B} \mathbf{B} \,.\, \nabla\phi \,\mathrm{d}V = -\int_{S_B} \mathbf{n} \,.\, \mathbf{B}\phi \,\mathrm{d}S = 0. \tag{6.33}$$

This result reduces to (6.32) in the particular case when a toroidal surface S_B shrinks onto a closed curve C; for then $\mathbf{B}\,\mathrm{d}V \to \Phi\,\mathrm{d}\mathbf{x}$ where Φ is the total flux of **B** round the torus.

Similarly if S_u (interior V_u) is a closed surface on which $\mathbf{n} \,.\, \mathbf{u} = 0$, and if $\nabla \,.\, \mathbf{u} = 0$ in V_u, then by the same reasoning, under steady conditions,

$$\int_{V_u} \lambda \mathbf{u} \,.\, (\nabla \wedge \mathbf{B}) \,\mathrm{d}V = 0. \tag{6.34}$$

The results (6.33) and (6.34) hold even if λ is non-uniform.

6.6. Some comments on the situation $\mathbf{B} \,.\, (\nabla \wedge \mathbf{B}) \equiv 0$

If **B** is an axisymmetric poloidal field, then clearly it satisfies

$$\mathbf{B} \,.\, \nabla \wedge \mathbf{B} \equiv 0. \tag{6.35}$$

Similarly an axisymmetric toroidal field satisfies (6.35). We know from § 6.4 that in either case dynamo action is impossible. It seems likely that dynamo action is impossible for *any* field satisfying (6.35), but this has not been proved.

The condition (6.35) is well known as the necessary and sufficient condition for the existence of a family of surfaces everywhere orthogonal to **B**, or equivalently for the existence of functions $\alpha(\mathbf{x})$ and $\beta(\mathbf{x})$ such that

$$\mathbf{B} = \beta(\mathbf{x})\nabla\alpha(\mathbf{x}), \qquad \nabla \wedge \mathbf{B} = \nabla\beta \wedge \nabla\alpha, \tag{6.36}$$

the orthogonal surfaces then being $\alpha = \text{cst.}$ In the external region $\hat{V}$ where $\nabla \wedge \mathbf{B} = 0$, we may clearly take $\beta = 1$.

It is easy to show that there is no field of the form (6.36) satisfying the conditions $\mathbf{B} = O(r^{-3})$ at infinity and $\mathbf{B}^2 > 0$ for all finite $\mathbf{x}$. For, in this case, $\beta > 0$ for all finite $\mathbf{x}$, and so $\beta^{-1}\mathbf{B}^2 > 0$ everywhere. But

$$\int_{V_\infty} \beta^{-1}\mathbf{B}^2 \, \mathrm{d}V = \int_{V_\infty} \mathbf{B} \cdot \nabla\alpha \, \mathrm{d}V = \int_{S_\infty} (\mathbf{n} \cdot \mathbf{B})\alpha \, dS = 0 \tag{6.37}$$

and we have a contradiction.

It follows that any field of the form (6.36) and at most dipole at infinity must vanish for at least one finite value of $\mathbf{x}$. The simplest topology (an arbitrary distortion of the axisymmetric poloidal case) is that in which every point of a closed curve C in V is an O-type neutral point of **B**. If there is just one such curve, then all the surfaces $\alpha = \text{cst.}$ intersect on C. In this situation, $\beta = 0$ on C and α is not single-valued, and the simple statement (6.37) is certainly not applicable.

Pichackchi (1966) has claimed to prove that dynamo action is impossible if $\mathbf{E} \equiv 0$, i.e. if $\lambda\nabla \wedge \mathbf{B} = \mathbf{u} \wedge \mathbf{B}$ in V ((6.35) being an immediate consequence). His argument however rests on the unjustified (and generally incorrect) assertion that the surfaces $\alpha = \text{cst.}$ do not intersect. The result claimed nevertheless seems plausible, and a correct proof would be of considerable interest.

6.7 The impossibility of dynamo action with purely toroidal motion

It is physically plausible that poloidal velocities are necessary to regenerate poloidal magnetic fields, and that a velocity field that is purely toroidal will therefore be incapable of sustained dynamo action. This result was discovered by Bullard & Gellman (1954) as a by-product of their treatment of the induction equation by spherical harmonic decomposition. A simpler direct proof was given by Backus (1958), whose method we follow here. We revert to the standard situation in which λ is assumed uniform in V, and we restrict attention to incompressible velocity fields for which $\nabla . \mathbf{u} = 0$. From the induction equation in the form

$$\mathrm{D}\mathbf{B}/\mathrm{D}t = \mathbf{B} . \nabla\mathbf{u} + \lambda \nabla^2 \mathbf{B}, \tag{6.38}$$

we may then immediately deduce an equation for the scalar $\mathbf{x} . \mathbf{B}$, viz.

$$\frac{\mathrm{D}}{\mathrm{D}t}(\mathbf{x} . \mathbf{B}) \equiv \mathbf{x} . \frac{\mathrm{D}\mathbf{B}}{\mathrm{D}t} + \mathbf{u} . \mathbf{B} = (\mathbf{B} . \nabla)(\mathbf{x} . \mathbf{u}) + \lambda \nabla^2 (\mathbf{x} . \mathbf{B}). \tag{6.39}$$

Hence if the motion is purely toroidal, so that $\mathbf{x} . \mathbf{u} = 0$, the quantity $Q = \mathbf{x} . \mathbf{B}$ satisfies the diffusion equation

$$\mathrm{D}Q/\mathrm{D}t = \lambda \nabla^2 Q \quad \text{in } V. \tag{6.40}$$

Moreover $\nabla^2 Q = 0$ in $\hat{V}$, and Q and $\partial Q/\partial n$ are continuous across the surface S of V. Standard manipulation then gives the result

$$\frac{\mathrm{d}}{\mathrm{d}t}\int_V Q^2 \,\mathrm{d}V = -2\lambda \int_{V_\infty} (\nabla Q)^2 \,\mathrm{d}V, \tag{6.41}$$

so that $Q \to 0$ everywhere.

Now, in the standard poloidal–toroidal decomposition of $\mathbf{B}$, we have $L^2 P = -\mathbf{x} . \mathbf{B}$ (2.38a); and the equation $L^2 P = 0$ with outer boundary condition $P = O(r^{-2})$ at infinity has only the trivial solution $P \equiv 0$. It follows that a steady state is possible only if $\mathbf{B}_P \equiv 0$.

The equation for the toroidal field $\mathbf{B}_T$ then becomes

$$\frac{\partial \mathbf{B}_T}{\partial t} = \nabla \wedge (\mathbf{u}_T \wedge \mathbf{B}_T) + \lambda \nabla^2 \mathbf{B}_T. \tag{6.42}$$

With $\mathbf{B}_T = -\mathbf{x} \wedge \nabla T$, and $\mathbf{x} . \mathbf{u}_T = 0$, we have

$$\mathbf{u}_T \wedge \mathbf{B}_T = -\mathbf{x}(\mathbf{u}_T . \nabla) T, \tag{6.43}$$

so that (6.42) becomes

$$-\mathbf{x} \wedge \nabla \frac{\partial T}{\partial t} = (\mathbf{x} \wedge \nabla)\mathbf{u}_T . \nabla T - \lambda (\mathbf{x} \wedge \nabla)\nabla^2 T, \tag{6.44}$$

using the commutativity of $\mathbf{x} \wedge \nabla$ and ∇^2. Hence

$$\frac{\mathrm{D}T}{\mathrm{D}t} \equiv \frac{\partial T}{\partial t} + \mathbf{u}_T \,.\, \nabla T = \lambda \nabla^2 T + f(r), \tag{6.45}$$

for some function $f(r)$. When we multiply by T and integrate (6.45) throughout V, the term involving $f(r)$ makes no contribution since T integrates to zero over each surface $r = \text{cst}$. Moreover $T = 0$ on the surface S of V, and so we obtain

$$\frac{\mathrm{d}}{\mathrm{d}t} \int_V T^2 \,\mathrm{d}V = -2\lambda \int_V (\nabla T)^2 \,\mathrm{d}V. \tag{6.46}$$

Hence T (and so $\mathbf{B}_T$ also) ultimately decays to zero.

By further manipulations of (6.39), Busse (1975*a*) has succeeded in obtaining an estimate for the magnitude of the radial velocity field that would be required to prevent the diffusive decay of the radial magnetic field. Comparison of the two terms on the right of (6.39) gives the preliminary estimate

$$u_r \sim \lambda B_r / |\boldsymbol{B}| R. \tag{6.47}$$

If the term involving $\mathbf{x} \,.\, \mathbf{u}$ in (6.39) is retained, then Busse shows that (6.41) may be replaced by the inequality

$$\frac{\mathrm{d}}{\mathrm{d}t} \int_V Q^2 \,\mathrm{d}V \leqslant -2\left[\lambda - \max\,(\mathbf{u} \,.\, \mathbf{x})\left(\frac{M}{2M_P}\right)^{1/2}\right] \int_{V_\infty} (\nabla Q)^2 \,\mathrm{d}V, \tag{6.48}$$

where M represents the total magnetic energy, and M_P the energy of the poloidal part of the magnetic field. Hence a necessary condition for amplification of $\int_V Q^2 \,\mathrm{d}V$, is

$$\max\,(\mathbf{u} \,.\, \mathbf{x}) > \lambda\,(2M_P/M)^{1/2}, \tag{6.49}$$

which may be compared with (6.47). For a given magnetic field distribution (and therefore a given value of the ratio M_P/M), (6.49) clearly provides a necessary (although again by no means sufficient) condition that a velocity field must satisfy if it is to maintain the magnetic field steadily against ohmic decay.

The condition (6.49) is of particular interest and relevance in the terrestrial context in view of the arguments in favour of a stably stratified core (§ 4.4). Stable stratification implies inhibition of radial (convective) velocity fields, but may nevertheless permit wave motions (e.g. internal gravity waves modified by Coriolis forces) if perturbing force fields are present. The condition (6.49) indicates a minimum level for radial fluctuation velocities to maintain a given level of poloidal field energy.

It is instructive to re-express (6.49) in terms of magnetic Reynolds numbers R_{mT} and R_{mP} characterising the toroidal and poloidal motions. We have seen that when $R_{mT} \gg 1$ the toroidal magnetic energy builds up to $O(R_{mT}^2)$ times the poloidal magnetic energy, so that $(M_P/M)^{1/2} = O(R_{mT}^{-1})$; hence (6.49) becomes simply (with $R_{mP} = \max(\mathbf{u} \,.\, \mathbf{x})/\lambda$)

$$R_{mP}R_{mT} \gtrsim 1. \tag{6.50}$$

It is interesting to note that the result that a purely toroidal flow is incapable of dynamo action has no analogue in a cylindrical geometry, i.e. when the velocity is confined to cylindrical (rather than spherical) surfaces. The reason is that diffusion in a cylindrical geometry with coordinates (s, φ, z) introduces a coupling between the radial component B_s and the azimuthal component B_φ. There is however an analogous result in a Cartesian configuration; this is obtained in the following section.

6.8 The impossibility of dynamo action with plane two-dimensional motion

The analogue of purely toroidal motion in a Cartesian configuration is a motion $\mathbf{u}$ such that $\mathbf{k} \,.\, \mathbf{u} = 0$ where $\mathbf{k}$ is a unit vector, say $(0, 0, 1)$. Under this condition the field component $\mathbf{B} \,.\, \mathbf{k}$ satisfies

$$\frac{\mathrm{D}}{\mathrm{D}t}(\mathbf{B} \,.\, \mathbf{k}) = \lambda \nabla^2(\mathbf{B} \,.\, \mathbf{k}), \tag{6.51}$$

and so $\mathbf{B} \,.\, \mathbf{k}$ tends everywhere to zero in the absence of external sources.

Suppose then that $\mathbf{B} = \nabla \wedge \mathbf{A}$ with $\mathbf{A} = A(x, y, z)\mathbf{k}$; note that this choice of $\mathbf{A}$ does not satisfy $\nabla \,.\, \mathbf{A} = 0$ (unless $\partial A/\partial z = 0$). Since $\mathbf{u} \wedge \mathbf{B} = -\mathbf{u} \wedge (\mathbf{k} \wedge \nabla A) = -(\mathbf{u} \,.\, \nabla A)\mathbf{k}$, the equation for $\mathbf{A}$ (2.118) becomes

$$\mathrm{D}\mathbf{A}/\mathrm{D}t = -\nabla\phi - \lambda \nabla \wedge (\nabla \wedge \mathbf{A}). \tag{6.52}$$

Suppose that the fluid extends to infinity[5] and that $\mathbf{A}=O(s^{-2})$, $\phi=O(s^{-1})$ as $s=(x^2+y^2)^{1/2}\to\infty$; then (6.52) gives (with S_∞ the 'cylinder at infinity')

$$\frac{\mathrm{d}}{\mathrm{d}t}\iint \mathbf{A}^2\,\mathrm{d}x\,\mathrm{d}y=-\int_{S_\infty}\mathbf{n}\,.\,(\mathbf{A}\phi+\lambda\mathbf{A}\wedge(\nabla\wedge\mathbf{A}))\,\mathrm{d}S$$

$$-\lambda\iint(\nabla\wedge\mathbf{A})^2\,\mathrm{d}x\,\mathrm{d}y. \quad (6.53)$$

Under the assumed conditions there is no contribution from the surface integral, and it follows immediately that $\nabla\wedge\mathbf{A}\to 0$ everywhere.

Note that this result holds whether $\mathbf{B}$ is z-dependent or not, provided solely that $\mathbf{k}\,.\,\mathbf{u}=0$. The result was proved in a weaker form (assuming $\partial\mathbf{B}/\partial z=0$) by Lortz (1968). An analogous result was first discussed in the context of two-dimensional turbulence by Zel'dovich (1957).

6.9. Rotor dynamos

We now turn to some examples of kinematically possible motions in a homogeneous conductor which *do* give rise to steady dynamo action. In order to avoid the consequences of the foregoing anti-dynamo theorems, such motions must necessarily be quite complicated, and the associated analysis is correspondingly complex. Nevertheless it is important to find at least one explicit example of successful dynamo action if only to be confident that there can be no all-embracing anti-dynamo theorem. The first such example, provided by Herzenberg (1958), was significant in that it provided unequivocal proof that steady motions $\mathbf{u}(\mathbf{x})$ do exist in a sphere of conducting fluid which can maintain a steady magnetic field $\mathbf{B}(\mathbf{x})$ against ohmic decay, and which give a non-zero dipole moment outside the sphere.

Herzenberg's velocity field consisted of two spherical rotors imbedded in a conducting sphere of fluid otherwise at rest; within each rotor the angular velocity was constant and the radius of the

[5] The proof may be easily adapted to the case when the fluid is confined to a finite domain in two or three dimensions.

rotors was small compared with the distance between their centres which in turn was small compared with the radius of the conducting sphere. Herzenberg's analysis has been greatly clarified by the discussion of Gibson (1968*a*, *b*), Gibson & Roberts (1967) and Roberts (1971); the following discussion is based largely on these papers.

Let $S_\alpha (\alpha = 1, 2, \ldots, n)$ denote the n spheres $|\mathbf{x}-\mathbf{x}_\alpha| = a$, and suppose that for each pair (α, β), $|\mathbf{x}_\alpha - \mathbf{x}_\beta| \gg a$, i.e. the spheres are all far apart relative to their radii. We define a velocity field

$$\mathbf{u}(\mathbf{x}) = \begin{cases} \boldsymbol{\omega}_\alpha \wedge (\mathbf{x}-\mathbf{x}_\alpha) & \text{when } |\mathbf{x}-\mathbf{x}_\alpha| < a \quad (\alpha = 1, 2, \ldots, n) \\ 0 & \text{otherwise} \end{cases} \tag{6.54}$$

where the $\boldsymbol{\omega}_\alpha (\alpha = 1, 2, \ldots, n)$ are constants; i.e. the fluid inside S_α rotates with uniform angular velocity $\boldsymbol{\omega}_\alpha$, and the fluid outside all n spheres is at rest. We enquire under what circumstances such a velocity field in a fluid of infinite extent and uniform conductivity can maintain a steady magnetic field $\mathbf{B}(\mathbf{x})$, at most $O(r^{-3})$ at infinity.

If such a field $\mathbf{B}(\mathbf{x})$ exists, then in the neighbourhood of the sphere S_α it may be decomposed into its poloidal and toroidal parts

$$\mathbf{B}_P^\alpha = \nabla \wedge \nabla \wedge \mathbf{r}_\alpha P(\mathbf{r}_\alpha), \qquad \mathbf{B}_T^\alpha = \nabla \wedge \mathbf{r}_\alpha T(\mathbf{r}_\alpha) \tag{6.55}$$

where $\mathbf{r}_\alpha = \mathbf{x} - \mathbf{x}_\alpha$. Let $P^s(\mathbf{r}_\alpha)$ and $T^s(\mathbf{r}_\alpha)$ be the average of $P(\mathbf{r}_\alpha)$ and $T(\mathbf{r}_\alpha)$ over the azimuth angle about the rotation vector $\boldsymbol{\omega}_\alpha$, and let $P^a(\mathbf{r}_\alpha)$, $T^a(\mathbf{r}_\alpha)$ be defined by

$$P = P^s + P^a, \qquad T = T^s + T^a, \tag{6.56}$$

(the superfixes s and a indicating symmetry and asymmetry about the axis of rotation). Let the corresponding decomposition of $\mathbf{B}_P^\alpha$ and $\mathbf{B}_T^\alpha$ be

$$\mathbf{B}_P^\alpha = \mathbf{B}_P^{\alpha s} + \mathbf{B}_P^{\alpha a}, \qquad \mathbf{B}_T^\alpha = \mathbf{B}_T^{\alpha s} + \mathbf{B}_T^{\alpha a}. \tag{6.57}$$

We know from § 6.4 that the toroidal motion of the sphere S_α has no direct regenerative effect on $\mathbf{B}_P^\alpha$, which is maintained entirely by the inductive effects of the other $n-1$ spheres. Hence $\mathbf{B}_P^\alpha$ can be regarded as an 'applied' field in the neighbourhood of S_α, and the rotation of S_α in the presence of this applied field determines the structure of the toroidal field $\mathbf{B}_T^\alpha$. The antisymmetric part of the total field $\mathbf{B}^\alpha$ is merely excluded from the rotating region essentially by the process of § 3.8. The symmetric part of $\mathbf{B}_P^\alpha$ on the other hand interacts with the differential rotation ($\omega'(r)$ is here concentrated on the spherical surface $r=|\mathbf{x}-\mathbf{x}_\alpha|=a$) to provide $\mathbf{B}_T^{\alpha s}$, which, by the arguments of § 3.11, is $O(R_{m\alpha})|\mathbf{B}_P^{\alpha s}|$, where $R_{m\alpha}=\omega_\alpha a^2/\lambda$. If $R_{m\alpha}\gg 1$, as we shall suppose, then $\mathbf{B}_T^{\alpha s}$ is the dominant part of the total field $\mathbf{B}^\alpha$ in a large neighbourhood of S_α.

Now $\mathbf{B}_P^{\alpha s}$ may be expanded about the point $\mathbf{x}=\mathbf{x}_\alpha$ in Taylor series:

$$\begin{aligned} B_{Pi}^{\alpha s}(\mathbf{x}) &= B_{Pi}^{\alpha s}(\mathbf{x}_\alpha)+r_{\alpha j}B_{Pi,j}^{\alpha s}(\mathbf{x}_\alpha)+O(r_\alpha^2) \\ &= B_{Pi}^{\alpha s}(\mathbf{x}_\alpha)+\tfrac{1}{2}r_{\alpha j}(B_{Pi,j}^{\alpha s}(\mathbf{x}_\alpha)+B_{Pj,i}^{\alpha s}(\mathbf{x}_\alpha))+O(r_\alpha^2), \end{aligned} \tag{6.58}$$

since $\nabla\wedge\mathbf{B}_P^\alpha$ is the toroidal current which vanishes at $\mathbf{x}_\alpha$; this may be expressed in the equivalent form

$$\mathbf{B}_P^{\alpha s}(\mathbf{x})=\nabla\Phi+O(r_\alpha^2), \tag{6.59}$$

where

$$\Phi=A_1^\alpha rP_1(\cos\theta)+a^{-1}A_2^\alpha r^2P_2(\cos\theta), \tag{6.60}$$

and (with the convention that repeated Greek suffices are *not* summed)

$$A_1^\alpha=\omega_\alpha^{-1}\boldsymbol{\omega}_\alpha\,.\,\mathbf{B}_P^{\alpha s}(\mathbf{x}_\alpha),\qquad A_2^\alpha=\tfrac{1}{2}a\omega_\alpha^{-2}\boldsymbol{\omega}_\alpha\,.\,\nabla\mathbf{B}_P^{\alpha s}(\mathbf{x}_\alpha)\,.\,\boldsymbol{\omega}_\alpha. \tag{6.61}$$

Hence from the results of § 3.11, for $r_\alpha\gg a$,

$$\begin{aligned} \mathbf{B}_T^{\alpha s}(\mathbf{x}) \sim & -\frac{1}{5}a^3A_1^\alpha R_{m\alpha}\frac{(\boldsymbol{\omega}_\alpha\,.\,\mathbf{r}_\alpha)(\boldsymbol{\omega}_\alpha\wedge\mathbf{r}_\alpha)}{\omega_\alpha^2 r_\alpha^5} \\ & +\frac{2}{5}a^2A_2^\alpha R_{m\alpha}\frac{\boldsymbol{\omega}_\alpha\wedge\mathbf{r}_\alpha}{\omega_\alpha r_\alpha^3}+O(r_\alpha^{-4}). \end{aligned} \tag{6.62}$$

The field in the neighbourhood of S_β is the sum of fields of the form (6.62) resulting from the presence of all the other spheres $S_\alpha\,(\alpha\neq\beta)$, i.e.

$$\mathbf{B}^\beta(\mathbf{x})=\sum_{\alpha(\neq\beta)}\mathbf{B}_T^{\alpha s}(\mathbf{x}). \tag{6.63}$$

For self-consistency, we must have, for $\beta=1,2,\ldots,n$,

$$A_1^\beta=\omega_\beta^{-1}\boldsymbol{\omega}_\beta\,.\,\mathbf{B}_P^{\beta s}(\mathbf{x}_\beta)=\sum_{\alpha(\neq\beta)}\omega_\beta^{-1}\boldsymbol{\omega}_\beta\,.\,\mathbf{B}_T^{\alpha s}(\mathbf{x}_\beta), \tag{6.64}$$

and, similarly,

$$A_2^\beta = \sum_{\alpha(\neq\beta)} \tfrac{1}{2} a\omega_\beta^{-2} \boldsymbol{\omega}_\beta \,.\, \nabla \mathbf{B}_T^{\alpha s}(\mathbf{x}_\beta) \,.\, \boldsymbol{\omega}_\beta. \tag{6.65}$$

If the terms $O(r_\alpha^{-4})$ in (6.62) are neglected, these provide $2n$ linear equations for A_1^α and $A_2^\alpha (\alpha = 1, \ldots, n)$, and the determinant of the coefficients must vanish for a non-trivial solution.

3-sphere dynamo

The procedure is most simply followed for the case of three spheres (Gibson, 1968*b*) in the configuration of fig. 6.3. Let

$$\mathbf{x}_1 = (d, 0, 0), \qquad \mathbf{x}_2 = (0, d, 0), \qquad \mathbf{x}_3 = (0, 0, \mathrm{d}), \tag{6.66}$$

and let

$$\boldsymbol{\omega}_1 = (0, 0, -\omega), \qquad \boldsymbol{\omega}_2 = (-\omega, 0, 0), \qquad \boldsymbol{\omega}_3 = (0, -\omega, 0), \tag{6.67}$$

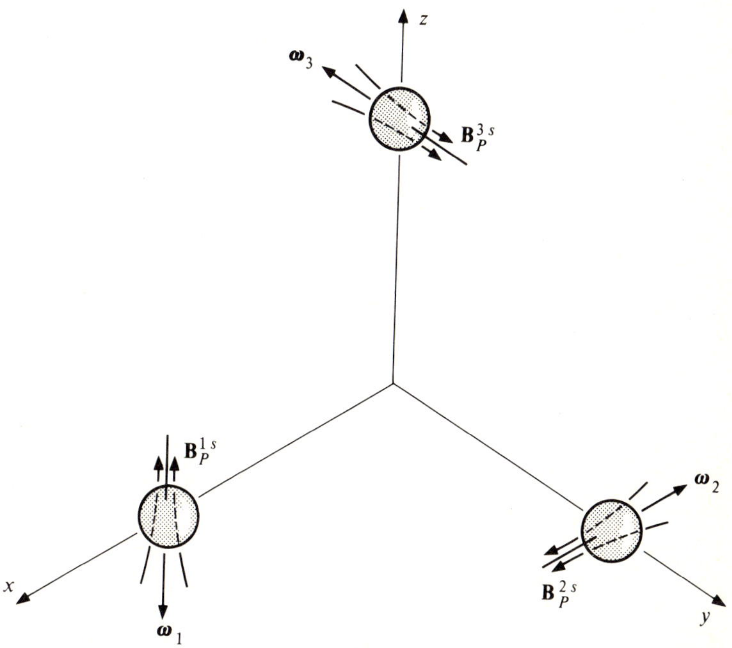

Fig. 6.3 The 3-rotor dynamo of Gibson (1968*b*) for the particular configuration that is invariant under rotations of $2\pi/3$ and $4\pi/3$ about the direction (1, 1, 1).

where $\omega > 0$. The reason for this choice of sign will emerge below. This configuration has a three-fold symmetry, in that it is invariant under rotations of $2\pi/3$ and $4\pi/3$ about the direction (1, 1, 1). Let us therefore look only for magnetic fields $\mathbf{B}(\mathbf{x})$ which exhibit this same degree of symmetry. In particular, in the above notation, we have

$$A_1^1 = A_1^2 = A_1^3 = A_1 \text{ say}, \qquad A_2^1 = A_2^2 = A_2^3 = A_2 \text{ say}. \tag{6.68}$$

There are then only two conditions deriving from (6.64) and (6.65) relating A_1 and A_2.

In the neighbourhood of the sphere S_3

$$\mathbf{r}_1 = \mathbf{x}_3 - \mathbf{x}_1 \approx (-d, 0, d), \quad \text{and} \quad \mathbf{r}_2 = \mathbf{x}_3 - \mathbf{x}_2 \approx (0, -d, d), \tag{6.69}$$

and so, after some vector manipulation, (6.64) and (6.65), with $\beta = 3$, become

$$A_1\left(1 + \frac{R_m}{10}\left(\frac{a}{R}\right)^3\right) = \frac{4}{15\sqrt{2}} R_m \left(\frac{a}{R}\right)^3 A_2 \frac{R}{a}, \tag{6.70}$$

and

$$A_2\left(1 - \frac{a^3 R_m}{5R^3}\right) = 0, \tag{6.71}$$

where $R_m = \omega a^2/\lambda$, and $R = d\sqrt{2}$ is the distance between the sphere centres. Hence (6.71) gives a critical magnetic Reynolds number

$$R_{mc} = 5(R/a)^3, \tag{6.72}$$

and (6.70) then becomes

$$A_2 = \frac{9}{4\sqrt{2}} \frac{A_1 a}{R}. \tag{6.73}$$

The configuration of the $\mathbf{B}_P^{\alpha s}$-lines in the neighbourhood of each sphere are as indicated in fig. 6.3. (It would be difficult to portray the full three-dimensional field pattern.)

Certain points in the above calculation deserve particular comment. First note that since $a \ll R$, A_2 is an order of magnitude smaller than A_1. This means that the field $\mathbf{B}_P^{\alpha s}$ is approximately uniform in the neighbourhood of S_α. It would be quite wrong however to treat it as exactly uniform; this would correspond to putting $A_2 = 0$ in (6.70) and (6.71) with the erroneous conclusion that dynamo action will occur if $R_m = -10(R/a)^3$ (requiring $\omega < 0$ in (6.67)). The small field gradient in the neighbourhood of each sphere is important because of the phenomenon noted in § 3.11 and evident in the expression (6.62) that terms in $\mathbf{B}_T^{\alpha s}$ arising from the gradient of $\mathbf{B}_P^{\alpha s}$ fall off more slowly with distance than do terms arising from the magnitude of $\mathbf{B}_P^{\alpha s}$ itself in the sphere neighbourhood. This effect compensates for the smaller value of the coefficient A_2.

Secondly, note that the directions of $\boldsymbol{\omega}_1$, $\boldsymbol{\omega}_2$ and $\boldsymbol{\omega}_3$ in fig. 6.3 were chosen for maximum simplicity, but the same method would work if $\boldsymbol{\omega}_1$ were taken in any direction and $\boldsymbol{\omega}_2$ and $\boldsymbol{\omega}_3$ were obtained from $\boldsymbol{\omega}_1$ by rotations of $2\pi/3$ and $4\pi/3$ about (1, 1, 1). If one such triad $(\boldsymbol{\omega}_1, \boldsymbol{\omega}_2, \boldsymbol{\omega}_3)$ gives dynamo action, however, then the triad $(-\boldsymbol{\omega}_1, -\boldsymbol{\omega}_2, -\boldsymbol{\omega}_3)$ cannot support a field $\mathbf{B}(\mathbf{x})$ having the same three-fold symmetry.

Thirdly, there is no real need for $\boldsymbol{\omega}$ to be uniform throughout each sphere. The results of § 3.11 indicate that if $\boldsymbol{\omega} = \omega(r)\mathbf{k}$, then, as far as the induced field far from the sphere is concerned, the only quantity that matters is $\bar{\omega}$ where

$$\tfrac{1}{5}a^5\bar{\omega} = \int_0^a r^4\omega(r)\,dr. \qquad (6.74)$$

If ω varies with radius within each sphere, but $\bar{\omega}$ is the same for each sphere, then the above results apply with $R_m = \bar{\omega}a^2/\lambda$.

2-sphere dynamo

If there are only two spherical rotors, then taking origin at the mid-point of the line joining their centres we may take

$$\mathbf{x}_1 = (-\tfrac{1}{2}R, 0, 0), \qquad \mathbf{x}_2 = (\tfrac{1}{2}R, 0, 0). \qquad (6.75)$$

Suppose (fig. 6.4) that

$$\boldsymbol{\omega}_1 = \omega(0, \cos\varphi/2, -\sin\varphi/2), \qquad \boldsymbol{\omega}_2 = \omega(0, \cos\varphi/2, \sin\varphi/2). \qquad (6.76)$$

The configuration is then invariant under a rotation of π about the axis Oy, and we may therefore look for a magnetic field exhibiting this same two-fold symmetry. Putting $A_1^1 = A_1^2 = A_1$ and

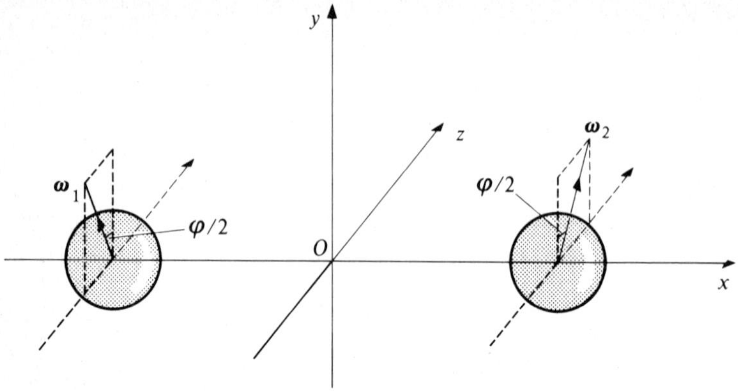

Fig. 6.4 Two-sphere dynamo configuration defined by (6.75) and (6.76).

$A_2^1 = A_2^2 = A_2$, the conditions (6.64) and (6.65) reduce to[6]

$$A_1 = -\frac{2a^2}{15R^2} A_2 R_m \sin\varphi, \tag{6.77}$$

and

$$A_2 = \frac{a^4}{10R^4} A_1 R_m \sin\varphi \cos\varphi. \tag{6.78}$$

Hence

$$(\tfrac{1}{5}R_m(a/R)^3 \sin\varphi)^2(-\cos\varphi) = 3, \tag{6.79}$$

so that dynamo action is possible if $\pi/2 < \varphi < 3\pi/2$ (excluding $\varphi = \pi$), and then from (6.77)

$$A_2 = \tfrac{1}{2}A_1(a/R)(-3\cos\varphi)^{1/2}. \tag{6.80}$$

The dynamo action is most efficient (i.e. the resulting value of R_m is least) when $\tan\varphi = \pm 2^{-1/2}$, i.e. $\varphi \approx 145°$ or $215°$, and then

$$A_2 \approx 0{\cdot}78 A_1(a/R). \tag{6.81}$$

[6] The antisymmetric possibility $A_1^1 = -A_1^2$ and $A_2^1 = -A_2^2$ leads to the same value of R_m (6.79) and a field structure which is transformed into its inverse ($\mathbf{B}(\mathbf{x}) \to -\mathbf{B}(\mathbf{x})$) under rotation through π about Oy.

The result (6.79) is a particular case of the result obtained by Herzenberg (1958) who studied the more general configuration for which

$$\begin{aligned}\boldsymbol{\omega}_1 &= \omega(\cos\theta_1, \sin\theta_1\cos\varphi/2, -\sin\theta_1\sin\varphi/2),\\ \boldsymbol{\omega}_2 &= \omega(\cos\theta_2, \sin\theta_2\cos\varphi/2, \sin\theta_2\sin\varphi/2).\end{aligned} \tag{6.82}$$

Unless $\theta_1 = \pm\theta_2$, this configuration does not exhibit two-fold symmetry about any axis, and there is therefore no *a priori* justification for putting $A_1^1 = A_1^2$ and $A_2^1 = A_2^2$ in general. There are therefore four equations linear in these quantities obtained from (6.64) and (6.65); vanishing of the determinant of the coefficients yields the condition

$$\begin{aligned}\lambda^2 \equiv \{[\tfrac{1}{5}R_m(a/R)^3\sin\theta_1\sin\theta_2\sin\varphi]^2(\cos\theta_1\cos\theta_2\\ -\sin\theta_1\sin\theta_2\cos\varphi) - 3\}^2 = 0.\end{aligned} \tag{6.83}$$

The condition $\lambda = 0$ reduces to (6.79) when $\theta_1 = \theta_2 = \pi/2$. However Herzenberg observed that terms neglected in the expansion scheme (those denoted $O(r_\alpha^{-4})$ in (6.62)) could conceivably, if included, give a negative contribution, $-\varepsilon^2$ say, on the right of (6.83). We would then have $\lambda = \pm i\varepsilon$, and the resulting magnetic Reynolds number would be complex indicating that steady dynamo action is not in fact possible for the given velocity field.

Herzenberg resolved this difficulty by taking account of fields reflected from the distant spherical boundary of the conductor. He found that these modified the relationship (6.83) to the form

$$(\lambda - \lambda_1)(\lambda - \lambda_2) = 0, \tag{6.84}$$

where λ_1 and λ_2 are small numbers determined by these distant boundary effects. In general $\lambda_1 \neq \lambda_2$, and the possibilities $\lambda = \lambda_1$ and $\lambda = \lambda_2$ give two distinct values of R_m at which steady dynamo action is possible, with two corresponding (but different) field structures.

When the conductor extends to infinity in all directions, the double root of (6.83) for R_m suggests that, even when the configuration has no two-fold symmetry about any axis, there is nevertheless a degeneracy of the eigenvalue problem, i.e. corresponding to the critical value R_m there are two linearly independent eigenfunctions $\mathbf{B}_1(\mathbf{x})$ and $\mathbf{B}_2(\mathbf{x})$; when the configuration has the two-fold symmetry of fig. 6.4, it seems altogether plausible that one of these should be symmetric and one anti-symmetric in the sense that $\mathbf{B}_1(\mathbf{x}) \to \mathbf{B}_1(\mathbf{x})$, $\mathbf{B}_2(\mathbf{x}) \to -\mathbf{B}_2(\mathbf{x})$ under a rotation of π about Oy. Gibson (1968a) has in fact shown that the degeneracy implicit in (6.83) is not removed by the retention of the term of order r_α^{-4} in (6.62); it seems likely that the degeneracy persists to all orders, at any rate for the symmetric case (6.76).

The rotor dynamo of Lowes & Wilkinson

Lowes & Wilkinson (1963, 1968) have constructed a laboratory dynamo consisting of two solid cylindrical rotors imbedded in a solid block of the same material, electrical contact between the rotors and the block being provided by a thin lubricating film of mercury. Fig. 6.5 shows a typical orientation of the cylinders. The

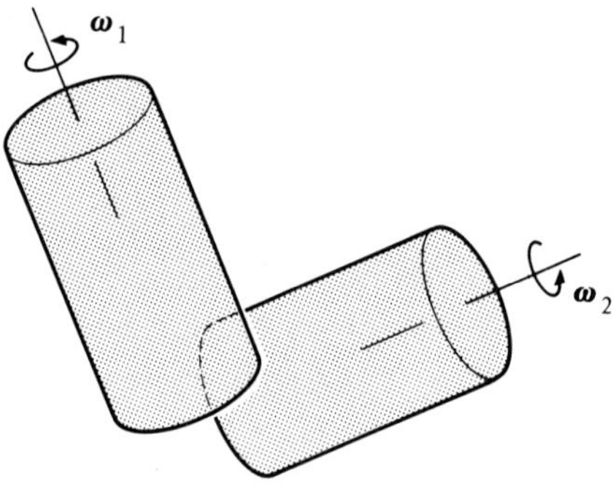

Fig. 6.5 Rotating cylinder configuration of Lowes & Wilkinson (1963, 1968).

principle on which the dynamo operates is essentially that of the Herzenberg model: the 'applied' poloidal field of rotor A is the induced toroidal field of rotor B, and vice versa. The use of cylinders rather than spheres was dictated by experimental expediency; but the interaction between two cylinder ends is in fact stronger than that between two spheres since, roughly speaking, the toroidal field lines generated by the rotation of a cylinder in a nearly uniform axial field all have the same sense in the neighbourhood of the cylinder ends (Herzenberg & Lowes, 1957) (contrast the case of the rotating sphere, where the toroidal field changes sign across the equatorial plane). A reasonably small value of the magnetic diffusivity $\lambda = (\mu\sigma)^{-1}$ was achieved through the use of ferromagnetic material for the rotors and block (the iron alloy 'Perminvar' in the earlier model, annealed mild steel in the later model). For the most favourable orientation of the cylinders, dynamo action was found to occur at a critical angular velocity of the rotors of about 400 rpm. corresponding to a critical magnetic Reynolds number $R_m \approx 200$. As the angular velocities are increased to the critical value, dynamo action manifests itself as a sudden increase of the magnetic field measured outside the block. The currents in the block (and the corresponding field) increase until the retarding torque on the rotors associated with the Lorentz force distribution is just sufficient to prevent the angular velocities from increasing above the critical value. In the steady state (Lowes & Wilkinson, 1963), the power supplied to the rotors equals the sum of the rate of ohmic dissipation and the rate of viscous dissipation in the lubricating films of mercury. In the later improved model (Lowes & Wilkinson, 1968), the rate of viscous dissipation was much reduced, and the system exhibited interesting oscillatory behaviour about the possible steady states. This type of behaviour, which may have a bearing on the question of reversals of the Earth's dipole field (§ 4.3), involves dynamical effects which will be considered in later chapters – see particularly § 12.4.

6.10. Dynamo action associated with a pair of ring vortices

A further ingenious example of dynamo action associated with a pair of rotors has been analysed by Gailitis (1970). The velocity field, as sketched in fig. 6.6, is axisymmetric, has circular streamlines, and is confined to the interior of two toroidal rings $\mathcal{T}_1$ and $\mathcal{T}_2$. We know from Cowling's theorem that such a velocity field cannot support a steady axisymmetric magnetic field vanishing at infinity; it can however, under certain circumstances, support a non-axisymmetric field proportional to $e^{im\varphi}$, where φ is the azimuth angle about the common axis of the toroids and m is an integer. The figure shows the field lines in the neighbourhood of each torus when $m = 1$ and indicates in a qualitative way how the rotation within each torus can generate a magnetic field which acts as the inducing field for the other torus.

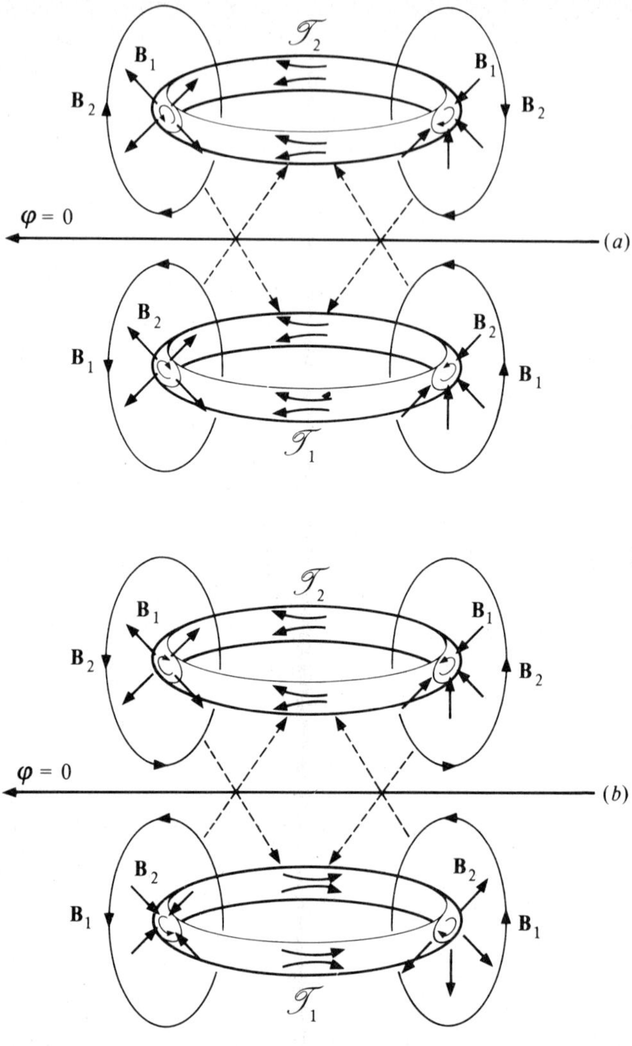

Fig. 6.6 Field maintenance by a pair of ring vortices (after Gailitis, 1970): (*a*) dipole configuration; (*b*) quadrupole configuration. In each case, rotation within each torus induces a field in the neighbourhood of the other torus which has the radial divergence indicated.

The analysis of Gailitis (1970) (with slight changes of notation) proceeds as follows. Let (z, s, φ) be the usual cylindrical polar coordinates, and let (τ, χ, φ) be displaced polar coordinates (Roberts, 1971) defined by

$$s = c - \tau \cos \chi, \qquad z = \tau \sin \chi. \tag{6.85}$$

Let $\mathcal{T}_1$ be the torus $\tau = a$ where $a \ll c$, and let $\mathcal{T}_2$ be the torus obtained by reflecting $\mathcal{T}_1$ in the plane $z = \frac{1}{2} z_0$ where $a \ll z_0$ also. Terms of order a/c and a/z_0 are neglected throughout. Let the velocity field be

$$\mathbf{u} = \mathbf{u}_1(\mathbf{x}) + \mathbf{u}_2(\mathbf{x}), \tag{6.86}$$

where

$$\mathbf{u}_1(\mathbf{x}) = \begin{cases} v_1(\tau)\mathbf{i}_\chi & \text{inside} \quad \mathcal{T}_1, \\ 0 & \text{outside} \quad \mathcal{T}_1, \end{cases} \tag{6.87}$$

and $\mathbf{u}_2(\mathbf{x})$ is similarly defined relative to $\mathcal{T}_2$. The total velocity field is zero except in the two toroids. Note that the assumption $a \ll c$ allows us to neglect the small variation of $v_1(\tau)$ with χ that would otherwise arise from the incompressibility condition $\nabla \cdot \mathbf{u}_1 = 0$.

The steady induction equation is

$$\lambda \nabla^2 \mathbf{B} + \nabla \wedge (\mathbf{u} \wedge \mathbf{B}) = 0, \tag{6.88}$$

and this is formally satisfied by $\mathbf{B} = \mathbf{B}_1 + \mathbf{B}_2$, where

and
$$\left.\begin{aligned} \lambda \nabla^2 \mathbf{B}_1 + \nabla \wedge (\mathbf{u}_1 \wedge \mathbf{B}_1) &= -\nabla \wedge (\mathbf{u}_1 \wedge \mathbf{B}_2), \\ \lambda \nabla^2 \mathbf{B}_2 + \nabla \wedge (\mathbf{u}_2 \wedge \mathbf{B}_2) &= -\nabla \wedge (\mathbf{u}_1 \wedge \mathbf{B}_1). \end{aligned}\right\} \tag{6.89}$$

We regard $\mathbf{B}_1$ as the field *induced* by the motion $\mathbf{u}_1$ *and* as the *inducing field* for the motion $\mathbf{u}_2$; similarly for $\mathbf{B}_2$.

The essential idea behind the Gailitis analysis is very similar to that applying in the Herzenberg model. Let $\mathbf{B}_2^s$ be the vector obtained by averaging the (τ, χ, φ) components of $\mathbf{B}_2$ over the angle χ, and let

$$\mathbf{B}_2^a = \mathbf{B}_2 - \mathbf{B}_2^s; \tag{6.90}$$

the components of $\mathbf{B}_2^\alpha$ average to zero over χ. The corresponding solution of (6.89*a*) may be denoted by $\mathbf{B}_1^a + \mathbf{B}_1^s$. We know from the analysis of § 3.10 that the effect of the closed streamline motion $\mathbf{u}_1$ is simply to expel the asymmetric field from the rotating region when the magnetic Reynolds number R_m, based on a and an appropriate average of $v(\tau)$ (see (6.105) below), is large, i.e.

$$\mathbf{B}_1^a+\mathbf{B}_2^a\approx 0 \quad \text{in } \mathscr{T}_1. \tag{6.91}$$

On the other hand, the motion $\mathbf{u}_1$ generates from $\mathbf{B}_2^s$ by the differential rotation mechanism of § 3.7 a field $\mathbf{B}_1^s$ satisfying

$$|\mathbf{B}_1^s|=O(R_m)|\mathbf{B}_2^s|. \tag{6.92}$$

Clearly for $R_m \gg 1$, this is the dominant contribution to $\mathbf{B}_1$, and so in calculating $\mathbf{B}_1$ from (6.89a) we may regard $\mathbf{B}_2$ as symmetric with respect to χ in the neighbourhood of $\mathscr{T}_1$.

Suppose then that in $\mathscr{T}_1$

$$B_{2\varphi}=B_0\,\mathrm{e}^{im\varphi}. \tag{6.93}$$

The condition $\nabla\,.\,\mathbf{B}_2=0$ then implies that in $\mathscr{T}_1$

$$B_{2\tau}=-(\mathrm{i}\tau m/2c)B_0\,\mathrm{e}^{im\varphi}, \tag{6.94}$$

and so, from (6.87),

$$\mathbf{u}_1\wedge\mathbf{B}_2=B_0 v_1\,\mathrm{e}^{im\varphi}(\mathbf{i}_\tau+(im\tau/2c)\mathbf{i}_\varphi) \quad \text{in } \mathscr{T}_1, \tag{6.95}$$

and so

$$\mathbf{g}\equiv\nabla\wedge(\mathbf{u}_1\wedge\mathbf{B}_2)=-\frac{imB_0\,\mathrm{e}^{im\varphi}}{2c}\tau^2\frac{\mathrm{d}}{\mathrm{d}\tau}\left(\frac{v}{\tau}\right)\mathbf{i}_\chi \quad \text{in } \mathscr{T}_1. \tag{6.96}$$

As expected, the motion $\mathbf{u}_1$ generates a field in the χ-direction as a result of differential rotation within $\mathscr{T}_1$. It follows that $\mathbf{u}_1\wedge\mathbf{B}_1=0$, and so the solution of (6.89a) (Poisson's equation) is

$$\mathbf{B}_1(\mathbf{x}')=\frac{1}{4\pi\lambda}\int_{\mathscr{T}_1}\frac{\mathbf{g}(\mathbf{x})}{|\mathbf{x}-\mathbf{x}'|}\,\mathrm{d}V. \tag{6.97}$$

We now wish to evaluate the φ-component of this integral on the curved axis of $\mathscr{T}_2$ to see whether $\mathbf{B}_1$ can act as the inducing field for $\mathscr{T}_2$. Let $\mathbf{x}_0$ be a point on the axis of $\mathscr{T}_1$ with cylindrical polar coordinates $(0, c, \varphi)$, $\mathbf{x}_0'$ a point on the axis of $\mathscr{T}_2$ with coordinates (z_0, c, φ'), and let $\psi=\varphi'-\varphi$. Then

$$|\mathbf{x}_0-\mathbf{x}_0'|=[z_0^2+2c^2(1-\cos\psi)]^{1/2}, \tag{6.98}$$

and, for $\mathbf{x}\in\mathscr{T}_1$,

$$|\mathbf{x}-\mathbf{x}_0'|^{-1}=|\mathbf{x}_0-\mathbf{x}_0'|^{-1}\left(1-\frac{\tau\mathbf{i}_\tau\,.\,(\mathbf{x}_0-\mathbf{x}_0')}{|\mathbf{x}_0-\mathbf{x}_0'|^3}+O(\tau^3)\right). \tag{6.99}$$

Moreover

$$\mathbf{i}_{\varphi'} \,.\, \mathbf{i}_{\chi} = -\sin \chi \sin \psi. \tag{6.100}$$

When (6.99) and (6.100) are substituted in the equation

$$\mathbf{B}_1(\mathbf{x}_0') \,.\, \mathbf{i}_{\varphi'} = \frac{1}{4\pi\lambda} \int_{\mathscr{T}_1} \frac{\mathbf{g}(\mathbf{x}) \,.\, \mathbf{i}_{\varphi'}}{|\mathbf{x}-\mathbf{x}_0|} c\tau \,\mathrm{d}\tau \,\mathrm{d}\chi \,\mathrm{d}\varphi, \tag{6.101}$$

only a term proportional to $\sin^2 \chi$ gives a non-zero contribution when integrated over χ. Using (6.96) the result simplifies to

$$\mathbf{B}_1(\mathbf{x}_0') \,.\, \mathbf{i}_{\varphi'} = B_0' \,\mathrm{e}^{im\varphi'},$$

where

$$B_0' = B_0 \left(\frac{a}{c}\right)^2 \frac{V_1 a}{\lambda} F_m\left(\frac{z_0}{c}\right), \qquad V_1 a^3 = \int_0^a \tau^2 v_1(\tau) \,\mathrm{d}\tau, \tag{6.102}$$

and

$$F_m(q) = \tfrac{1}{2} mq \int_0^{2\pi} \frac{\sin \psi \sin m\psi \,\mathrm{d}\psi}{(q^2+2-2\cos \psi)^3} \quad (>0 \text{ for } q>0). \tag{6.103}$$

Similarly, by analysing the inductive effect of the motion in $\mathscr{T}_2$, we obtain

$$B_0 = -B_0' \left(\frac{a}{c}\right)^2 \frac{V_2 a}{\lambda} F_m\left(\frac{z_0}{c}\right), \tag{6.104}$$

where V_2 is defined like V_1 but in relation to the motion in $\mathscr{T}_2$. The results (6.102*a*) and (6.104) are compatible only if V_1 and V_2 have opposite signs, so that the net circulations (weighted according to (6.102*b*)) must be opposite in the two toroids. Defining

$$R_m = +(-V_1 V_2)^{1/2} a/\lambda, \tag{6.105}$$

we then have the condition for steady dynamo action in the form

$$R_m = \left(\frac{c}{a}\right)^2 T_m\left(\frac{z_0}{c}\right), \tag{6.106}$$

where $T_m = F_m^{-1}$. When $m=1$ the possibility $V_1 = -V_2 < 0$ corresponds to the field configuration of fig. 6.6(*a*) for which the field has a steady dipole moment perpendicular to Oz and in the plane $\varphi = 0$, and the possibility $V_1 = -V_2 > 0$ corresponds to the field configuration of fig 6.6(b) for which the dipole moment is evidently zero and the far field is that of a quadrupole.

The functions $T_m(q)$ as computed by Gailitis for $m = 1, 2, \ldots, 10$, are reproduced in fig. 6.7. For $z_0/c > 1{\cdot}16$, $T_m(z_0/c)$

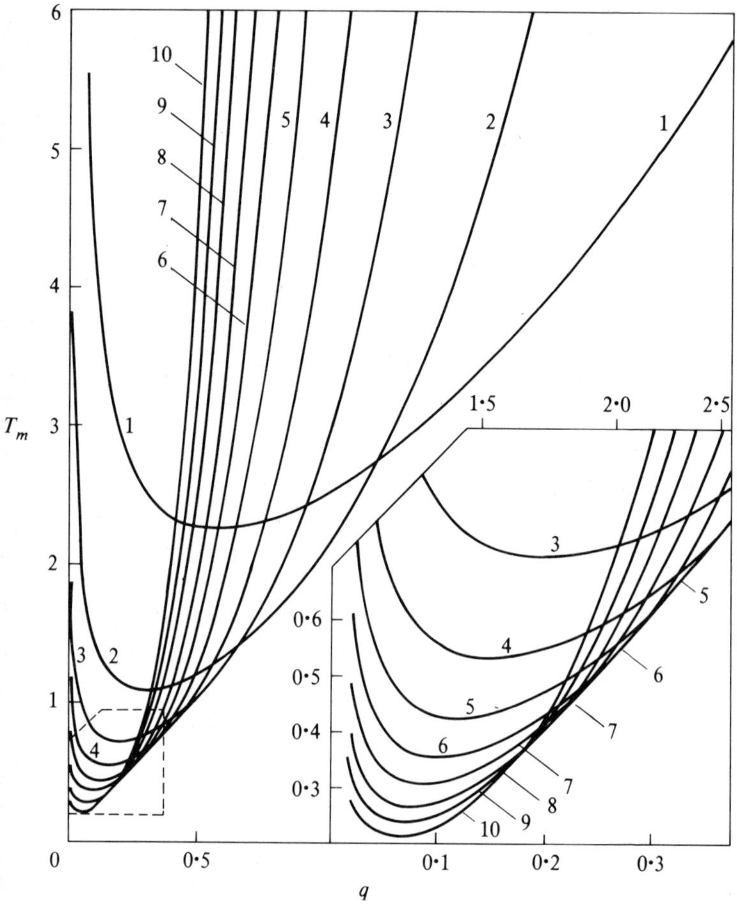

Fig. 6.7 The functions $T_m(q) = (F_m(q))^{-1}$ as computed from (6.103). The numbers by the curves indicate the value of m. The portion surrounded by a dashed line is shown in four-fold magnification. (From Gailitis, 1970.)

is least (corresponding to the most easily excited magnetic mode) for $m=1$. As z_0/c is decreased, the value of m corresponding to the most easily excited mode increases; this is physically plausible in that as the rings approach each other, each becomes more sensitive to the detailed field structure within the other. The analysis of course breaks down if z_0 is decreased to values of order a.

The above analysis is only approximate in that terms of order a/c and a/z_0 are neglected throughout. Strictly a formal perturbation procedure in terms of these small parameters is required, and a rigorous proof of dynamo action would require strict upper bounds to be put on the neglected terms of the Gailitis analysis. There seems little doubt however that such a procedure (paralleling the procedure followed by Herzenberg, 1958) would confirm the validity of the 'zero-order' analysis presented above.

It might be thought that the Gailitis dynamo is nearer to physical reality than the Herzenberg dynamo, in that vortex rings are a well-known dynamically realisable phenomenon in nearly inviscid fluids, whereas spherical rotors are not. There is perhaps some truth in this; however it must be recognised that the velocity field $\mathbf{u}_1(\mathbf{x})$ given by (6.87) is unlike that of a real ring vortex in that the net flux of vorticity around $\mathcal{T}_1$ (including a possible surface contribution) is zero; if it were non-zero (as in a real vortex ring) then the vortex would necessarily be accompanied by an irrotational flow outside $\mathcal{T}_1$. Two real vortices oriented as in fig. 6.6 would as a result of this irrotational flow either separate and contract (case (a)) or approach each other and expand (case (b)). There can be no question of maintenance of a *steady* magnetic field by an unsteady motion of either kind.

6.11. The Bullard–Gellman formalism

Suppose now that V is the sphere $r<R$, and that $\mathbf{u}(\mathbf{x})$ is a given steady solenoidal velocity field, satisfying $\mathbf{u}\,.\,\mathbf{n}=0$ on $r=R$. It is convenient to use R as the unit of length, $u_m=\max|\mathbf{u}|$ as the unit of velocity[7], and R^2/λ as the unit of time, and to define $R_m=u_mR/\lambda$.

[7] Different authors adopt different conventions here and care is needed in making detailed comparisons.

The problem (6.4) then takes the dimensionless form

$$\left.\begin{aligned} \partial \mathbf{B}/\partial t &= R_m \nabla \wedge (\mathbf{u} \wedge \mathbf{B}) + \nabla^2 \mathbf{B} && \text{for } r<1, \\ \nabla \wedge \boldsymbol{B} &= 0 && \text{for } r>1, \\ [\mathbf{B}] &= 0 && \text{across } r=1. \end{aligned}\right\} \tag{6.107}$$

As in § 2.7, the problem admits solutions of the form

$$\mathbf{B}(\mathbf{x}, t) = \mathbf{B}(\mathbf{x})\, \mathrm{e}^{pt}, \tag{6.108}$$

where

$$\left.\begin{aligned} (p-\nabla^2)\mathbf{B} &= R_m \nabla \wedge (\mathbf{u} \wedge \mathbf{B}) && \text{for } r<1, \\ \nabla \wedge \mathbf{B} &= 0 && \text{for } r>1, \\ [\mathbf{B}] &= 0 && \text{across } r=1. \end{aligned}\right\} \tag{6.109}$$

This may be regarded as an eigenvalue problem for the parameter p, the eigenvalues $p_1, p_2, \ldots,$ being functions of R_m (as well as depending of course on the structural properties of the **u**-field). When $R_m = 0$, the eigenvalues p_α are given by the free decay mode theory of § 2.7, and they are all real and negative. As R_m increases from zero (for a given structure $\mathbf{u}(\mathbf{x})$), each may be expected to vary continuously and may become complex. If $\operatorname{Re} p_\alpha$ becomes positive for some finite value of R_m, then the corresponding field structure $\mathbf{B}^{(\alpha)}(\mathbf{x})$ is associated with an exponential growth factor in (6.108), and dynamo action occurs. If all the $\operatorname{Re} p_\alpha$ remain negative (as would happen if for example **u** were purely toroidal) then dynamo action does not occur for any value of R_m.

The natural procedure for solving (6.109) is a direct extension of that adopted in § 2.7 (when $R_m = 0$). Let P and T be the defining scalars of $\mathbf{B}_P$ and $\mathbf{B}_T$, and let

$$P(r, \theta, \varphi) = \sum_{n,m} (P_n^{mc}(r) \cos m\varphi + P_n^{ms}(r) \sin m\varphi) P_n^m(\cos \theta), \tag{6.110}$$

and similarly for T. Then, as in § 2.7, the conditions in $r \geqslant 1$ may be replaced by boundary conditions

$$T_n^m(r) = 0, \qquad \partial P_n^m/\partial r + (n+1)P_n^m = 0 \quad \text{on } r=1, \tag{6.111}$$

where $T_n^m(r)$ denotes either $T_n^{mc}(r)$ or $T_n^{ms}(r)$, and similarly for

$P_n^m(r)$. The equations for $T_n^m(r)$, $P_n^m(r)$ in $r<1$ take the form

$$\left(p-\frac{1}{r^2}\frac{\mathrm{d}}{\mathrm{d}r}r^2\frac{\mathrm{d}}{\mathrm{d}r}-\frac{n(n+1)}{r^2}\right)\begin{Bmatrix}T_n^{ms/c}\\ P_n^{ms/c}\end{Bmatrix}=R_m\begin{Bmatrix}I_n^{ms/c}\\ J_n^{ms/c}\end{Bmatrix}, \quad (6.112)$$

where $I_n^m(r)$, $J_n^m(r)$ are terms that arise through the interaction of **u** and **B**. The determination of these interaction terms requires detailed prescription of **u**, and is in general an intricate matter. It is clear however that, in view of the linearity of the induction equation (in **B**), the terms I_n^m, J_n^m can each (for given **u**) be expressed as a sum of terms linear in $T_{n'}^{m'}$, $P_{n'}^{m'}$, over a range of values of m', n' dependent on the particular choice of **u**. In general we thus obtain an infinite set of coupled linear second-order ordinary differential equations for the functions T_n^m, P_n^m.

From a purely analytical point of view, there is little more that can be done, and recourse must be had to numerical methods to make further progress. From a numerical point of view also, the problem is quite formidable. The method usually adopted is to truncate the system (6.112) by ignoring all harmonics having $n>N$, where N is some fixed integer (m is of course limited by $0\leq m\leq n$). The radial derivatives are then replaced by finite differences[8], the range $0<r<1$, being divided into say M segments. Determination of p is then reduced to a numerical search for the roots of the discriminant of the resulting set of linear algebraic equations. Interest centres on the value p_1 having largest real part. This value depends on N and M, and the method can be deemed successful only when N and M are sufficiently large that further increase in N and/or M induces negligible change in p_1. Actually, as demonstrated by Gubbins (1973), the eigenfunction $\mathbf{B}^{(1)}(\mathbf{x})$ is much more sensitive than the eigenvalue p_1 to changes in N and M, and a more convincing criterion for convergence of the procedure is provided by requiring that this eigenfunction show negligible variation with increasing N and M.

The velocity field $\mathbf{u}(\mathbf{x})$ can, like **B**, be expresssed as the sum of poloidal and toroidal parts, each of which may be expanded in surface harmonics. The motion selected for detailed study by

[8] There are other possibilities here, e.g. functions of r may be expanded as a series of spherical Bessel functions (the free decay modes) with truncation after m terms, as suggested by Elsasser (1946). This procedure has been followed by Pekeris *et al.* (1973).

Bullard & Gellman (1954) was of the form

$$\mathbf{u} = \varepsilon \nabla \wedge (\mathbf{x} Q_T(r) P_1(\cos\theta)) + \nabla \wedge \nabla \wedge (\mathbf{x} Q_P(r) P_2^2(\cos\theta)\cos 2\varphi), \tag{6.113}$$

where Q_P and Q_T had simple forms, e.g.

$$Q_P \propto r^3(1-r)^2, \qquad Q_T \propto r^2(1-r). \tag{6.114}$$

We may use the notation $\mathbf{u} = \{\varepsilon \mathbf{T}_1 + \mathbf{P}_2^{2c}\}$ as a convenient abbreviation for (6.113). Interaction of $\mathbf{u}$ and $\mathbf{B}$ for this choice of motion is depicted diagrammatically in fig. 6.8, in which the small circles

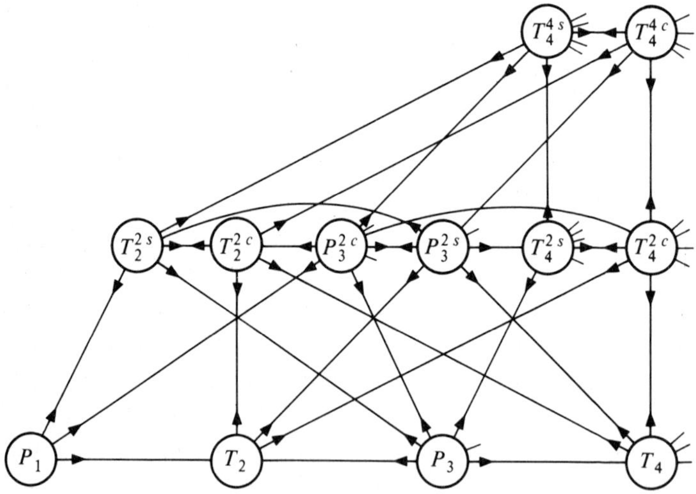

Fig. 6.8 Diagrammatic representation of the interaction of harmonics of velocity and magnetic fields when the velocity field consists of a $\mathbf{T}_1$ ingredient and a $\mathbf{P}_2^{2c}$ ingredient. Each circle indicates an excited magnetic mode; coupling along the rows is provided by the $\mathbf{T}_1$-motion and coupling between the rows by the $\mathbf{P}_2^{2c}$-motion. (From Bullard & Gellman, 1954.)

represent excited magnetic modes. The $\mathbf{T}_1$-motion introduces the coupling along the rows: for example interaction of a $\mathbf{T}_1$-motion with a $\mathbf{P}_1$-field generates a $\mathbf{T}_2$-field (this is just the process of generation of toroidal field by differential rotation analysed in § 3.11). Similarly the $\mathbf{P}_2^{2c}$-motion introduces the coupling between the rows: for example, interaction of a $\mathbf{P}_2^{2c}$-motion with a $\mathbf{P}_1$-field generates a $\{\mathbf{T}_2^{2s}+\mathbf{P}_3^{2c}\}$ field. The figure is 'truncated' for $n \geq 5$; even at this low level of truncation, the complexity of the interactions is impressive! The particular shape of the interaction diagram is entirely determined by the choice of **u**: to each possible choice of **u** there corresponds one and only one such diagram.

It was shown by Gibson & Roberts (1969) that the Bullard & Gellman velocity field (6.113) could not in fact sustain a dynamo: the procedure outlined above failed to converge as M was increased. A similar convergence failure has been demonstrated by Gubbins (1973) in respect of the more complex motion $\{\mathbf{T}_1+\mathbf{P}_2^{2c}+\mathbf{P}_2^{2s}\}$ proposed as a dynamo model by Lilley (1970). Positive results have however been obtained by Gubbins (1973) who considered axisymmetric velocity fields of the form $\varepsilon\mathbf{P}_n+\mathbf{T}_n$ with $n=6, 4, 2$ and $\varepsilon=\frac{1}{30}, \frac{1}{30}, \frac{1}{10}$ respectively. Such a motion cannot maintain a field axisymmetric about the same axis (by Cowling's theorem), but may conceivably maintain a non-axisymmetric field (cf. the Gailitis dynamo discussed in § 6.9). The dependence of p_1 on R_m as obtained numerically by Gubbins for the three cases is shown in fig. 6.9; in each case p_1 remains real as R_m increases and changes sign at

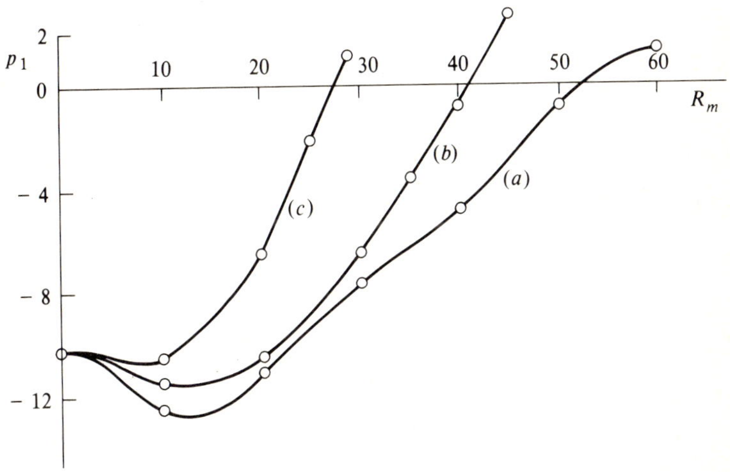

Fig. 6.9 Dependence of growth rate p_1 on R_m for a motion of the form $\mathbf{T}_n+\varepsilon\mathbf{P}_n$; (*a*) $n=2$, $\varepsilon=\frac{1}{10}$; (*b*) $n=4$, $\varepsilon=\frac{1}{30}$; (*c*) $n=6$, $\varepsilon=\frac{1}{30}$ (replotted from Gubbins, 1973). It may be shown (Gubbins, private communication) that $\mathrm{d}p_1/\mathrm{d}R_m=0$ at $R_m=0$.

a critical value R_{mc} of R_m ($R_{mc}=27, 41, 53$ in the three cases). It is noteworthy that p_1 first decreases slightly in all three cases as R_m increases from zero (indicating accelerated decay) before increasing to zero; the reason for this behaviour is by no means clear.

Positive results have also been obtained by Pekeris *et al.* (1973), who studied kinematic dynamo action associated with velocity fields satisfying the 'maximal helicity' condition $\nabla \wedge \mathbf{u}=k\mathbf{u}$ in the sphere $r<1$. Such motions are interesting in that they can be made to satisfy the equations of inviscid incompressible fluid mechanics (the Euler equations) and the condition $\mathbf{u}\,.\,\mathbf{n}=0$ on $r=1$ (Moffatt, 1969; Pekeris, 1972). The defining scalars for both poloidal and toroidal ingredients of the velocity field are then both proportional to $r^{1/2}J_{n+\frac{1}{2}}(kr)S_n(\theta, \varphi)$ where k satisfies $J_{n+\frac{1}{2}}(k)=0$ (see table 2.1 on p. 39). Pekeris *et al.* studied particularly the case $n=2$, with $S_n(\theta, \varphi)\propto \sin^2\theta \cos 2\varphi$ and found steady dynamo action for each $k=x_{2q}$ $(q=1, 2, \ldots, 20)$ in the notation of table 2.1. The corresponding critical magnetic Reynolds number R_m decreased with increasing q (i.e. with increasing radial structure in the velocity field) from 99·2 (when $q=1$) to 26·9 when $q=6$, and 26.4 when $q=20$. The values of N and M used in the numerical calculations were $N=10$, $M=100$.

The results of Pekeris *et al.* have been independently confirmed by Kumar & Roberts (1975), who also studied the numerical convergence of eigenvalues and eigenfunctions for a range of motions of the form $\{\mathbf{T}_1+\varepsilon_1\mathbf{P}_2+\varepsilon_2\mathbf{P}_2^{2c}+\varepsilon_3\mathbf{P}_2^{2s}\}$. The helicity density $\mathbf{u}\,.\,\nabla\wedge\mathbf{u}$ of such motions is antisymmetric about the equatorial plane (unlike the motions studied by Pekeris *et al.*) and in this respect are more relevant in the geophysical context[9] – see the later discussion in § 9.5.

6.12. The stasis dynamo of Backus (1958)

As noted above, it is difficult in general to justify truncation of the spherical harmonic expansion of $\mathbf{B}$, and erroneous conclusions can

[9] In this context, see also Bullard & Gubbins (1976) who have computed eigenvalues and eigenfunctions for further velocity fields having an axisymmetric structure similar to that driven by thermal convection in a rotating sphere (Weir 1976); the fields maintained by such motions are of course non-axisymmetric.

result from such truncation. There are two situations however in which truncation (of one form or another) *can* be rigorously justified. The first of these is the rotor dynamo situation of Herzenberg (1958) considered above in § 6.9, in which the radius of each rotor is small compared with the distance between rotors: in this situation the spatial attenuation of higher harmonics of the field induced by each rotor permits the imposition of a rigorous upper bound on the influence of these higher harmonics and hence permits rigorous justification of the process described (without due respect for rigour!) in § 6.9.

The second situation in which 'unwanted' higher harmonics may be dropped without violation of mathematical rigour was conceived by Backus (1958), and invokes temporal rather than spatial attenuation. We know from the theory of free decay modes that higher harmonics (i.e. those corresponding to higher values of n and q in the notation of § 2.7) decay faster than lower harmonics when $\mathbf{u} \equiv 0$, and that the fundamental harmonic (i.e. that corresponding to the lowest available values of n and q) will survive the longest and will ultimately dominate during a period of free decay (when $\mathbf{u} \equiv 0$). Suppose then that at some initial instant we start with a poloidal field $\mathbf{B}_{P11}$ where the suffix 11 indicates that only the fundamental ('dipole') harmonic $n = q = 1$ is present. Suppose that we subject this field to the influence of the following time-dependent velocity field (devised and justifiable in terms of mathematical expediency rather than physical plausibility): (i) a short period of intense differential rotation $\mathbf{u}_T$, thus generating a strong toroidal field $\mathbf{B}_T$ by the mechanism analysed in § 3.11; (ii) a period of 'stasis' ($\mathbf{u} \equiv 0$) so that all but the fundamental harmonic $\mathbf{B}_{T11}$ of $\mathbf{B}_T$ decay to a negligible level; (iii) a short period of intense non-axisymmetric poloidal motion $\mathbf{u}_P$ generating a poloidal field $\mathbf{B}_P^*$ from $\mathbf{B}_T$ through the mechanism described (at least in part) by (6.39); (iv) a second period of stasis to allow all but the fundamental harmonic $\mathbf{B}_{P11}^*$ of $\mathbf{B}_P^*$ to decay to a negligible level; (v) a rapid rigid body rotation $\boldsymbol{\omega} \wedge \mathbf{x}$ to bring $\mathbf{B}_{P11}^*$ (plus whatever remains of $\mathbf{B}_{P11}$) into alignment with the original direction of $\mathbf{B}_{P11}$. If the fields $\mathbf{u}_T$, $\mathbf{u}_P$ and $\boldsymbol{\omega} \wedge \mathbf{x}$, and the durations $t_1, t_2, \ldots, t_5$ of the phases (i), . . . , (v) are suitably chosen, then a net field amplification (with arbitrarily little change of structure) can be guaranteed. The interaction diagram corresponding to the particular velocity fields $\mathbf{u}_T$, $\mathbf{u}_P$ and $\boldsymbol{\omega} \wedge \mathbf{x}$ chosen by Backus to give substance to this skeleton procedure is indicated in fig. 6.10, in which the evanescent (and impotent) higher harmonics suppressed (relatively) by the periods of stasis are indicated by dotted lines.

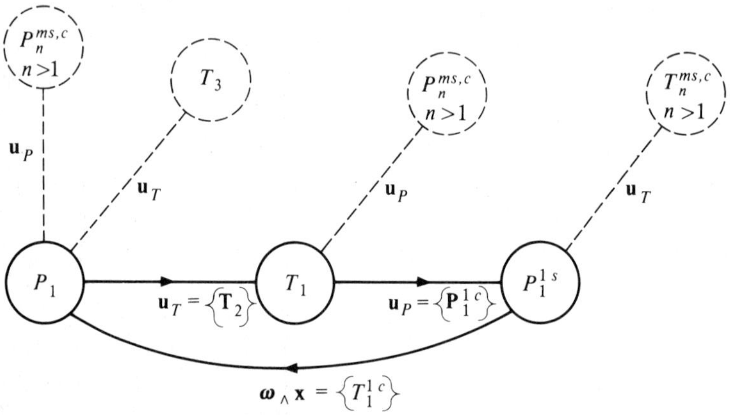

Fig. 6.10 Interaction diagram for the stasis dynamo of Backus (1958). The motion considered was of the form $\mathbf{u} = \{\mathbf{T}_2 + \mathbf{P}_1^{1c} + \mathbf{T}_1^{1c}\}$ in the notation of § 6.11, with periods of stasis between separate phases of application of the three ingredients. The dotted lines and circles indicate excitations which are evanescent due to their relatively rapid decay during the periods of stasis. The field ingredients represented by $\mathbf{P}_1$, $\mathbf{T}_1$ and $\mathbf{P}_1^{1s}$ then provide the dominant contributions in the closed dynamo cycle.

The Backus dynamo, like the Herzenberg dynamo, makes no claims to dynamic (as opposed to merely kinematic) plausibility. Although artificial (from a purely dynamical viewpoint), the enduring interest of dynamo models incorporating spatial or temporal decay (of which Herzenberg and Backus almost simultaneously devised the respective prototypes) resides in the fact that by either technique the existence of velocity fields capable of dynamo action (as defined in § 6.1) can be rigorously demonstrated. In this respect, the Herzenberg and Backus dynamos provide corner-stones (i.e. reliable positive results) that can act both as a test and a basis for subsequent developments in which mathematical rigour must necessarily give way to the more pressing demands of physical plausibility.

CHAPTER 7

THE MEAN ELECTROMOTIVE FORCE GENERATED BY RANDOM MOTIONS

7.1. Turbulence and random waves

We have so far treated the velocity field $\mathbf{U}(\mathbf{x}, t)$ as a known function of position and time[1]. In this chapter we consider the situation of greater relevance in both solar and terrestrial contexts when $\mathbf{U}(\mathbf{x}, t)$ includes a random ingredient whose statistical (i.e. average) properties are assumed known, but whose detailed (unaveraged) properties are too complicated for either analytical description or observational determination. Such a velocity field generates random perturbations of electric current and magnetic field, and our aim is to determine the evolution of the statistical properties of the magnetic field (and in particular of its local mean value) in terms of the ('given') statistical properties of the **U**-field.

The random velocity field may be a turbulent velocity field as normally understood, or it may consist of a random superposition of interacting wave motions. The distinction can be most easily appreciated for the case of a thermally stratified fluid. If the stratification is unstable (i.e. if the fluid is strongly heated from below) then thermal turbulence will ensue, the net upward transport of heat being then predominantly due to turbulent convection. If the stratification is stable (i.e. if the temperature either increases with height, or decreases at a rate less than the adiabatic rate) then turbulence will not occur, but the medium may support internal gravity waves which will be present to a greater or lesser extent, in proportion to any random influences that may be present, distributed either throughout the fluid or on its boundaries. For example, if the outer core of the Earth is stably stratified (as maintained by Higgins & Kennedy, 1971 – see § 4.4), random inertial waves may be excited either by sedimentation of iron-rich material released from the mantle across the core–mantle interface or by flotation of

[1] From now on, we shall use **U** to represent the total velocity field, reserving **u** for its random ingredient.

light compounds (rich in silicon or sulphur) released by chemical separation at the interface between inner core and outer core, or possibly by shear-induced instability in the boundary layers and shear layers formed as a result of the slow precession of the Earth's angular velocity vector. Such effects may generate radial perturbation velocities whose amplitude is limited by the stabilising buoyancy forces; the flow fields will then have the character of a field of forced weakly interacting internal waves, rather than of strongly non-linear turbulence of the 'conventional' type.

We shall throughout this chapter suppose that the random ingredient of the motion is characterised by a length-scale l_0 which is small compared with the 'global' scale L of variation of mean quantities (fig. 7.1). L will in general be of the same order of

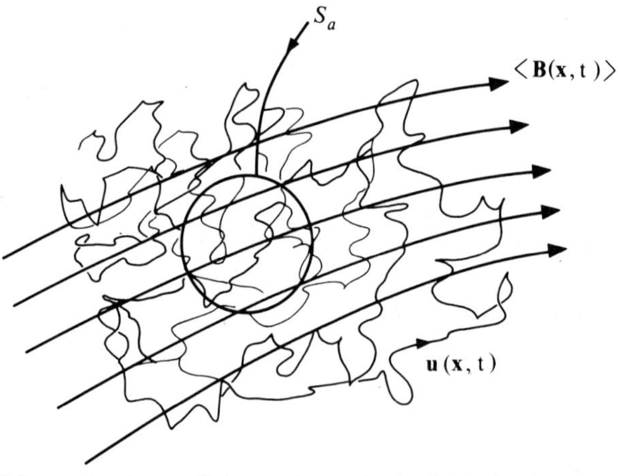

Fig. 7.1 Schematic picture of the random velocity field $\mathbf{u}(\mathbf{x}, t)$ varying on the small length-scale l_0 and the mean magnetic field varying on the large scale L. The mean is defined as an average over the sphere S_a of radius a where $l_0 \ll a \ll L$.

magnitude as the linear dimension of the region occupied by the conducting fluid, i.e. $L = O(R)$ when the fluid is confined to a sphere of radius R. In the case of turbulence, l_0 may be loosely defined as the scale of the energy-containing eddies (see e.g. Batchelor, 1953). Likewise, in the case of random waves, l_0 may be identified in order of magnitude with the wavelength of the constituent waves of maximum energy. On any intermediate scale a

satisfying the double inequality

$$l_0 \ll a \ll L, \tag{7.1}$$

the global variables (e.g. mean velocity and mean magnetic field) may be supposed nearly uniform; here the 'mean', which we shall denote by angular brackets, may reasonably be defined as an average over a sphere of intermediate radius a: i.e. for any $\psi(\mathbf{x}, t)$, we define

$$\langle\psi(\mathbf{x}, t)\rangle_a = \frac{3}{4\pi a^3}\int_{|\boldsymbol{\xi}|<a} \psi(\mathbf{x}+\boldsymbol{\xi}, t)\,\mathrm{d}^3\boldsymbol{\xi}, \tag{7.2}$$

with the expectation that this average is insensitive to the precise value of a provided merely that (7.1) is satisfied. The statistical (i.e. mean) properties of the **U**-field are weakly varying on the scale a; the methods of the theory of homogeneous turbulence (Batchelor, 1953) may therefore be employed in calculating effects on these intermediate scales.

We could equally use time-scales rather than spatial scales in defining mean quantities. If T is the time-scale of variation of global fields and t_0 is the time-scale characteristic of the fluctuating part of the **U**-field, then we shall require as a matter of consistency that T be large compared with t_0. If analysis reveals that in any situation T and t_0 are comparable, then the general approach (as applied to that situation) must be regarded as suspect. When $T \gg t_0$, then for any intermediate time-scale τ satisfying

$$t_0 \ll \tau \ll T, \tag{7.3}$$

we can define

$$\langle\psi(\mathbf{x}, t)\rangle_\tau = \frac{1}{2\tau}\int_{-\tau}^{\tau} \psi(\mathbf{x}, t+\tau')\,\mathrm{d}\tau', \tag{7.4}$$

with again a reasonable expectation that this is insensitive to the value of τ provided (7.3) is satisfied. We shall use the notation $\langle\psi(\mathbf{x}, t)\rangle$ without suffix a or τ to denote either average (7.2) or (7.4), which from a purely mathematical point of view may both be identified with an 'ensemble' average (in the asymptotic limits $l_0/L \to 0$, $t_0/T \to 0$, respectively).

Having thus defined a mean, the velocity and magnetic fields may be separated into mean and fluctuating parts:

$$\mathbf{U}(\mathbf{x}, t) = \mathbf{U}_0(\mathbf{x}, t) + \mathbf{u}(\mathbf{x}, t), \quad \langle \mathbf{u} \rangle = 0, \tag{7.5}$$

$$\mathbf{B}(\mathbf{x}, t) = \mathbf{B}_0(\mathbf{x}, t) + \mathbf{b}(\mathbf{x}, t), \quad \langle \mathbf{b} \rangle = 0. \tag{7.6}$$

Likewise the induction equation (4.10) may be separated into *its* mean and fluctuating parts:

$$\partial \mathbf{B}_0/\partial t = \nabla \wedge (\mathbf{U}_0 \wedge \mathbf{B}_0) + \nabla \wedge \mathscr{E} + \lambda \nabla^2 \mathbf{B}_0, \tag{7.7}$$

$$\partial \mathbf{b}/\partial t = \nabla \wedge (\mathbf{U}_0 \wedge \mathbf{b}) + \nabla \wedge (\mathbf{u} \wedge \mathbf{B}_0) + \nabla \wedge \mathbf{G} + \lambda \nabla^2 \mathbf{b}, \tag{7.8}$$

where

$$\mathscr{E} = \langle \mathbf{u} \wedge \mathbf{b} \rangle, \qquad \mathbf{G} = \mathbf{u} \wedge \mathbf{b} - \langle \mathbf{u} \wedge \mathbf{b} \rangle. \tag{7.9}$$

Note that in (7.7) there now appears a term associated with a product of random fluctuations. The mean electromotive force $\mathscr{E}$ is a quantity of central importance in the theory: the aim must clearly be to find a way to express $\mathscr{E}$ in terms of the mean fields $\mathbf{U}_0$ and $\mathbf{B}_0$ so that, for given $\mathbf{U}_0$, (7.7) may be integrated.

The idea of averaging equations involving random fluctuations is of course well-known in the context of conventional turbulence theory: averaging of the Navier–Stokes equations likewise leads to the appearance of the important quadratic mean $-\langle u_i u_j \rangle$ (the Reynolds stress tensor) which is the counterpart of the $\langle \mathbf{u} \wedge \mathbf{b} \rangle$ of the present context. There is no satisfactory theory of turbulence that succeeds in expressing $\langle u_i u_j \rangle$ in terms of the mean field $\mathbf{U}_0$. By contrast, there *is* now a satisfactory body of theory for the determination of $\mathscr{E}$. The reason for this (comparative) degree of success can be attributed to the linearity (in $\mathbf{B}$) of the induction equation. There is no counterpart of this linearity in the dynamics of turbulence.

The two-scale approach in the context of the induction equation was first introduced by Steenbeck, Krause & Rädler (1966), and many ideas of the present chapter can be traced either to this pioneering paper, or to the series of papers by the same authors that followed; these papers, originally published in German, are available in English translation – Roberts & Stix (1971).

7.2. The linear relation between $\mathscr{E}$ and $\mathbf{B}_0$

The term $\nabla \wedge (\mathbf{u} \wedge \mathbf{B}_0)$ in (7.8) acts as a source term generating the fluctuating field $\mathbf{b}$. If we suppose that $\mathbf{b} = 0$ at some initial instant $t = 0$, then the linearity of (7.8) guarantees that the fields $\mathbf{b}$ and $\mathbf{B}_0$ are linearly related. It follows that the fields $\mathscr{E} = \langle \mathbf{u} \wedge \mathbf{b} \rangle$ and $\mathbf{B}_0$ are likewise linearly related, and since the spatial scale of $\mathbf{B}_0$ is by assumption large (compared with scales involved in the detailed solution of (7.8)) we may reasonably anticipate that this relationship may be developed as a rapidly convergent series of the form[2]

$$\mathscr{E}_i = \alpha_{ij} B_{0j} + \beta_{ijk} \frac{\partial B_{0j}}{\partial x_k} + \gamma_{ijkl} \frac{\partial^2 B_{0j}}{\partial x_k\, \partial x_l} + \ldots, \tag{7.10}$$

where the coefficients $\alpha_{ij}, \beta_{ijk}, \ldots$, are pseudo-tensors ('pseudo' because $\mathscr{E}$ is a polar vector whereas $\mathbf{B}_0$ is an axial vector). It is clear that, since the solution $\mathbf{b}(\mathbf{x}, t)$ of (7.8) depends on $\mathbf{U}_0$, $\mathbf{u}$ and λ, the pseudo-tensors $\alpha_{ij}, \beta_{ijk}, \ldots$, may be expected to depend on, and indeed are totally determined by, (i) the mean field $\mathbf{U}_0$, (ii) the statistical properties of the random field $\mathbf{u}(\mathbf{x}, t)$, and (iii) the value of the parameter λ. These pseudo-tensors will in general vary on the 'macroscale' L, but in the conceptual limit $L/l_0 \to \infty$, when $\mathbf{U}_0$ becomes uniform and $\mathbf{u}$ becomes statistically strictly homogeneous, the pseudo-tensors $\alpha_{ij}, \beta_{ijk}, \ldots$, which may themselves be regarded as 'statistical properties of the $\mathbf{u}$-field', also become strictly uniform.

If $\mathbf{U}_0$ is uniform, then it is natural to take axes moving with velocity $\mathbf{U}_0$, and to redefine $\mathbf{u}(\mathbf{x}, t)$ as the random velocity in this frame of reference. With this convention, (7.7) becomes

$$\frac{\partial B_{0i}}{\partial t} = \varepsilon_{ijk} \frac{\partial}{\partial x_j}\left(\alpha_{kl} B_{0l} + \beta_{klm} \frac{\partial B_{0l}}{\partial x_m} + \ldots\right) + \lambda \nabla^2 B_{0i}. \tag{7.11}$$

It is immediately evident that if $\mathbf{B}_0$ is also uniform (and if $\mathbf{u}$ is statistically homogeneous so that $\alpha_{kl}, \beta_{klm}, \ldots$ are uniform) then $\partial \mathbf{B}_0/\partial t = 0$, i.e. infinite length-scale for $\mathbf{B}_0$ implies infinite time-scale. We may therefore in general anticipate that as $L/l_0 \to \infty$,

[2] Terms involving time derivatives $\partial B_{0j}/\partial t$, $\partial^2 B_{0j}/\partial t^2, \ldots$, may also appear in the expression for $\mathscr{E}$. Such terms may however always be replaced by terms involving only space derivatives by means of (7.7).

$T/t_0 \to \infty$ also, in the notation of § 7.1, and the notions of spatial and temporal means are compatible.

When $\mathbf{B}_0$ is weakly non-uniform it is evident from (7.11) that the first term on the right (incorporating α_{kl}) is likely to be of dominant importance when $\alpha_{kl} \neq 0$, since it involves the lowest derivative of $\mathbf{B}_0$; the second term (incorporating β_{klm}) is also of potential importance, and cannot in general be discarded, since like the natural diffusion term $\lambda \nabla^2 \mathbf{B}_0$, it involves second spatial derivatives of $\mathbf{B}_0$. Subsequent terms indicated by ... in (7.11) should however be negligible provided the scale of $\mathbf{B}_0$ is sufficiently large for the series (7.10) to be rapidly convergent. We shall in the following two sections consider some general properties of the α- and β-terms of (7.11), and we shall go on in subsequent sections of this chapter to evaluate α_{ij} and β_{ijk} explicitly in certain limiting situations.

7.3. The α-effect

Let us now focus attention on the leading term of the series (7.10), viz.

$$\mathscr{E}_i^{(0)} = \alpha_{ij} B_{0j}. \qquad (7.12)$$

The pseudo-tensor α_{ij} (which is uniform insofar as the **u**-field is statistically homogeneous) may be decomposed into symmetric and antisymmetric parts:

$$\alpha_{ij} = \alpha_{ij}^{(s)} - \varepsilon_{ijk} a_k \quad \text{where } a_k = -\tfrac{1}{2}\varepsilon_{ijk}\alpha_{ij}, \qquad (7.13)$$

and correspondingly, from (7.12),

$$\mathscr{E}_i^{(0)} = \alpha_{ij}^{(s)} B_{0j} + (\mathbf{a} \wedge \mathbf{B}_0)_i. \qquad (7.14)$$

It is clear that the effect of the antisymmetric part is merely to provide an additional ingredient **a** (evidently a polar vector) to the 'effective' mean velocity which acts upon the mean magnetic field: if $\mathbf{U}_0$ is the actual mean velocity, then $\mathbf{U}_0 + \mathbf{a}$ is the *effective mean velocity* (as far as the field $\mathbf{B}_0$ is concerned).

The nature of the symmetric part $\alpha_{ij}^{(s)}$ is most simply understood in the important special situation when the **u**-field is (statistically) isotropic[3] as well as homogeneous. In this situation, by definition, all statistical properties of the **u**-field are invariant under rotations (as well as translations) of the frame of reference, and in particular α_{ij} must be isotropic, i.e.

$$\alpha_{ij} = \alpha\delta_{ij}, \tag{7.15}$$

and of course in this situation $\mathbf{a} = 0$.

The parameter α is a *pseudo-scalar* (cf. the mean helicity $\langle \mathbf{u} \cdot \boldsymbol{\omega} \rangle$) and it must therefore change sign under any transformation from a right-handed to a left-handed frame of reference ('parity transformations'). Since α is a statistical property of the **u**-field, it can be non-zero only if the **u**-field itself is not statistically invariant under such a transformation. The simplest such transformation is reflexion in the origin $\mathbf{x}' = -\mathbf{x}$, and we shall say that the **u**-field is *reflexionally symmetric* if all its statistical properties are invariant under this transformation[4]. Otherwise the **u**-field *lacks reflexional symmetry*. Only in this latter case can α be non-zero.

Combination of (7.12) and (7.15) gives the very simple result

$$\mathscr{E}^{(0)} = \alpha \mathbf{B}_0, \tag{7.16}$$

and, from Ohm's law (2.117), we have a corresponding contribution to the mean *current* density

$$\mathbf{J}^{(0)} = \sigma \mathscr{E}^{(0)} = \sigma\alpha \mathbf{B}_0. \tag{7.17}$$

This possible appearance of a current parallel to the local mean field $\mathbf{B}_0$ is in striking contrast to the conventional situation in which the induced current $\sigma \mathbf{U} \wedge \mathbf{B}$ is perpendicular to the field $\mathbf{B}$. It may appear paradoxical that two fields ($\mathbf{B}$ and $\mathbf{U} \wedge \mathbf{B}$) that are everywhere perpendicular may nevertheless have mean parts that

[3] We shall use the word 'isotropic' in the weak sense to indicate 'invariant under rotations but not necessarily under reflexions' of the frame of reference.

[4] Alternatively, parity transformations of the kind $x' = -x$, $y' = y$, $z' = z$ representing mirror reflexion in the plane $x = 0$ could be adopted as the basis of a definition of 'mirror symmetry' (a term in frequent use in many published papers). Care is however needed: the mirror transformation can be regarded as a superposition of reflexion in the origin followed by a rotation; a **u**-field that is reflexionally symmetric but anisotropic will then not in general be mirror symmetric since its statistical properties will be invariant under the reflexion in the origin but not under the subsequent rotation.

are not perpendicular (and that may even be parallel); and to demonstrate beyond doubt that this is in fact a real possibility it is necessary to obtain an explicit expression for the parameter α and to show that it can indeed be non-zero (see § 7.8). The appearance of an electromotive force of the form (7.16) was described by Steenbeck & Krause (1966) as the 'α-effect', a terminology that, although arbitrary[5], is now well-established. It is this effect that is at the heart of all modern dynamo theory. The reason essentially is that it provides an obvious means whereby the 'dynamo cycle' $\mathbf{B}_P \rightleftarrows \mathbf{B}_T$ may be completed. We have seen that toroidal field $\mathbf{B}_T$ may very easily be generated from poloidal field $\mathbf{B}_P$ by the process of differential rotation. If we think now in terms of mean fields, then (7.17) indicates that the α-effect will generate toroidal current (and hence poloidal field) from the toroidal field. It is this latter step $\mathbf{B}_T \to \mathbf{B}_P$ that is so hard to describe in laminar dynamo theory; in the turbulent (or random wave) context it is brought to the same simple level as the much more elementary process $\mathbf{B}_P \to \mathbf{B}_T$. Cowling's anti-dynamo theorem no longer applies to *mean* fields, and axisymmetric analysis of mean-field evolution is then both possible and promising (see chapter 9).

As noted above, α can be non-zero only if the $\mathbf{u}$-field lacks reflexional symmetry, and in this situation the mean helicity $\langle \mathbf{u} \cdot \boldsymbol{\omega} \rangle$ will in general be non-zero also. To understand the physical nature of the α-effect, consider the situation depicted in fig. 7.2 (as conceived essentially by Parker, 1955*b*). Following Parker, we define a 'cyclonic event' as a velocity field $\mathbf{u}(\mathbf{x}, t)$ that is localised in space and time and for which the helicity $I = \int (\mathbf{u} \cdot \nabla \wedge \mathbf{u})\, dV$ is non-zero. For definiteness suppose that (in a right-handed frame of reference) $I > 0$. Such an event tends to distort a line of force of an initial field $\mathbf{B}_0$ in the manner indicated in fig. 7.2, the process of distortion being resisted more or less by diffusion. The normal $\mathbf{n}$ to the field loop generated has a component parallel to $\mathbf{B}_0$, with $\mathbf{n} \cdot \mathbf{B}_0$ less than or greater than zero depending on the net angle of twist of the loop, the former being certainly more likely if diffusion is strong or if the events are very short-lived.

[5] The effect was in fact first isolated by Parker (1955b) who introduced, on the basis of physical arguments, a parameter Γ which may be identified (almost) with the α of Steenbeck & Krause (1966).

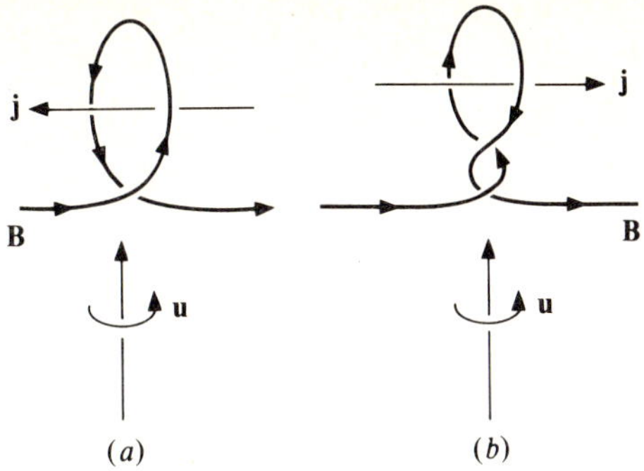

Fig. 7.2 Field distortion by a localised helical disturbance (a 'cyclonic event' in the terminology of Parker, 1970). In (*a*) the loop is twisted through an angle $\pi/2$ and the associated current is anti-parallel to **B**; in (*b*) the twist is $3\pi/2$, and the associated current is parallel to **B**.

Suppose now that cyclonic events all with $I>0$ are randomly distributed in space and time (a possible idealisation of a turbulent velocity field with positive mean helicity). Each field loop generated can be associated with an elemental perturbation current in the direction **n**, and the spatial mean of these elemental currents will have the form $\mathbf{J}^{(0)} = \sigma\alpha\mathbf{B}_0$, where (if the case $\mathbf{n}\,.\,\mathbf{B}_0 < 0$ dominates) the coefficient α will be negative. We shall in fact find below that, in the diffusion dominated situation, $\alpha\langle\mathbf{u}\,.\,\boldsymbol{\omega}\rangle < 0$ consistent with this picture.

If the **u**-field is not isotropic, then the simple relationship (7.15) of course does not hold. The symmetric pseudo-tensor $\alpha_{ij}^{(s)}$ may however be referred to its principal axes:

$$\alpha_{ij}^{(s)} = \begin{pmatrix} \alpha^{(1)} & \cdot & \cdot \\ \cdot & \alpha^{(2)} & \cdot \\ \cdot & \cdot & \alpha^{(3)} \end{pmatrix}, \tag{7.18}$$

and the corresponding contribution to $\mathscr{E}^{(0)}$ is, from (7.14),

$$\mathscr{E}^{(0s)} = (\alpha^{(1)}B_{01}, \alpha^{(2)}B_{02}, \alpha^{(3)}B_{03}). \tag{7.19}$$

Again under the reflexion $\mathbf{x}' = -\mathbf{x}$, $\alpha^{(1)}$, $\alpha^{(2)}$ and $\alpha^{(3)}$ must change sign; and so in general $\alpha_{ij}^{(s)}$ vanishes unless the **u**-field lacks reflexional symmetry.

7.4. Effects associated with the coefficient β_{ijk}

Consider now the second term of the series (7.10), viz.

$$\mathscr{E}_i^{(1)} = \beta_{ijk}\, \partial B_{0j}/\partial x_k. \tag{7.20}$$

In the simplest situation, in which the **u**-field is isotropic, β_{ijk} is also isotropic, and so

$$\beta_{ijk} = \beta\varepsilon_{ijk}, \tag{7.21}$$

where β is a pure scalar. Equation (7.20) then becomes

$$\boldsymbol{\mathscr{E}}^{(1)} = -\beta\nabla\wedge\mathbf{B}_0 = -\beta\mu_0\mathbf{J}_0, \tag{7.22}$$

where $\mathbf{J}_0$ is the mean current. Hence also (if β is uniform)

$$\nabla\wedge\boldsymbol{\mathscr{E}}^{(1)} = \beta\nabla^2\mathbf{B}_0, \tag{7.23}$$

and it is evident from (7.7) that the net effect of the emf $\boldsymbol{\mathscr{E}}^{(1)}$ is simply to alter the value of the effective magnetic diffusivity, which becomes $\lambda+\beta$ rather than simply λ. We shall find that in nearly all circumstances in which β can be calculated explicitly, it is positive, consistent with the simple physical notion of an 'eddy diffusivity': one would expect random mixing (due to the **u**-field) to enhance (rather than detract from) the process of molecular diffusion that gives rise to a positive value of λ. However there is no general proof that β must inevitably and in all circumstances be positive, and there are some indications (see § 7.11) that it may in fact in some extreme circumstances be negative; if $\beta<-\lambda$, there would of course be dramatic consequences as far as solutions of (7.7) are concerned.

If the **u**-field is not isotropic, then departures from the simple relationship (7.21) are to be expected. Suppose for example that the **u**-field is (statistically) invariant under rotations about an axis defined by the unit (polar) vector **e**, but not under general rotations, i.e. **e** defines a 'preferred direction'. Turbulence with this property is described as axisymmetric about the direction of **e**. (The situation is of course of potential importance in the context of turbulence that is strongly influenced by Coriolis forces in a system rotating with angular velocity $\mathbf{\Omega}$: in this situation $\mathbf{e} = \pm\mathbf{\Omega}/\Omega$.) The pseudo-tensor β_{ijk}, which is then also axisymmetric about the direction of $\mathbf{e}$, may be expressed in the form

$$\begin{aligned}\beta_{ijk} = \beta_0\varepsilon_{ijk} + \tilde{\beta}_1 e_i\delta_{jk} + \tilde{\beta}_2 e_j\delta_{ki} + \tilde{\beta}_3(e_k\delta_{ij} - e_j\delta_{ki})\\ + \tilde{\beta}_0 e_i e_j e_k + \beta_1\varepsilon_{mjk}e_m e_i + \beta_2\varepsilon_{imk}e_m e_j + \beta_3\varepsilon_{ijm}e_m e_k,\end{aligned} \tag{7.24}$$

where $\beta_0, \ldots, \beta_3$ are pure scalars and $\tilde{\beta}_0, \ldots, \tilde{\beta}_3$ are pseudo-scalars, which can be non-zero only if the $\mathbf{u}$-field lacks reflexional symmetry. The corresponding expression for $\mathscr{E}^{(1)}$ from (7.20) is

$$\begin{aligned}\mathscr{E}^{(1)} = -\beta_0\nabla \wedge \mathbf{B}_0\\ + \beta_1\mathbf{e}\,\nabla \,.\, (\mathbf{e} \wedge \mathbf{B}_0) + \beta_2(\mathbf{e} \wedge \nabla)(\mathbf{e} \,.\, \mathbf{B}_0) - \beta_3(\mathbf{e} \,.\, \nabla)(\mathbf{e} \wedge \mathbf{B}_0)\\ + \tilde{\beta}_0\mathbf{e}(\mathbf{e} \,.\, \nabla)(\mathbf{e} \,.\, \mathbf{B}_0) + \tilde{\beta}_2\nabla(\mathbf{e} \,.\, \mathbf{B}_0) + \tilde{\beta}_3\mathbf{e} \wedge (\nabla \wedge \mathbf{B}_0).\end{aligned} \tag{7.25}$$

The complexity of this type of expression as compared with the simple isotropic relationship (7.22) is noteworthy (and it is not hard to see that if the assumption of axisymmetry is relaxed and *two* preferred directions $\mathbf{e}^{(1)}$ and $\mathbf{e}^{(2)}$ are introduced, the relevant expression for $\mathscr{E}^{(1)}$ becomes still more complex with a dramatic increase in the number of scalar and pseudo-scalar coefficients).

It seems likely that the terms of (7.25) involving β_1, β_2 and β_3 admit interpretation in terms of contributions to a non-isotropic (eddy) diffusivity for the mean magnetic field. These terms do not however appear to have been given detailed study, and it may be that more interesting effects may be concealed in their structure.

As for the terms involving the pseudo-scalars $\tilde{\beta}_0$, $\tilde{\beta}_2$ and $\tilde{\beta}_3$, that involving $\tilde{\beta}_3$ has been singled out for detailed examination by Rädler (1969*a*,*b*). This term indicates the possible generation of a mean emf perpendicular to the mean current $\mathbf{J}_0 = \mu_0^{-1}\nabla \wedge \mathbf{B}_0$, and is of particular significance again in the context of the closure of the dynamo cycle: through this effect (the 'Rädler-effect') a toroidal emf (and so a toroidal current) may be directly generated from a poloidal current, or equivalently a poloidal field may be generated from a toroidal field. In conjunction with the complementary effect of differential rotation, a closed dynamo cycle may be envisaged, and has indeed been demonstrated by Rädler (1969*b*).

It must be emphasised however that $\tilde{\beta}_3$, being a pseudo-scalar, can be non-zero only if the **u**-field lacks reflexional symmetry[6], and in this situation the pseudo-tensor α_{ij} will in general also be non-zero. Since the dominant term of the series (7.10) involves α_{ij}, it seems almost inevitable that whenever the Rädler-effect is operative, it will be dominated by the α-effect.

7.5. First-order smoothing[7]

We now turn to the detailed solution of (7.8) and the subsequent derivation of $\mathscr{E} = \langle \mathbf{u} \wedge \mathbf{b} \rangle$. We suppose for the remainder of this chapter that $\mathbf{U}_0 = 0$, and that the **u**-field is statistically homogeneous. Effects associated with non-zero $\mathbf{U}_0$ (in particular with strong differential rotation in a spherical geometry) will be deferred to chapter 8. The difficulty in solving (7.8) in general arises from the term $\nabla \wedge \mathbf{G}$ involving the interaction of the fluctuating fields **u** and **b**, and it is natural first to consider circumstances in which this awkward term may be neglected. There are two distinct circumstances when this neglect (*the first-order smoothing approximation*) would appear to be justified. The order of magnitude of the terms in (7.8) (with $\mathbf{U}_0 = 0$) is indicated in (7.26):

$$\begin{array}{ccccccc} \partial \mathbf{b}/\partial t & = \nabla \wedge (\mathbf{u} \wedge \mathbf{B}_0) & + & \nabla \wedge \mathbf{G} & + & \lambda \nabla^2 \mathbf{b}. \\ O(b_0/t_0) & O(B_0 u_0/l_0) & & O(u_0 b_0/l_0) & & O(\lambda b_0/l_0^2) \end{array} \qquad (7.26)$$

Here, as usual, l_0 and t_0 are length- and time-scales characteristic of the **u**-field, and u_0 and b_0 are, say, the root mean square values of **u** and **b**:

$$u_0 = \langle \mathbf{u}^2 \rangle^{1/2}, \qquad b_0 = \langle \mathbf{b}^2 \rangle^{1/2}. \qquad (7.27)$$

Here we must distinguish between two situations:

conventional turbulence: $u_0 t_0 / l_0 = O(1)$, (7.28)

random waves: $u_0 t_0 / l_0 \ll 1$. (7.29)

[6] This conclusion is at variance with that of Rädler (1969*a*) who expressed the argument throughout in terms of an axial vector $\mathbf{\Omega}$ rather than a polar vector **e**, a procedure that (from a purely kinematic point of view) is hard to justify.

[7] The first-order smoothing approximation described in this section is analogous to the Born approximation in scattering theory; some authors use alternative terms, e.g. the 'quasi-linear approximation' (Kraichnan, 1976*a*).

If (7.29) is satisfied, then it is immediately clear from (7.26) that $|\nabla \wedge \mathbf{G}| \ll |\partial \mathbf{b}/\partial t|$, and that a good first approximation is given by

$$\partial \mathbf{b}/\partial t = \nabla \wedge (\mathbf{u} \wedge \mathbf{B}_0) + \lambda \nabla^2 \mathbf{b}, \tag{7.30}$$

an equation first studied (with $\mathbf{B}_0$ uniform and $\mathbf{u}$ random) by Liepmann (1952).

If on the other hand (7.28) is satisfied, then $|\partial \mathbf{b}/\partial t|$ and $|\nabla \wedge \mathbf{G}|$ are of the same order of magnitude, and both are negligible compared with $|\lambda \nabla^2 \mathbf{b}|$ provided

$$R_m = u_0 l_0/\lambda = O(l_0^2/\lambda t_0) \ll 1. \tag{7.31}$$

Under this assumption of small (turbulent) magnetic Reynolds number, a legitimate first approximation to (7.26) is

$$0 = \nabla \wedge (\mathbf{u} \wedge \mathbf{B}_0) + \lambda \nabla^2 \mathbf{b}. \tag{7.32}$$

Although the physical situations described by (7.30) and (7.32) are rather different, both equations say essentially the same thing: fluctuations $\mathbf{b}(\mathbf{x}, t)$ are generated by the interaction of $\mathbf{u}$ with the local mean field $\mathbf{B}_0$. In (7.32) this process is instantaneous because of the dominant influence of diffusion, whereas in (7.30) $\mathbf{b}(\mathbf{x}, t)$ can evidently depend on the previous history of $\mathbf{u}(\mathbf{x}, t)$ (i.e. on $\mathbf{u}(\mathbf{x}, t')$ for all $t' \leq t$). It may be anticipated that solutions of (7.30) will approximate to solutions of (7.32) when (7.31) is satisfied. We may therefore focus attention on the more general equation (7.30), bearing in mind that, in application to the turbulent (as opposed to random wave) situation, the study is relevant only if the additional condition (7.31) is satisfied.

7.6. Spectrum tensor of a stationary random vector field

Before considering the consequences of (7.30), we must digress briefly to recall certain basic properties of a random velocity field $\mathbf{u}(\mathbf{x}, t)$ that is statistically homogeneous in $\mathbf{x}$ and stationary in t. We may define (in the sense of generalised functions – see for example Lighthill, 1959) the Fourier transform (with $d\mathbf{x} \equiv d^3\mathbf{x}$)

$$\tilde{\mathbf{u}}(\mathbf{k}, \omega) = \frac{1}{(2\pi)^4} \int\int \mathbf{u}(\mathbf{x}, t)\, e^{-i(\mathbf{k}.\mathbf{x} - \omega t)}\, d\mathbf{x}\, dt, \tag{7.33}$$

which satisfies the inverse relation

$$\mathbf{u}(\mathbf{x}, t) = \iint \tilde{\mathbf{u}}(\mathbf{k}, \omega)\, e^{i(\mathbf{k}.\mathbf{x}-\omega t)}\, d\mathbf{k}\, d\omega. \tag{7.34}$$

Since $\mathbf{u}$ is real, we have for all $\mathbf{k}, \omega$,

$$\tilde{\mathbf{u}}(-\mathbf{k}, -\omega) = \tilde{\mathbf{u}}^*(\mathbf{k}, \omega), \tag{7.35}$$

where the star denotes a complex conjugate. Moreover, if $\mathbf{u}$ satisfies $\nabla . \mathbf{u} = 0$, then

$$\mathbf{k} . \tilde{\mathbf{u}}(\mathbf{k}, \omega) = 0 \quad \text{for all } \mathbf{k}. \tag{7.36}$$

Now consider the mean quantity

$$\langle \tilde{u}_i(\mathbf{k}, \omega)\tilde{u}_j^*(\mathbf{k}', \omega')\rangle$$
$$= \frac{1}{(2\pi)^8} \iiiint \langle u_i(\mathbf{x}, t)u_j(\mathbf{x}', t')\rangle\, e^{-i(\mathbf{k}.\mathbf{x}-\mathbf{k}'.\mathbf{x}'-\omega t+\omega' t')}\, d\mathbf{x}\, d\mathbf{x}'\, dt\, dt'. \tag{7.37}$$

If $\mathbf{x}' = \mathbf{x} + \mathbf{r}$ and $t' = t + \tau$, then, under the assumption of homogeneity and stationarity,

$$\langle u_i(\mathbf{x}, t)u_j(\mathbf{x}', t')\rangle = R_{ij}(\mathbf{r}, \tau), \tag{7.38}$$

the correlation tensor of the field $\mathbf{u}$. Using the basic property of the δ-function

$$\iint e^{-i(\mathbf{k}-\mathbf{k}').\mathbf{x}}\, e^{i(\omega-\omega')t}\, d\mathbf{x}\, dt = (2\pi)^4 \delta(\mathbf{k}-\mathbf{k}')\delta(\omega-\omega'), \tag{7.40}$$

(7.37) then takes the form

$$\langle \tilde{u}_i(\mathbf{k}, \omega)\tilde{u}_j^*(\mathbf{k}', \omega')\rangle = \Phi_{ij}(\mathbf{k}, \omega)\delta(\mathbf{k}-\mathbf{k}')\delta(\omega-\omega'), \tag{7.41}$$

where

$$\Phi_{ij}(\mathbf{k}, \omega) = \frac{1}{(2\pi)^4} \iint R_{ij}(\mathbf{r}, \tau)\, e^{-i(\mathbf{k}.\mathbf{r}-\omega\tau)}\, d\mathbf{r}\, d\tau. \tag{7.42}$$

The relation inverse to (7.42) is

$$R_{ij}(\mathbf{r}, \tau) = \iint \Phi_{ij}(\mathbf{k}, \omega)\, e^{i(\mathbf{k}.\mathbf{r}-\omega\tau)}\, d\mathbf{k}\, d\omega. \tag{7.43}$$

The tensor $\Phi_{ij}(\mathbf{k}, \omega)$ is the *spectrum tensor* of the field $\mathbf{u}(\mathbf{x}, t)$, and it plays a fundamental role in the subsequent theory. From (7.35), it satisfies the condition of Hermitian symmetry

$$\Phi_{ij}(\mathbf{k}, \omega) = \Phi_{ji}(-\mathbf{k}, -\omega) = \Phi^*_{ji}(\mathbf{k}, \omega), \tag{7.44}$$

while, from (7.36), if $\nabla \cdot \mathbf{u} = 0$ then, for all $\mathbf{k}$,

$$k_j \Phi_{ij}(\mathbf{k}, \omega) = 0, \qquad k_i \Phi_{ij}(\mathbf{k}, \omega) = 0. \tag{7.45}$$

The energy spectrum function $E(k, \omega)$ is defined by

$$E(k, \omega) = \tfrac{1}{2} \int_{S_k} \Phi_{ii}(\mathbf{k}, \omega)\, \mathrm{d}S, \tag{7.46}$$

where the integration is over the surface of the sphere S_k of radius k in $\mathbf{k}$-space. Note that

$$\begin{aligned} \tfrac{1}{2}\langle \mathbf{u}^2 \rangle &= \tfrac{1}{2} R_{ii}(0, 0) \\ &= \tfrac{1}{2} \iint \Phi_{ii}(\mathbf{k}, \omega)\, \mathrm{d}\mathbf{k}\, \mathrm{d}\omega = \iint E(k, \omega)\, \mathrm{d}k\, \mathrm{d}\omega, \end{aligned} \tag{7.47}$$

where the integral over $k = |\mathbf{k}|$ naturally runs from 0 to ∞. Hence $\rho E(k, \omega)\, \mathrm{d}k\, \mathrm{d}\omega$ is the contribution to the kinetic energy density from the wave-number range $(k, k + \mathrm{d}k)$ and frequency range $(\omega, \omega + \mathrm{d}\omega)$. Note that the scalar quantity $\Phi_{ii}(\mathbf{k}, \omega)$ is non-negative for all $\mathbf{k}, \omega$. (If it could be negative, integration of (7.41) over an infinitesimal neighbourhood of $(\mathbf{k}, \omega)$ would give a contradiction.) Hence also

$$E(k, \omega) \geqslant 0 \quad \text{for all } k, \omega. \tag{7.48}$$

The vorticity field $\boldsymbol{\omega} = \nabla \wedge \mathbf{u}$ evidently has Fourier transform $\tilde{\boldsymbol{\omega}} = \mathrm{i}\mathbf{k} \wedge \tilde{\mathbf{u}}$, and spectrum tensor

$$\Omega_{ij}(\mathbf{k}, \omega) = \varepsilon_{imn} \varepsilon_{jpq} k_m k_p \Phi_{nq}(\mathbf{k}, \omega). \tag{7.49}$$

In particular, using (7.45),

$$\Omega_{ii}(\mathbf{k}, \omega) = k^2 \Phi_{ii}(\mathbf{k}, \omega), \tag{7.50}$$

and as an immediate consequence

$$\tfrac{1}{2}\langle \boldsymbol{\omega}^2 \rangle = \iint k^2 E(k, \omega)\, \mathrm{d}k\, \mathrm{d}\omega. \tag{7.51}$$

By analogy with the definition of $E(k, \omega)$, we define the *helicity spectrum function* $F(k, \omega)$ by

$$F(k, \omega) = \mathrm{i} \int_{S_k} \varepsilon_{ikl} k_k \Phi_{il}(\mathbf{k}, \omega)\, \mathrm{d}S, \tag{7.52}$$

so that, with $\boldsymbol{\omega} = \nabla \wedge \mathbf{u}$,

$$\langle \mathbf{u} \cdot \boldsymbol{\omega} \rangle = \mathrm{i}\varepsilon_{ikl} \iint k_k \Phi_{il}(\mathbf{k}, \omega)\, \mathrm{d}\mathbf{k}\, \mathrm{d}\omega = \iint F(k, \omega)\, \mathrm{d}k\, \mathrm{d}\omega. \tag{7.53}$$

The function $F(k, \omega)$ is real (by virtue of (7.44)) and is a pseudo-scalar, and so vanishes if the **u**-field is reflexionally symmetric. We have seen however in the previous sections that lack of reflexional symmetry is likely to be of crucial importance in the dynamo context, and it is important therefore to consider situations in which $F(k, \omega)$ may be non-zero. The mean helicity $\langle \mathbf{u} \cdot \boldsymbol{\omega} \rangle$ is the simplest (although by no means the only) measure of the lack of reflexional symmetry of a random **u**-field.

Unlike $E(k, \omega)$, $F(k, \omega)$ can be positive or negative. It is however limited in magnitude; in fact from the Schwarz inequality in the form

$$\left| \int_{S_k} \langle \tilde{\mathbf{u}} \cdot \tilde{\boldsymbol{\omega}}^* + \tilde{\mathbf{u}}^* \cdot \tilde{\boldsymbol{\omega}} \rangle\, \mathrm{d}S \right|^2 \leqslant 4 \int_{S_k} \langle |\tilde{\mathbf{u}}|^2 \rangle\, \mathrm{d}S \int_{S_k} \langle |\tilde{\boldsymbol{\omega}}|^2 \rangle\, \mathrm{d}S, \tag{7.54}$$

together with (7.47), (7.50) and (7.53), we may deduce that[8]

$$|F(k, \omega)| \leqslant 2kE(k, \omega), \quad \text{for all } k, \omega. \tag{7.55}$$

If the **u**-field is statistically isotropic[9], as well as homogeneous, then the functions $E(k, \omega)$ and $F(k, \omega)$ are sufficient to completely specify $\Phi_{ij}(\mathbf{k}, \omega)$. In fact the most general isotropic form for $\Phi_{ij}(\mathbf{k}, \omega)$ consistent with (7.45), (7.46) and (7.52) is

$$\Phi_{ij}(\mathbf{k}, \omega) = \frac{E(k, \omega)}{4\pi k^4}(k^2\delta_{ij} - k_i k_j) + \frac{\mathrm{i}F(k, \omega)}{8\pi k^4}\varepsilon_{ijk} k_k. \tag{7.56}$$

[8] The results (7.48) and (7.55) are particular consequences of the fact that $X_i X_j^* \Phi_{ij}(\mathbf{k}, \omega) \geqslant 0$ for arbitrary complex vectors $\mathbf{X}$ (Cramer's theorem); in the isotropic case, when $\Phi_{ij}(\mathbf{k}, \omega)$ is given by (7.56), choosing $\mathbf{X}$ real gives (7.48), and choosing $\mathbf{X} = \mathbf{p} + \mathrm{i}\mathbf{q}$, where $\mathbf{p}$ and $\mathbf{q}$ are unit orthogonal vectors both orthogonal to $\mathbf{k}$, gives (7.55).

[9] See footnote 3 on p. 151.

The assumption of isotropy can lead to dramatic simplifications in the mathematical analysis. Since however turbulence (or a random wave field) that lacks reflexional symmetry can arise in a natural way only in a rotating system in which there is necessarily a preferred direction (the direction of the rotation vector $\mathbf{\Omega}$), it is perhaps unrealistic to place too much emphasis on the isotropic situation. There are however 'unnatural' ways of generating isotropic non-reflexionally symmetric turbulence, and it may be useful to describe one such 'thought experiment' if only to fix ideas. Suppose that the fluid is contained in a large sphere whose surface S is perforated by a large number of small holes placed randomly. Suppose that a small right-handed screw propeller is freely mounted at the centre of each hole, and suppose that fluid is injected at high velocity through a random subset of the holes, an equal mass flux then emerging from the complement of this subset. In a neighbourhood of the centre of the sphere the turbulent velocity field that results from the interaction of the incoming swirling jets will be approximately homogeneous and isotropic (there is clearly no preferred direction at the centre.) The turbulence nevertheless certainly lacks reflexional symmetry since each fluid particle entering the sphere follows a right-handed helical path at the start of its trajectory, so that (presumably) $\langle \mathbf{u} \cdot \boldsymbol{\omega} \rangle$ will be positive throughout the sphere. Note that the net angular momentum generated will be zero since the torques exerted on the fluid by propellers at opposite ends of a diameter will tend to cancel, and the cancellation will be complete when the injection statistics are uniform over the surface of the sphere.

If the $\mathbf{u}$-field is not isotropic, but is nevertheless statistically axisymmetric about the direction of a unit vector $\mathbf{e}$, then the most general form for $\Phi_{ij}(\mathbf{k}, \omega)$ compatible with (7.44) and (7.45) is (with $\mathbf{k} \cdot \mathbf{e} = k\mu$)

$$\begin{aligned} \Phi_{ij}(\mathbf{k}, \omega) = {} & \varphi_1(k^2\delta_{ij} - k_i k_j) + \varphi_2(k e_i e_j + k\mu^2\delta_{ij} - \mu k_i e_j - \mu k_j e_i) \\ & + i\tilde{\varphi}_3 \varepsilon_{ijk} k_k + i\tilde{\varphi}_4 \varepsilon_{ijk} e_k \\ & + \tilde{\varphi}_5 (\mathbf{k} \wedge \mathbf{e})_i k_j + \tilde{\varphi}_5^* (\mathbf{k} \wedge \mathbf{e})_j k_i \\ & + \tilde{\varphi}_6 (\mathbf{k} \wedge \mathbf{e})_i e_j + \tilde{\varphi}_6^* (\mathbf{k} \wedge \mathbf{e})_j e_i, \end{aligned} \qquad (7.57)$$

where

$$i\tilde{\varphi}_4 + k^2\tilde{\varphi}_5 + \mu k\tilde{\varphi}_6 = 0. \qquad (7.58)$$

Here, as usual, the star denotes a complex conjugate and the tilde denotes a pseudo-scalar. The functions $\varphi_1, \ldots, \tilde{\varphi}_6$ are functions of k, μ and ω; φ_1, φ_2, $\tilde{\varphi}_3$ and $\tilde{\varphi}_4$ are real, while $\tilde{\varphi}_5$ and $\tilde{\varphi}_6$ are complex. The energy and helicity spectrum functions defined by (7.46) and (7.52) are related to these functions by

$$E(k,\omega) = \pi \int_0^1 (2k^2\varphi_1 + k^3(1+\mu^2)\varphi_2)\, d\mu, \qquad (7.59)$$

$$F(k,\omega) = 4\pi \int_0^1 (k^4\tilde{\varphi}_3 + k^3\mu\tilde{\varphi}_4 + k^4(\mu^2-1)\,\mathrm{Im}\,\tilde{\varphi}_6)\, d\mu. \qquad (7.60)$$

It may of course happen that the **u**-field exhibits in its statistical properties more than one preferred direction. For example if both Coriolis forces and buoyancy forces act on the fluid, the rotation vector $\mathbf{\Omega}$ and the gravity vector $\mathbf{g}$ provide independent preferred directions which will influence the statistics of any turbulence present. The general formulae for Φ_{ij}, E and F corresponding to (7.57)–(7.60) can be readily obtained, and again (as is to be expected) only that part of Φ_{ij} involving pure scalar functions contributes to E, while only the part involving pseudo-scalar functions contributes to F. We shall meet such situations in later chapters (see particularly § 10.3).

7.7. Determination of α_{ij} for a helical wave motion

Since the expansion (7.10) is valid for *any* field distribution $\mathbf{B}_0$ of sufficiently large length-scale, we may in calculating α_{ij} suppose that $\mathbf{B}_0$ is uniform (and therefore time-independent). Restricting attention to incompressible motions for which $\nabla \cdot \mathbf{u} = 0$, (7.30) then becomes

$$\partial\mathbf{b}/\partial t - \lambda\nabla^2\mathbf{b} = (\mathbf{B}_0 \cdot \nabla)\mathbf{u}. \qquad (7.61)$$

Before treating the general random field $\mathbf{u}(\mathbf{x}, t)$, it is illuminating to consider first the effect of a single 'helical wave' given by

$$\mathbf{u}(\mathbf{x}, t) = u_0(\sin(kz - \omega t), \cos(kz - \omega t), 0) = \mathrm{Re}\, \mathbf{u}_0\, \mathrm{e}^{\mathrm{i}(\mathbf{k}.\mathbf{x} - \omega t)}, \tag{7.62}$$

where

$$\mathbf{u}_0 = u_0(-\mathrm{i}, 1, 0), \qquad \mathbf{k} = (0, 0, k), \tag{7.63}$$

and where for definiteness we suppose $k > 0$, $\omega > 0$. Note that for this motion

$$\left.\begin{aligned} \nabla \wedge \mathbf{u} = k\mathbf{u}, \qquad \mathbf{u}\,.\,(\nabla \wedge \mathbf{u}) = ku_0^2,\\ \text{and} \qquad\qquad \\ \mathrm{i}\mathbf{u}_0 \wedge \mathbf{u}_0^* = 2u_0^2(0, 0, 1). \end{aligned}\right\} \tag{7.64}$$

The helicity density is evidently uniform and positive. With this choice of $\mathbf{u}$, the corresponding periodic solution of (7.61) has the form

$$\mathbf{b}(\mathbf{x}, t) = \mathrm{Re}\, \mathbf{b}_0\, \mathrm{e}^{\mathrm{i}(\mathbf{k}.\mathbf{x} - \omega t)}, \tag{7.65}$$

where

$$\mathbf{b}_0 = \frac{\mathrm{i}\mathbf{B}_0\,.\,\mathbf{k}}{-\mathrm{i}\omega + \lambda k^2}\mathbf{u}_0. \tag{7.66}$$

Hence

$$\mathbf{b} = \frac{\mathbf{B}_0\,.\,\mathbf{k}}{\omega^2 + \lambda^2 k^4}(-\omega\mathbf{u} + \lambda k^2 \mathbf{v}), \tag{7.67}$$

where

$$\mathbf{v} = u_0(\cos(kz - \omega t), -\sin(kz - \omega t), 0) = \mathrm{Re}\, \mathrm{i}\mathbf{u}_0\, \mathrm{e}^{\mathrm{i}(\mathbf{k}.\mathbf{x} - \omega t)}, \tag{7.68}$$

and we can immediately obtain

$$\mathscr{E} = \langle \mathbf{u} \wedge \mathbf{b} \rangle = \frac{\lambda(\mathbf{B}_0\,.\,\mathbf{k})k^2}{\omega^2 + \lambda^2 k^4}\langle \mathbf{u} \wedge \mathbf{v} \rangle = -\frac{\lambda u_0^2(\mathbf{B}_0\,.\,\mathbf{k})k^2}{\omega^2 + \lambda^2 k^4}(0, 0, 1). \tag{7.69}$$

Hence in this case $\mathscr{E}_i = \alpha_{ij} B_{0j}$, where

$$\alpha_{ij} = \alpha^{(3)}\delta_{i3}\delta_{j3}, \qquad \alpha^{(3)} = -\frac{\lambda u_0^2 k^3}{\omega^2 + \lambda^2 k^4}. \tag{7.70}$$

The α-effect is clearly non-isotropic because the velocity field (7.62) exhibits the preferred direction (0, 0, 1). More significantly, note that $\alpha_{ij} \to 0$ as $\lambda \to 0$, i.e. *some diffusion is essential to generate an α-effect.* The role of diffusion is evidently (from (7.67)) to shift the phase of $\mathbf{b}$ relative to that of $\mathbf{u}$, a process that is crucial in producing a non-zero value of $\mathcal{E}$.

Note further that $\mathbf{u} \wedge \mathbf{b}$ is in fact uniform in the above situation, so that $\mathbf{G} = \mathbf{u} \wedge \mathbf{b} - \langle \mathbf{u} \wedge \mathbf{b} \rangle \equiv 0$. This means that the first-order smoothing approximation (in which the $\mathbf{G}$-term in (7.26) is ignored) is exact when the wave field contains only one Fourier component like (7.62). If more than one Fourier component is present, then $\mathbf{G}$ is no longer zero. We can however give slightly more precision to the condition for the validity of first-order smoothing. Suppose that the wave spectrum (discrete or continuous) is sharply peaked around a wave-number k_0 and a frequency ω_0, and that $u_0 = \langle \mathbf{u}^2 \rangle^{1/2}$. Then from (7.29) and (7.31), the effects of the $\mathbf{G}$-term in (7.26) should be negligible provided either

$$u_0/\lambda k_0 \ll 1 \quad \text{or} \quad u_0 k_0/\omega_0 \ll 1. \tag{7.71}$$

Conversely, if $\lambda \ll u_0/k_0$, then first-order smoothing must be regarded as a dubious approximation for all pairs $(\mathbf{k}, \omega)$ strongly represented in the wave spectrum for which $|\omega| \lesssim u_0 k$.

Note finally that the solution (7.65) does not of course satisfy an initial condition $\mathbf{b}(\mathbf{x}, 0) = 0$. If we insist on this condition, we must simply add to (7.65) the transient term

$$\mathbf{b}_1 = -\mathrm{Re}\, \mathbf{b}_0\, e^{i\mathbf{k}.\mathbf{x}}\, e^{-\lambda k^2 t}. \tag{7.72}$$

This makes an additional contribution to $\mathcal{E}$ which however decays to zero in a time of order $(\lambda k^2)^{-1}$. The limit $\lambda \to 0$ again poses problems: the influence of initial conditions is 'forgotten' only for $t \gtrsim O(\lambda k^2)^{-1}$, and the result obtained for $\mathcal{E}$ will depend on the ordering of the limiting processes $\lambda \to 0$ and $t \to \infty$ (cf. the problem discussed in § 3.8). If we first let $t \to \infty$ (with $\lambda > 0$) so that the transient effect disappears, then we obtain the result (7.69). Alternatively, if we first let $\lambda \to 0$, then we obtain (with $\mathbf{b}(\mathbf{x}, 0) = 0$)

$$\mathbf{b}(\mathbf{x}, t) = -\omega^{-1} u_0 k\,(\sin(kz - \omega t) - \sin kz,\ \cos(kz - \omega t) - \cos kz,\ 0), \tag{7.73}$$

and so

$$\mathscr{E} = \langle \mathbf{u} \wedge \mathbf{b} \rangle = -\omega^{-1} u_0^2 k \sin \omega t (0, 0, 1), \tag{7.74}$$

and $\mathscr{E}$ does not settle down to any steady value as $t \to \infty$.

7.8. Determination of α_{ij} for a random u-field under first-order smoothing

Suppose now that **u** is a stationary random function of **x** and t with Fourier transform (7.33). The Fourier transform of (7.61) is

$$(-\mathrm{i}\omega + \lambda k^2)\tilde{\mathbf{b}} = \mathrm{i}(\mathbf{B}_0 \cdot \mathbf{k})\tilde{\mathbf{u}}, \tag{7.75}$$

and we may immediately calculate

$$\mathscr{E} = \langle \mathbf{u} \wedge \mathbf{b} \rangle = \iiiint \frac{\langle \tilde{\mathbf{u}}(\mathbf{k}, \omega) \wedge \tilde{\mathbf{u}}^*(\mathbf{k}', \omega')\rangle \mathrm{i}\mathbf{B}_0 \cdot \mathbf{k}}{-\mathrm{i}\omega + \lambda k^2} \times \exp\{\mathrm{i}(\mathbf{k}-\mathbf{k}') \cdot \mathbf{x} - \mathrm{i}(\omega - \omega')t\}\, \mathrm{d}\mathbf{k}\, \mathrm{d}\mathbf{k}'\, \mathrm{d}\omega\, \mathrm{d}\omega'. \tag{7.76}$$

Using (7.41) and noting that $\mathrm{i}\varepsilon_{ikl}\Phi_{kl}(\mathbf{k}, \omega)$ is real by virtue of (7.44), and that the imaginary part of (7.76) must vanish (since $\mathscr{E}$ is real), we obtain $\mathscr{E}_i = \alpha_{ij} B_{0j}$ where

$$\alpha_{ij} = \mathrm{i}\lambda\varepsilon_{ikl} \iint (\omega^2 + \lambda^2 k^4)^{-1} k^2 k_j \Phi_{kl}(\mathbf{k}, \omega)\, \mathrm{d}\mathbf{k}\, \mathrm{d}\omega, \tag{7.77}$$

essentially a superposition of contributions like (7.70). Note that if we define $\alpha = \frac{1}{3}\alpha_{ii}$ (consistent with $\alpha_{ij} = \alpha\delta_{ij}$ in the isotropic situation) then from (7.52) and (7.77) we have

$$\alpha = -\tfrac{1}{3}\lambda \iint \frac{k^2 F(k, \omega)}{\omega^2 + \lambda^2 k^4}\, \mathrm{d}k\, \mathrm{d}\omega, \tag{7.78}$$

a result that holds irrespective of whether the **u**-field is isotropic or not. It is here that the relation between α and helicity is at its most transparent: α is simply a weighted integral of the helicity spectrum function. As remarked earlier, $F(k, \omega)$ can take positive or negative values; if however $F(k, \omega)$ is non-negative for all k, ω (and not identically zero, so that $\langle \mathbf{u} \cdot \boldsymbol{\omega} \rangle > 0$) then evidently, from (7.78), $\alpha < 0$; likewise if $F(k, \omega) \leq 0$ for all k, ω (but not identically zero) then $\alpha > 0$.

In the case of turbulence, first-order smoothing is valid only if $\lambda k^2 \gg |\omega|$ for all $(\mathbf{k}, \omega)$ for which a significant contribution is made to the integral (7.77). Hence in this situation the factor $(\omega^2+\lambda^2k^4)^{-1}$ may be replaced by $\lambda^{-2}k^{-4}$, giving

$$\alpha_{ij}\approx \mathrm{i}\lambda^{-1}\varepsilon_{ikl}\int k^{-2}k_j\Phi_{kl}(\mathbf{k})\,\mathrm{d}\mathbf{k}, \tag{7.79}$$

where

$$\Phi_{kl}(\mathbf{k})=\int \Phi_{kl}(\mathbf{k},\omega)\,\mathrm{d}\omega=\frac{1}{(2\pi)^3}\int R_{kl}(\mathbf{r},0)\,\mathrm{e}^{-\mathrm{i}\mathbf{k}.\mathbf{r}}\,\mathrm{d}\mathbf{r}. \tag{7.80}$$

Correspondingly (7.78) becomes

$$\alpha\approx-\frac{1}{3\lambda}\int k^{-2}F(k)\,\mathrm{d}k,\qquad F(k)=\int F(k,\omega)\,\mathrm{d}\omega. \tag{7.81}$$

$\Phi_{kl}(\mathbf{k})$ is the conventional zero-time-delay spectrum tensor of homogeneous turbulence (see e.g. Batchelor 1953). The results (7.79) and (7.81) may be most simply obtained directly from (7.32) (Moffatt, 1970*a*).

In the random wave situation, the full expression (7.77) must be retained. Note again the property that if there are no 'zero frequency' waves in the wave spectrum, or more precisely if

$$\Phi_{kl}(\mathbf{k},\omega)=O(\omega^2)\quad \text{as}\ \omega\to 0, \tag{7.82}$$

then

$$\alpha_{ij}=O(\lambda)\quad \text{as}\ \lambda\to 0. \tag{7.83}$$

If on the other hand $\Phi_{kl}(\mathbf{k},0)\neq 0$ then, since

$$\int_{-\infty}^{\infty}\frac{\mathrm{d}\omega}{\omega^2+\lambda^2k^4}=\frac{2\pi}{\lambda k^2}, \tag{7.84}$$

we obtain formally from (7.77)

$$\alpha_{ij}\sim 2\pi\mathrm{i}\varepsilon_{ikl}\int k_j\Phi_{kl}(\mathbf{k},0)\,\mathrm{d}\mathbf{k}\quad \text{as}\ \lambda\to 0. \tag{7.85}$$

Here however we must bear in mind the limitations of the first-order smoothing approximation. As indicated in the remark following (7.71), this approximation is suspect when λ is small and $|\omega|\lesssim u_0k$; since the asymptotic expression (7.85) is determined entirely by the spectral density at $\omega=0$, the limiting procedure that yields (7.85) is in fact incompatible with the first-order smoothing approximation. The limiting result (7.85) is therefore of dubious validity.

If the **u**-field is isotropic, then from (7.56)

$$\mathrm{i}\varepsilon_{ikl}\Phi_{kl}(\mathbf{k},\omega)=-(4\pi k^4)^{-1}k_iF(k,\omega), \tag{7.86}$$

and so (7.77) becomes simply $\alpha_{ij}=\alpha\delta_{ij}$ where α is given by (7.78).

Under the weaker symmetry condition that the **u**-field is statistically axisymmetric, (7.57) gives

$$\mathrm{i}\varepsilon_{ikl}\Phi_{kl}(\mathbf{k},\omega)=-2[k_i\tilde{\varphi}_3+e_i\tilde{\varphi}_4+(k^2e_i-k\mu k_i)\operatorname{Im}\tilde{\varphi}_5 \\ +(k\mu e_i-k_i)\operatorname{Im}\tilde{\varphi}_6], \tag{7.87}$$

and (7.77) then reduces to the axisymmetric form

$$\alpha_{ij}=\alpha\delta_{ij}+\alpha_1(\delta_{ij}-3e_ie_j), \tag{7.88}$$

where $\alpha=\frac{1}{3}\alpha_{ii}$ is still given by (7.78), and

$$\alpha_1=\tfrac{1}{2}\alpha+\lambda\iiint\frac{k^3\mu}{\omega^2+\lambda^2k^4}(k\mu\tilde{\varphi}_3+\tilde{\varphi}_4 \\ +k^2(1-\mu^2)\operatorname{Im}\tilde{\varphi}_5)\,\mathrm{d}k\,\mathrm{d}\mu\,\mathrm{d}\omega. \tag{7.89}$$

It is relevant to note here that the α-effect has been detected in the laboratory in an ingenious experiment carried out by Steenbeck *et al.* (1967). A velocity field having a deliberately contrived negative mean helicity was generated in liquid sodium by driving the liquid through two linked copper ducts (fig. 7.3(*a*)). The effective magnetic Reynolds number was low, and the result (7.81) is therefore relevant. The streamline linkage was left-handed, and, assuming maximum helicity, $\langle\mathbf{u}\,.\,\boldsymbol{\omega}\rangle\approx-u_0^2/l_0$; from (7.81) (or (7.70) with $\omega=0$, $k=l_0^{-1}$), an order of magnitude estimate for α is given by

$$\alpha\sim l_0u_0^2/\lambda, \tag{7.90}$$

where u_0 is the mean velocity through either duct. The total potential drop between electrodes at the points X and Y is then

$$\Delta\phi\sim(nl_0u_0^2/\lambda)B_0, \tag{7.91}$$

where n is the number of duct sections between X and Y ($n=28$ in the experiment) and B_0 is the field applied (by external windings) parallel to the 'axis' XY. The measured values of $\Delta\phi$ ranged from zero up to 60 millivolts as u_0 and B_0 were varied. Fig. 7.3(*b*) and (*c*) shows the measured variation of $\Delta\phi$ with u_0^2 and with B_0. The linear relation between $\Delta\phi$ and u_0^2 is strikingly verified by these measure-

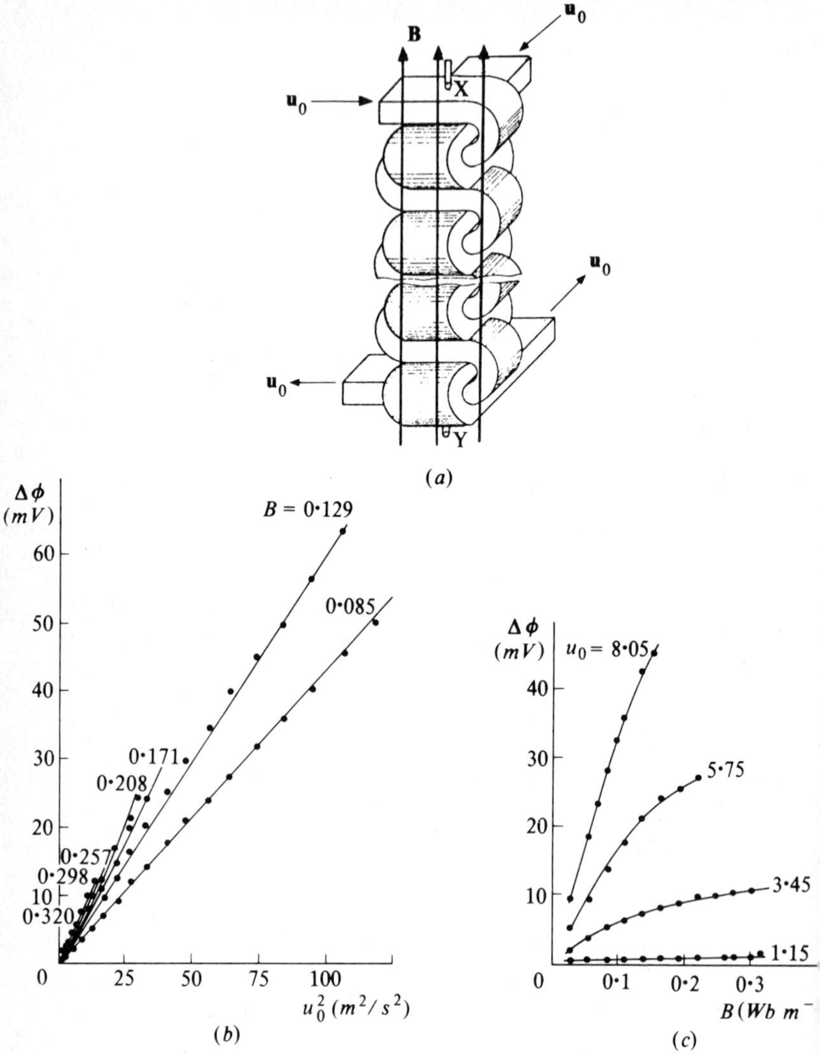

Fig. 7.3 Experimental verification of the α-effect. (*a*) Duct configuration; (*b*) potential difference $\Delta\phi$ measured between the electrodes X and Y as a function of u_0^2 for various values of the applied field $\mathbf{B}$; (*c*) $\Delta\phi$ as a function of B for various values of u_0. (Steenbeck *et al.*, 1967.)

ments. On the other hand the linear relation between $\Delta\phi$ and B_0 is evidently valid only when B_0 is weak ($\lesssim 0{\cdot}1$ Wb m^{-2}); reasons for the non-linear dependence of $\Delta\phi$ on B_0 when B_0 is strong must no doubt be sought in dynamical modifications of the (turbulent) velocity distribution in the duçts due to non-negligible Lorentz forces.

7.9. Determination of β_{ijk} under first-order smoothing

To determine β_{ijk}, we suppose now that the field $\mathbf{B}_0(\mathbf{x})$ in the expansion (7.10) is of the form

$$B_{0j}(\mathbf{x}) = x_k(\partial B_{0j}/\partial x_k), \tag{7.92}$$

where the field gradient $\partial B_{0j}/\partial x_k$ is uniform. Equation (7.30) then becomes

$$\left(\frac{\partial}{\partial t} - \lambda \nabla^2\right) b_i = x_k \frac{\partial B_{0j}}{\partial x_k} \frac{\partial u_i}{\partial x_j} - u_j \frac{\partial B_{0i}}{\partial x_j}, \tag{7.93}$$

with Fourier transform

$$(-\mathrm{i}\omega + \lambda k^2)\tilde{b}_i = \frac{\partial}{\partial k_k}(k_j \tilde{u}_i) \frac{\partial B_{0j}}{\partial x_k} - \tilde{u}_j \frac{\partial B_{0i}}{\partial x_j}. \tag{7.94}$$

Construction of $\langle \mathbf{u} \wedge \mathbf{b} \rangle_i$ now leads to an expression of the form $\beta_{ijk} \partial B_{0j}/\partial x_k$, where (after some manipulation)

$$\beta_{ijk} = \mathrm{Re}\, \varepsilon_{iml} \iint \frac{\mathrm{i}\omega + \lambda k^2}{\omega^2 + \lambda^2 k^4} \left\{ \frac{\partial}{\partial k_j} k_k \Phi_{lm}(\mathbf{k}, \omega) - \Phi_{jm}(\mathbf{k}, \omega) \delta_{kl} \right\} \mathrm{d}\mathbf{k}\, \mathrm{d}\omega. \tag{7.95}$$

Note the appearance in this expression of the gradient in **k**-space of the spectrum tensor. Again in the turbulence situation (7.95) must be replaced by

$$\beta_{ijk} \approx \mathrm{Re}\, \varepsilon_{iml} \int \lambda^{-1} k^{-2} \left\{ \frac{\partial}{\partial k_j} k_k \Phi_{lm}(\mathbf{k}) - \Phi_{jm}(\mathbf{k}) \delta_{kl} \right\} \mathrm{d}\mathbf{k}. \tag{7.96}$$

Now $\varepsilon_{iml} \Phi_{lm}(\mathbf{k})$ is pure imaginary (by virtue of the Hermitian symmetry of Φ_{lm}), and so (7.96) reduces to

$$\beta_{ijk} \approx -\mathrm{Re}\, \varepsilon_{imk} \int \lambda^{-1} k^{-2} \Phi_{jm}(\mathbf{k})\, \mathrm{d}\mathbf{k}. \tag{7.97}$$

In the case of an isotropic **u**-field, with $\Phi_{ij}(\mathbf{k}, \omega)$ given by (7.56), it is again only the second term under the integral (7.95) that makes a non-zero contribution, and we find $\beta_{ijk} = \beta\varepsilon_{ijk}$ where

$$\beta = \tfrac{1}{6}\varepsilon_{ijk}\beta_{ijk} = \tfrac{2}{3}\lambda \iint \frac{k^2 E(k,\omega)}{\omega^2+\lambda^2 k^4}\,dk\,d\omega, \tag{7.98}$$

the corresponding expression in the turbulence limit being

$$\beta \approx \tfrac{2}{3}\lambda^{-1}\int_0^\infty k^{-2}E(k)\,dk. \tag{7.99}$$

Similarly, for the case of rotational invariance about a direction **e**, substitution of (7.57) in (7.96) leads to explicit expressions for the coefficients in the axisymmetric form (7.24) of β_{ijk}. It is tedious to calculate these coefficients and the expressions will not be given here; it is enough to note that the scalar coefficients $\beta_0, \ldots, \beta_3$ emerge as linear functionals of $\varphi_1(k, \mu, \omega)$ and $\varphi_2(k, \mu, \omega)$ while the pseudo-scalar coefficients $\tilde{\beta}_0, \ldots, \tilde{\beta}_3$ emerge (like α and α_1 in (7.78) and (7.89)) as linear functionals of $\tilde{\varphi}_3, \ldots, \tilde{\varphi}_6$; and that, as commented earlier, in any circumstance in which $\tilde{\beta}_0, \ldots, \tilde{\beta}_3$ are non-zero, α and α_1 are generally non-zero also.

7.10. Lagrangian approach to the weak diffusion limit

For a turbulent velocity field $\mathbf{u}(\mathbf{x}, t)$ with $u_0t_0/l_0 = O(1)$, and in the weak diffusion limit $R_m = u_0l_0/\lambda \gg 1$, the first-order smoothing approach described in the previous sections is certainly not applicable. Even in the random wave situation ($u_0t_0/l_0 \ll 1$), we have seen that first-order smoothing may break down in the limit $\lambda \to 0$ if the wave spectral density at $\omega = 0$ is non-zero. An alternative approach that retains the influence of the interaction term $\nabla \wedge \mathbf{G}$ in (7.8) is therefore desirable. The following approach (Parker, 1971*a*; Moffatt, 1974) is analogous to the traditional treatment of turbulent diffusion of a passive scalar field in the limit of vanishing molecular diffusivity (Taylor, 1921).

The starting point is the Lagrangian solution of the induction equation, which is exact in the limit $\lambda = 0$, viz.

$$B_i(\mathbf{x}, t) = B_j(\mathbf{a}, 0)\,\partial x_i/\partial a_j, \tag{7.100}$$

in the Lagrangian notation of § 2.5. Hence we have immediately

$$\mathscr{E}_i(\mathbf{x}, t) = \langle \mathbf{u} \wedge \mathbf{b} \rangle_i = \langle \mathbf{u} \wedge \mathbf{B} \rangle_i$$
$$= \varepsilon_{ijk} \langle u_j^L(\mathbf{a}, t) B_l(\mathbf{a}, 0)\, \partial x_k / \partial a_l \rangle. \qquad (7.101)$$

Evaluation of α_{ij}

As in § 7.8, we may most simply obtain an expression for α_{ij} on the assumption that $\mathbf{B}_0$ is uniform (and therefore constant). If $\mathbf{b}(\mathbf{x}, 0) = 0$, then $\mathbf{B}(\mathbf{a}, 0) = \mathbf{B}_0$, and (7.101) then has the expected form $\mathscr{E}_i = \alpha_{il} B_{0l}$, where however α_{il} is a function of t:

$$\alpha_{il}(t) = \varepsilon_{ijk} \langle u_j^L(\mathbf{a}, t)\, \partial x_k(\mathbf{a}, t) / \partial a_l \rangle. \qquad (7.102)$$

Now the displacement of a fluid particle is simply the time integral of its Lagrangian velocity, i.e.

$$\mathbf{x}(\mathbf{a}, t) - \mathbf{a} = \int_0^t \mathbf{u}^L(\mathbf{a}, \tau)\, \mathrm{d}\tau. \qquad (7.103)$$

Hence, since $\langle \mathbf{u}^L(\mathbf{a}, t) \rangle = 0$, (7.102) becomes

$$\alpha_{il}(t) = \varepsilon_{ijk} \int_0^t \langle u_j^L(\mathbf{a}, t)\, \partial u_k^L(\mathbf{a}, \tau) / \partial a_l \rangle\, \mathrm{d}\tau. \qquad (7.104)$$

The time-dependence here is of course associated with the imposition of the initial condition $\mathbf{b}(\mathbf{x}, 0) = 0$ which trivially implies that $\alpha_{il} = 0$ when $t = 0$. When $t \gg t_1$, where t_1 is a typical turbulence correlation time, one would expect the influence of initial conditions to be 'forgotten'; equivalently one would expect $\alpha_{il}(t)$ to settle down asymptotically to a constant value given by

$$\alpha_{il} = \varepsilon_{ijk} \int_0^\infty \langle u_j^L(\mathbf{a}, t)\, \partial u_k^L(\mathbf{a}, \tau) / \partial a_l \rangle\, \mathrm{d}\tau. \qquad (7.105)$$

There is however some doubt concerning the convergence of the integral (7.104) as $t \to \infty$ for a general stationary random field of turbulence, and the delicate question of whether the influence of initial conditions is ever forgotten (when $\lambda = 0$) is still to some extent unanswered.[10]

Some light may be thrown on this question through comparison of (7.104) with the expression for the diffusion tensor of a passive scalar field, viz. (Taylor, 1921)

$$D_{ij}(t) = \int_0^t \langle u_i^L(\mathbf{a}, t) u_j^L(\mathbf{a}, \tau)\rangle \, d\tau. \qquad (7.106)$$

For a statistically stationary field of turbulence, the integrand here (the Lagrangian correlation tensor) is a function of the time difference $t - \tau$ only:

$$\langle u_i^L(\mathbf{a}, t) u_j^L(\mathbf{a}, \tau)\rangle = R_{ij}^{(L)}(t-\tau), \qquad (7.107)$$

and for $t \gg t_1$, provided simply that

$$t^{1+\mu} R_{ij}^{(L)}(t) \to 0 \quad \text{as } t \to \infty \quad \text{for some } \mu > 0, \qquad (7.108)$$

(7.104) gives

$$D_{ij} \sim \int_0^\infty R_{ij}^{(L)}(t) \, dt. \qquad (7.109)$$

The condition (7.108) is a very mild requirement on the statistics of the turbulence.

There is however a crucial difference between (7.106) and (7.104) in that the latter contains the novel type of derivative

$$\partial u_k^L / \partial a_l = (\partial u_k / \partial x_m)(\partial x_m / \partial a_l). \qquad (7.110)$$

Although $\partial u_k/\partial x_m$ is statistically stationary in time, $\partial x_m / \partial a_l$ in general is not, since any two particles $(\mathbf{a}, \mathbf{a}+\delta\mathbf{a})$ initially adjacent tend to wander further and further apart – in fact $|\delta\mathbf{x}|/|\delta\mathbf{a}| \sim t^{1/2}$ as $t \to \infty$; it follows that $\partial u_k^L/\partial a_l$ is not statistically stationary in time in general, and so the integrand in (7.104) depends on t and τ independently and not merely on the difference $t - \tau$.

[10] Computer simulations have recently been carried out by Kraichnan (1976*b*) with the aim of evaluating the integrals (7.115) and (7.116) below for the case of a statistically isotropic field with Gaussian statistics. The results show that the integrals for $\alpha(t)$ and $\beta(t)$ do in general converge as $t \to \infty$ to values of order u_0 and $u_0 l_0$ respectively; whether this is true for non-Gaussian statistics is not yet clear.

As in the discussion following (7.85), it is apparently the 'zero frequency' ingredients of the velocity field which are responsible for the possible divergence of (7.104). In any time-periodic motion (with zero mean) any two particles that are initially adjacent do not drift apart but remain permanently adjacent; it is spectral contributions in the neighbourhood of $\omega = 0$ which are responsible for the relative dispersion of particles in turbulent flow, and it is these same contributions that make it hard to justify the step between (7.104) and (7.105).

Evaluation of β_{ijk}

Suppose now that the mean field gradient $\partial B_{0i}/\partial x_j$ is uniform at time $t = 0$. From (7.11), we have

$$\frac{\partial}{\partial t}\frac{\partial B_{0i}}{\partial x_p} = \varepsilon_{ijk}\frac{\partial}{\partial x_j}\left(\alpha_{kl}\frac{\partial B_{0l}}{\partial x_p} + \beta_{klm}\frac{\partial^2 B_{0l}}{\partial x_m\,\partial x_p} + \ldots\right) + \lambda\nabla^2\frac{\partial B_{0i}}{\partial x_p}, \qquad (7.111)$$

so that (with α_{kl}, β_{klm} uniform in space), $\partial B_{0i}/\partial x_p$ then remains constant in time. We may therefore integrate (7.11) to give

$$B_{0i}(\mathbf{x}, t) = B_{0i}(\mathbf{x}, 0) + \varepsilon_{ijk}\frac{\partial B_{0l}}{\partial x_j}\int_0^t \alpha_{kl}(\tau)\,\mathrm{d}\tau. \qquad (7.112)$$

From (7.101), we now obtain

$$\mathscr{E}_i(\mathbf{x}, t) = \varepsilon_{ijk}\langle u_j^L(\mathbf{a}, t)\,\partial x_k/\partial a_l(B_{0l}(\mathbf{x}, 0) - (\mathbf{x}-\mathbf{a})_m\,\partial B_{0l}/\partial x_m)\rangle, \qquad (7.113)$$

and, using (7.112), this is of the form

$$\mathscr{E}_i(\mathbf{x}, t) = \alpha_{il}(t)B_{0l}(\mathbf{x}, t) + \beta_{ilm}(t)\,\partial B_{0l}/\partial x_m,$$

where $\alpha_{il}(t)$ is as given by (7.104), and

$$\beta_{ilm}(t) = -\varepsilon_{ijl}D_{jm}(t) - \varepsilon_{ijk}\int_0^t\int_0^t \left\langle u_j^L(\mathbf{a}, t)\frac{\partial u_k^L(\mathbf{a}, \tau_1)}{\partial a_l}u_m^L(\mathbf{a}, \tau_2)\right\rangle \mathrm{d}\tau_1\mathrm{d}\tau_2$$
$$-\varepsilon_{pmk}\int_0^t \alpha_{kl}(\tau)\alpha_{ip}(t)\,\mathrm{d}\tau, \qquad (7.114)$$

where $D_{jm}(t)$ is given by (7.106). This expression now involves the double integral of a triple Lagrangian correlation (whose convergence as $t\to\infty$ is open to the same doubts as expressed for the integral (7.104)). Moreover in the case of turbulence that lacks reflexional symmetry for which $\alpha_{ij}(t)\neq 0$, if $\alpha_{ij}(t)$ tends to a non-zero constant value as $t\to\infty$, then the final term of (7.114) is certainly unbounded as $t\to\infty$.[11] This result is of course due to the total neglect of molecular diffusivity effects; it is possible that inclusion of weak diffusion effects (i.e. small but non-zero λ) will guarantee the convergence of $\alpha_{ij}(t)$ and $\beta_{ijk}(t)$ to constant values as $t\to\infty$, but this is something that has yet to be proved.

The isotropic situation

From (7.104)

$$\alpha(t)=\tfrac{1}{3}\alpha_{ii}(t)=-\tfrac{1}{3}\int_0^t \langle \mathbf{u}^L(\mathbf{a},t)\,.\,\nabla_{\mathbf{a}}\wedge\mathbf{u}^L(\mathbf{a},\tau)\rangle\,\mathrm{d}\tau, \quad (7.115)$$

and if $\mathbf{u}(\mathbf{x},t)$ is isotropic then $\alpha_{ij}=\alpha\delta_{ij}$. The operator $\nabla_{\mathbf{a}}$ indicates differentiation with respect to $\mathbf{a}$. The integrand here contains a type of Lagrangian helicity correlation; note again the appearance of the minus sign in (7.115).

Similarly in the isotropic case, $\beta_{ijk}=\beta(t)\varepsilon_{ijk}$ where, from (7.114),

$$\begin{aligned}\beta(t)=\tfrac{1}{6}\varepsilon_{ijk}\beta_{ijk}&=\tfrac{1}{3}\int_0^t\langle\mathbf{u}^L(t)\,.\,\mathbf{u}^L(\tau)\rangle\,\mathrm{d}\tau+\int_0^t\alpha(t)\alpha(\tau)\,\mathrm{d}\tau\\&+\tfrac{1}{6}\int_0^t\int_0^t\langle\mathbf{u}^L(t)\,.\,\mathbf{u}^L(\tau_2)\nabla_{\mathbf{a}}\,.\,\mathbf{u}^L(\tau_1)\\&\qquad-(\mathbf{u}^L(t)\,.\,\nabla_{\mathbf{a}}\mathbf{u}^L(\tau_1))\,.\,\mathbf{u}^L(\tau_2)\rangle\,\mathrm{d}\tau_1\,\mathrm{d}\tau_2\end{aligned} \quad (7.116)$$

where the dependence of $\mathbf{u}^L$ on $\mathbf{a}$ is understood throughout. The first term here is the effective turbulent diffusivity for a scalar field,

[11] The fact that the computer simulations of Kraichnan (1976*b*) give a finite value for $\beta(t)=\frac{1}{6}\varepsilon_{ijk}\beta_{ijk}(t)$ as $t\to\infty$ (see footnote on p. 172) implies that this divergence must be compensated by simultaneous divergence of the second term of (7.114) involving the double integral. This fortuitous occurrence can hardly be of general validity, given the very different structure of the two terms, and may well be associated with the particular form of Gaussian statistics adopted by Kraichnan in the numerical specification of the velocity field. It may be noted that if the second and third terms on the right of (7.116) *exactly* compensate each other, then $\beta(t)=\frac{1}{3}D_{ii}(t)$, i.e. the magnetic turbulent diffusivity is equal to the scalar turbulent diffusivity. This was claimed as an exact result by Parker (1971*b*), but by an argument which has been questioned by Moffatt (1974). The results of Kraichnan (1976*a*,*b*) indicate that although $\beta(t)$ and $D(t)=\frac{1}{3}D_{ii}(t)$ may be of the same order of magnitude as $t\to\infty$, they are *not* in general identically equal.

and the second and third terms describe effects that are exclusively associated with the vector character of **B**. The structure of the second term, involving the product of values of α at different instants of time, suggests that *fluctuations* in helicity may have an important effect on the effective magnetic diffusivity. This suggestion, advanced by Kraichnan (1976*a*), will be examined further in the following section.

7.11. Effect of helicity fluctuations on effective turbulent diffusivity

When $\alpha_{kl} = \alpha\delta_{kl}$ and $\beta_{klm} = \beta\varepsilon_{klm}$, (7.11) takes the form

$$\partial \mathbf{B}_0/\partial t = \nabla \wedge (\alpha \mathbf{B}_0) + \lambda_1 \nabla^2 \mathbf{B}_0, \tag{7.117}$$

where $\lambda_1 = \lambda + \beta$, and β is assumed uniform. Let us now (following Kraichnan, 1976*a*) consider the effect of spatial and temporal fluctuations in α on scales l_α, t_α satisfying

$$l_0 \ll l_\alpha \ll L, \qquad t_0 \ll t_\alpha \ll T. \tag{7.118}$$

In order to handle such a situation, we need to define a double averaging process over scales a_1 and a_2 satisfying

$$l_0 \ll a_1 \ll l_\alpha \ll a_2 \ll L. \tag{7.119}$$

Preliminary averaging over the scale a_1 yields (7.117) as described in the foregoing sections. Now we treat $\alpha(\mathbf{x}, t)$ as a random function and examine the effect of averaging (7.117) over the scale a_2. (The process may also be interpreted in terms of an 'ensemble of ensembles': in each sub-ensemble α is constant, but it varies randomly from one sub-ensemble to another.) We shall use the notation $\langle\!\langle \ldots \rangle\!\rangle$ to denote averaging over the scale a_2 of quantities already averaged over the scale a_1. We shall suppose further that the **u**-field is globally reflexionally symmetric so that in particular $\langle\!\langle \alpha \rangle\!\rangle = 0$.

Spatial fluctuations in α will presumably occur in the presence of corresponding fluctuations in background helicity $\langle \mathbf{u} \cdot \boldsymbol{\omega} \rangle$. It is easy to conceive of a kinematically possible random velocity field exhibiting such fluctuations. A pair of vortex rings linked as in fig. 2.1(*a*) has an associated positive helicity; reversing the sign of one of the arrows gives a similar 'flow element' with negative helicity.

We can imagine such elements distributed at random in space in such a way as to give a velocity field that is homogeneous and isotropic, and reflexionally symmetric if elements of opposite parity occur with equal probability. Clustering of right-handed and left-handed elements will however give spatial fluctuations in helicity on the scale of the clusters.

From a dynamic point of view, there may seem little justification for consideration of somewhat arbitrary models of this kind. The reason for doing so is the following. When α is constant, (7.117) has solutions that grow exponentially when the length-scale is sufficiently large (see § 9.2 for details). In turbulence that is reflexionally symmetric, α is zero, and there then seems no possibility of growth of $\mathbf{B}_0$ according to (7.117). We have however encountered grave difficulties in calculating β (and so λ_1) in any circumstances which are not covered by the simple first-order smoothing approximation, and it is difficult to exclude the possibility that the effective diffusivity may even be negative in some circumstances. Kraichnan's (1976*a*) investigation was motivated by a desire to shed light on this question.

Let us then (following the same procedure as applied to the induction equation in § 7.1) split (7.117) into mean and fluctuating parts. Defining

$$\mathbf{B}_0 = \langle\langle \mathbf{B} \rangle\rangle + \mathbf{b}_1(\mathbf{x}, t) \quad \text{with } \langle\langle \mathbf{b}_1 \rangle\rangle = 0, \tag{7.120}$$

we obtain

$$\partial \langle\langle \mathbf{B} \rangle\rangle / \partial t = \nabla \wedge \langle\langle \alpha \mathbf{b}_1 \rangle\rangle + \lambda_1 \nabla^2 \langle\langle \mathbf{B} \rangle\rangle, \tag{7.121}$$

and

$$\partial \mathbf{b}_1 / \partial t = -\langle\langle \mathbf{B} \rangle\rangle \wedge \nabla \alpha + \alpha \nabla \wedge \langle\langle \mathbf{B} \rangle\rangle + \nabla \wedge \mathbf{G}_1 + \lambda_1 \nabla^2 \mathbf{b}_1, \tag{7.122}$$

where

$$\mathbf{G}_1 = \alpha \mathbf{b}_1 - \langle\langle \alpha \mathbf{b}_1 \rangle\rangle. \tag{7.123}$$

Let us now apply the first-order smoothing method to (7.122). The term $\nabla \wedge \mathbf{G}_1$ is negligible provided either

$$\alpha_0 t_\alpha / l_\alpha \ll 1 \qquad \text{or} \qquad \alpha_0 l_\alpha / \lambda_1 \ll 1, \tag{7.124}$$

where $\alpha_0^2 = \langle\langle \alpha^2 \rangle\rangle$, the mean square of the fluctuation field $\alpha(\mathbf{x}, t)$. The Fourier transforms of (7.122) (treating $\langle\langle \mathbf{B} \rangle\rangle$ and $\nabla \wedge \langle\langle \mathbf{B} \rangle\rangle$ as uniform) is then

$$(-\mathrm{i}\omega + \lambda_1 k^2)\tilde{\mathbf{b}}_1 = -\mathrm{i}\langle\langle \mathbf{B} \rangle\rangle \wedge \mathbf{k}\tilde{\alpha} + \tilde{\alpha}\nabla \wedge \langle\langle \mathbf{B} \rangle\rangle, \qquad (7.125)$$

from which we may readily obtain $\langle\langle \alpha \mathbf{b}_1 \rangle\rangle$ in the form

$$\langle\langle \alpha \mathbf{b}_1 \rangle\rangle = -\langle\langle \mathbf{B} \rangle\rangle \wedge \mathbf{Y} + X\nabla \wedge \langle\langle \mathbf{B} \rangle\rangle \qquad (7.126)$$

where

$$X = \iint \frac{\lambda_1 k^2 \Phi_\alpha(\mathbf{k}, \omega)}{\omega^2 + \lambda_1^2 k^4} \, \mathrm{d}\mathbf{k} \, \mathrm{d}\omega, \qquad \mathbf{Y} = \iint \frac{\mathbf{k}\Phi_\alpha(\mathbf{k}, \omega)}{\omega^2 + \lambda_1^2 k^4} \, \mathrm{d}\mathbf{k} \, \mathrm{d}\omega, \qquad (7.127)$$

where $\Phi_\alpha(\mathbf{k}, \omega)$ is the spectrum function of the field $\alpha(\mathbf{x}, t)$. Substitution of (7.127) in (7.121) now gives

$$\frac{\partial}{\partial t}\langle\langle \mathbf{B} \rangle\rangle = \nabla \wedge (\mathbf{Y} \wedge \langle\langle \mathbf{B} \rangle\rangle) + (\lambda_1 - X)\nabla^2 \langle\langle \mathbf{B} \rangle\rangle. \qquad (7.128)$$

The term involving $\mathbf{Y}$ here is not of great interest: it implies a uniform effective convection velocity $\mathbf{Y}$ of the field $\langle\langle \mathbf{B} \rangle\rangle$ relative to the fluid. If the α-field is statistically isotropic, then of course $\mathbf{Y} = 0$[12], since there is then no preferred direction.

The term involving X is of greater potential interest. It is evident from (7.127*a*) that $X > 0$, so that the helicity fluctuations do in fact make a negative contribution to the new effective diffusivity $\lambda_2 = \lambda_1 - X$. Let us estimate X when l_α is large enough for the following inequalities to be satisfied:

$$\varepsilon_1 = \alpha_0 t_\alpha / l_\alpha \ll 1, \qquad \varepsilon_2 = \lambda_1 t_\alpha / l_\alpha^2 \ll 1. \qquad (7.129)$$

The first justifies the use of first-order smoothing. The second allows asymptotic evaluation of (7.127*a*) (cf. the process leading to (7.85)) in the form

$$X \sim 2\pi \int \Phi_\alpha(\mathbf{k}, 0) \, \mathrm{d}\mathbf{k} = O(\alpha_0^2 t_\alpha). \qquad (7.130)$$

[12] Kraichnan (1976*a*) obtains only the X-term in (7.126) by (in effect) restricting attention to α-fields which, though possibly anisotropic, do have a particular statistical property that makes the integral (7.127*b*) vanish.

Hence

$$X/\lambda_1 = O(\alpha_0^2 t_\alpha/\lambda_1) = O(\varepsilon_1^2/\varepsilon_2), \tag{7.131}$$

and so apparently X *can* become of the same order as λ_1 or greater provided $\varepsilon_1^2 \gtrsim \varepsilon_2$, or equivalently provided t_α (as well as l_α) is sufficiently large.

The above argument, resting as it does on order of magnitude estimates, cannot be regarded as conclusive, but it is certainly suggestive, and the double ensemble technique merits further close study. A negative diffusivity $\lambda_2 = \lambda_1 - X$ in (7.128) implies on the one hand that all Fourier components of the field $\langle\langle \mathbf{B} \rangle\rangle$ grow exponentially in intensity, and on the other hand that the length-scale L of characteristic field structures will tend to *decrease* with time (the converse of the usual positive diffusion process). This type of result has to be seen in the context of the original assumption (7.118*a*) concerning scale separations: if L is reduced by negative diffusion to $O(l_\alpha)$ then the picture based on spatial averages is no longer meaningful.

Finally, it may be observed that (7.128) has the same general structure as the original induction equation (3.10). This means that if we push the Kraichnan (1976*a*) argument one stage further and consider random variations of $\mathbf{Y}(\mathbf{x}, t)$ on a length-scale l_Y satisfying $L \gg l_Y \gg l_\alpha \gg l_0$ (or via an 'ensemble of ensembles of ensembles'!) we are just back with the original problem but on a much larger length-scale, and no further new physical phenomenon can emerge from such a treatment.

CHAPTER 8

BRAGINSKII'S THEORY FOR WEAKLY ASYMMETRIC SYSTEMS

8.1. Introduction

The general arguments of §§ 7.1 and 7.2 indicate that when a velocity field consists of a steady mean part $\mathbf{U}_0(\mathbf{x})$ together with a fluctuating part $\mathbf{u}(\mathbf{x}, t)$, the mean magnetic field evolves according to the equation

$$\partial \mathbf{B}_0/\partial t = \nabla \wedge (\mathbf{U}_0 \wedge \mathbf{B}_0) + \nabla \wedge \boldsymbol{\mathscr{E}} + \lambda \nabla^2 \mathbf{B}_0, \tag{8.1}$$

where

$$\mathscr{E}_i = \alpha_{ij} B_{0j} + \beta_{ijk}\, \partial B_{0j}/\partial x_k + \dots. \tag{8.2}$$

In §§ 7.5–7.10, we obtained explicit expressions for α_{ij} and β_{ijk} as quadratic functionals of the $\mathbf{u}$-field, on the assumption that $\mathbf{U}_0 = 0$ (or cst.) and that the statistical properties of the $\mathbf{u}$-field are uniform in space and time. The approach was a two-scale approach, involving the definition of a mean as an average over the smaller length (or time) scale.

A very similar approach was developed by Braginskii (1964*a*, *b*) in an investigation of the effects of weak departures from axisymmetry in a spherical dynamo system. The idea motivating the study was that, although Cowling's theorem eliminates the possibility of axisymmetric dynamo action, if diffusive effects are weak (i.e. λ small) then weak departures from axisymmetry in the velocity field (and consequently in the magnetic field) may provide a regenerative electromotive force of the kind required to compensate ohmic decay.

In this situation it is natural to define the mean fields in terms of averages over the azimuth angle φ. For any scalar $\psi(s, \varphi, z)$, we therefore define its *azimuthal mean* by

$$\psi_0(s, z) = \langle \psi(s, \varphi, z) \rangle_{az} = \frac{1}{2\pi} \int_0^{2\pi} \psi(s, \varphi, z)\, \mathrm{d}\varphi, \tag{8.3}$$

and for any vector $\mathbf{f}(s, \varphi, z)$ we define similarly

$$\mathbf{f}_0(s, z) = \langle \mathbf{f} \rangle_{az} = \langle f_s \rangle \mathbf{i}_s + \langle f_\varphi \rangle \mathbf{i}_\varphi + \langle f_z \rangle \mathbf{i}_z. \tag{8.4}$$

We shall also use the notation

$$\mathbf{f}_M = \mathbf{f} - (\mathbf{f} \,.\, \mathbf{i}_\varphi)\mathbf{i}_\varphi = f_s \mathbf{i}_s + f_z \mathbf{i}_z \tag{8.5}$$

for the meridional projection of $\mathbf{f}$. Of course if $\mathbf{f}$ is axisymmetric, then $\mathbf{f}_0 = \mathbf{f}$; if in addition $\mathbf{f}$ is solenoidal, then $\mathbf{f}_M = \mathbf{f}_P$, the poloidal ingredient of $\mathbf{f}$.

Let us now consider a velocity field expressible in the form

$$\mathbf{U}_0(s, z) + \varepsilon \mathbf{u}'(s, \varphi, z, t), \tag{8.6}$$

where $\varepsilon \ll 1$ and $\langle \mathbf{u}' \rangle_{az} = 0$. Any magnetic field convected and distorted by this velocity field must exhibit at least the same degree of asymmetry about the axis Oz, and it is consistent to restrict attention to magnetic fields expressible in the form

$$\mathbf{B}_0(s, z) + \varepsilon \mathbf{b}'(s, \varphi, z, t), \qquad \langle \mathbf{b}' \rangle_{az} = 0. \tag{8.7}$$

(It is implicit in the notation of (8.6) and (8.7) that $\mathbf{u}'$ and $\mathbf{b}'$ are $O(1)$ as $\varepsilon \to 0$.) The azimuthal average of the induction equation is then (8.1) with

$$\mathscr{E}(s, z) = \varepsilon^2 \langle \mathbf{u}' \wedge \mathbf{b}' \rangle_{az}. \tag{8.8}$$

If we separate (8.1) into its toroidal and poloidal parts, with the notation

$$\mathbf{B}_0 = B(s, z)\mathbf{i}_\varphi + \mathbf{B}_P(s, z), \qquad \mathbf{U}_0 = U(s, z)\mathbf{i}_\varphi + \mathbf{U}_P(s, z), \tag{8.9}$$

we have for the toroidal part (cf. (3.43))

$$\frac{\partial B}{\partial t} + s(\mathbf{U}_P \,.\, \nabla)\frac{B}{s} = s(\mathbf{B}_P \,.\, \nabla)\frac{U}{s} + (\nabla \wedge \mathscr{E})_\varphi + \lambda\left(\nabla^2 - \frac{1}{s^2}\right)B. \tag{8.10}$$

Also, as in § 3.6, writing $\mathbf{B}_P = \nabla \wedge (A(s, z)\mathbf{i}_\varphi)$, the poloidal part may be 'uncurled' to give

$$\frac{\partial A}{\partial t} + s(\mathbf{U}_P \,.\, \nabla)\frac{A}{s} = \mathscr{E}_\varphi + \lambda\left(\nabla^2 - \frac{1}{s^2}\right)A. \tag{8.11}$$

We have seen in § 3.11 that the term $s(\mathbf{B}_P\,.\,\nabla)(U/s)$ in (8.10) can act as an adequate source term for the production of toroidal field by the process of differential rotation. We now envisage a situation in which the term $\mathscr{E}_\varphi$ in (8.11) acts as the complementary source term for the production of poloidal field $\mathbf{B}_P$ via its vector potential $A\,\mathbf{i}_\varphi$. If the rate of production of toroidal field is to be adequate to compensate ohmic dissipation, then a simple comparison of the terms on the right of (8.11) suggests that λ must be no greater than $O(\varepsilon^2)$, and we shall assume this to be the case in the following sections; i.e. we put

$$\lambda = \lambda_0 \varepsilon^2, \tag{8.12}$$

and assume that $\lambda_0 = O(1)$ in the limit $\varepsilon \to 0$. Equivalently if U_0 is a typical order of magnitude of the toroidal velocity $U(s, z)$ and L an overall length-scale, then

$$R_m^{-1} = \lambda / U_0 L = O(\varepsilon^2), \tag{8.13}$$

or equivalently, $\varepsilon = O(R_m^{-1/2})$.

The Braginskii approach is further based on the assumption that the dominant ingredient of the mean velocity $\mathbf{U}$ is the toroidal ingredient $U\mathbf{i}_\varphi$, and more specifically that

$$\mathbf{U}_P = \varepsilon^2 \mathbf{u}_P, \qquad \mathbf{u}_P = O(1). \tag{8.14}$$

This means that a magnetic Reynolds number based on say $|\mathbf{U}_P|_{\max}$ will be $O(1)$; this poloidal velocity will then redistribute toroidal and poloidal field, but will *not* be sufficiently intense to expel poloidal field from regions of closed $\mathbf{U}_P$-lines.

Since differential rotation tends to generate a toroidal field that is a factor $O(R_m)$ larger than the poloidal field, it is natural to introduce the further scaling (in parallel with (8.14))

$$\mathbf{B}_P = \varepsilon^2 \mathbf{b}_P, \qquad \mathbf{b}_P = O(1), \tag{8.15}$$

the implication then being that $|\mathbf{b}_P|$ and B are of the same order of magnitude, as $\varepsilon \to 0$.

With the total velocity field now of the form

$$U\mathbf{i}_\varphi + \varepsilon\,\mathbf{u}' + \varepsilon^2 \mathbf{u}_P, \tag{8.16}$$

it is clear that any fluid particle follows a nearly circular path about the axis Oz; in these circumstances, there is good reason to anticipate that $\mathscr{E} = \varepsilon^2\langle \mathbf{u}' \wedge \mathbf{b}' \rangle_{az}$ will be determined by *local* values of $B(s, z)$, $U(s, z)$ and azimuthal average properties of $\mathbf{u}'$. The situation is closely analogous to the two-scale approach of chapter 7, the smaller scale l_0 now being the mean departure of a fluid particle from a circular path in its trajectory round the axis Oz. A local expansion of the form (8.2) is therefore to be expected, but now with $\mathscr{E}(s, z)$, $\mathbf{B}_0(s, z)$ axisymmetric by definition. The feature that most distinguishes the Braginskii approach from the previous two-scale approach is the presence of the dominant azimuthal flow $U(s, z)\mathbf{i}_\varphi$; and a possible influence of local shear, given by $s\nabla(U/s)$, on $\mathscr{E}$ cannot be ruled out. It turns out however, as described in the following sections, that effects of the flow $U\mathbf{i}_\varphi$ on $\mathscr{E}$ are very simply accommodated through replacement of the fields $\mathbf{u}_P$ and $\mathbf{b}_P$ by 'effective fields' $\mathbf{u}_{eP}$ and $\mathbf{b}_{eP}$, the structure of equations (8.10) and (8.11) remaining otherwise unchanged. When this effect has been accounted for, the residual mean electromotive force (which is wholly diffusive in origin) is, with very minor modification, identifiable with that given by the first-order smoothing theory of § 7.8.

The theory that follows (although initiated by Braginskii, 1964*a*, *b*, before the mean-field electrodynamics of Steenbeck, Krause & Rädler, 1966) can best be regarded as a branch of mean-field electrodynamics that takes account of spatial inhomogeneity of mean velocity and of mean properties of the fluctuating ingredient of the velocity field. The close relationship between the two approaches was emphasised by Soward (1972), whose line of argument we follow in subsequent sections.

8.2. Lagrangian transformation of the induction equation when $\lambda = 0$

Soward's (1972) approach is based on a simple property of invariance of the induction equation in the limit of zero diffusion. For reasons that emerge, it is useful to modify the notation slightly: let $\tilde{\mathbf{B}}(\tilde{\mathbf{x}}, t)$, $\tilde{\mathbf{U}}(\tilde{\mathbf{x}}, t)$ represent field and velocity at $(\tilde{\mathbf{x}}, t)$; when $\lambda = 0$, we have

$$\partial\tilde{\mathbf{B}}/\partial t = \tilde{\nabla} \wedge (\tilde{\mathbf{U}} \wedge \tilde{\mathbf{B}}). \tag{8.17}$$

From § 2.5, we know that (when $\tilde{\nabla} \cdot \tilde{\mathbf{U}} = 0$) the Lagrangian solution of this equation is

$$\tilde{B}_i(\tilde{\mathbf{x}}, t) = \tilde{B}_j(\mathbf{a}, 0)\, \partial\tilde{x}_i/\partial a_j, \tag{8.18}$$

where $\tilde{\mathbf{x}}(\mathbf{a}, t)$ is the position of a fluid particle satisfying $\tilde{\mathbf{x}}(\mathbf{a}, 0) = \mathbf{a}$. Equivalently, the Lagrangian form of (8.17) is

$$\frac{\mathrm{D}}{\mathrm{D}t}\left(\tilde{B}_i(\tilde{\mathbf{x}}, t)\frac{\partial a_j}{\partial \tilde{x}_i}\right) = 0, \tag{8.19}$$

where $\mathrm{D}/\mathrm{D}t$ represents differentiation keeping $\mathbf{a}$ constant.

Consider now an 'incompressible' change of variable

$$\mathbf{x} = \mathbf{x}(\tilde{\mathbf{x}}, t) = \mathbf{x}(\tilde{\mathbf{x}}(\mathbf{a}, t), t) = \mathbf{X}(\mathbf{a}, t), \text{ say,} \tag{8.20}$$

the determinant of the transformation $\|\partial x_i/\partial \tilde{x}_j\|$ being equal to unity, a condition that may be expressed in the form

$$\varepsilon_{ijk}\frac{\partial x_i}{\partial \tilde{x}_p}\frac{\partial x_j}{\partial \tilde{x}_q}\frac{\partial x_k}{\partial \tilde{x}_r} = \varepsilon_{pqr}. \tag{8.21}$$

Equation (8.19) immediately transforms to

$$\frac{\mathrm{D}}{\mathrm{D}t}\left(B_k(\mathbf{x}, t)\frac{\partial a_j}{\partial x_k}\right) = 0, \tag{8.22}$$

where

$$B_k(\mathbf{x}, t) = \tilde{B}_i(\tilde{\mathbf{x}}, t)\, \partial x_k/\partial \tilde{x}_i, \tag{8.23}$$

and, reversing the process that led from (8.17) to (8.19), we see that the Eulerian equivalent of (8.22) is

$$\partial\mathbf{B}/\partial t = \nabla \wedge (\mathbf{U} \wedge \mathbf{B}), \tag{8.24}$$

where

$$U_k(\mathbf{x}, t) = \left(\frac{\partial X_k}{\partial t}\right)_{\mathbf{a}} = \frac{\partial x_k}{\partial t} + \tilde{U}_i\frac{\partial x_k}{\partial \tilde{x}_i}. \tag{8.25}$$

Equation (8.17) is therefore invariant under the transformation defined by (8.20), (8.23) and (8.25). $\mathbf{B}(\mathbf{x}, t)$ is (physically) the field that would result from $\tilde{\mathbf{B}}(\tilde{\mathbf{x}}, t)$ under an instantaneous frozen-field distortion of the medium defined by (8.20); $\mathbf{U}(\mathbf{x}, t)$ is the incompressible velocity field associated with the (hypothetical) Lagrangian displacement $\mathbf{X}(\mathbf{a}, t)$.

In view of the discussion of § 8.1, it is of particular interest to consider the effect of a mapping $\mathbf{x} \leftrightarrow \tilde{\mathbf{x}}$ which is nearly the identity. Such a mapping may be considered as the result of a steady Eulerian velocity field $\mathbf{v}(\mathbf{x})$ applied over some short time interval $\varepsilon\tau_0$ say. Then the particle path $\tilde{\mathbf{x}}(\mathbf{x}, \tau)$ is given by

$$d\tilde{\mathbf{x}}/d\tau = \mathbf{v}(\tilde{\mathbf{x}}), \qquad \tilde{\mathbf{x}}(\mathbf{x}, 0) = \mathbf{x}, \tag{8.26}$$

and, at time $\tau = \varepsilon\tau_0$, the net displacement $\tilde{\mathbf{x}}$ is given by

$$\tilde{\mathbf{x}} = \mathbf{x} + \varepsilon\boldsymbol{\eta}(\mathbf{x}) + \tfrac{1}{2}\varepsilon^2(\boldsymbol{\eta}\,.\,\nabla)\boldsymbol{\eta} + \frac{1}{3!}\varepsilon^3(\boldsymbol{\eta}\,.\,\nabla)^2\boldsymbol{\eta} + \ldots, \tag{8.27}$$

where $\boldsymbol{\eta}(\mathbf{x}) = \tau_0\mathbf{v}(\mathbf{x})$. If $\mathbf{v}(\mathbf{x})$ is incompressible, then $\nabla\,.\,\boldsymbol{\eta} = 0$, and, under this condition, the form of displacement given by (8.27) must automatically satisfy (8.21) to all orders in ε.

When $\mathbf{x}$ and $\tilde{\mathbf{x}}$ are related by (8.27), the instantaneous relation between $\mathbf{B}(\mathbf{x})$ and $\tilde{\mathbf{B}}(\mathbf{x})$ can be obtained as follows. Let $\tilde{\mathbf{B}}_\tau(\mathbf{x})$ be defined by

$$\partial\tilde{\mathbf{B}}_\tau/\partial\tau = \nabla \wedge (\mathbf{v}(\mathbf{x}) \wedge \tilde{\mathbf{B}}_\tau), \qquad \tilde{\mathbf{B}}_0(\mathbf{x}) = \mathbf{B}(\mathbf{x}), \tag{8.28}$$

so that evidently $\tilde{\mathbf{B}}(\mathbf{x}) = \tilde{\mathbf{B}}_{\varepsilon\tau_0}(\mathbf{x})$. We may write (8.28) in the equivalent integral form

$$\tilde{\mathbf{B}}_\tau(\mathbf{x}) = \mathbf{B}(\mathbf{x}) + \int_0^\tau \nabla \wedge (\mathbf{v}(\mathbf{x}) \wedge \tilde{\mathbf{B}}_{\tau'}(\mathbf{x}))\, d\tau', \tag{8.29}$$

and solve iteratively to obtain

$$\tilde{\mathbf{B}}_\tau(\mathbf{x}) = \mathbf{B}(\mathbf{x}) + \tau\nabla \wedge (\mathbf{v} \wedge \mathbf{B}) + \tfrac{1}{2}\tau^2\nabla \wedge [\mathbf{v} \wedge \nabla \wedge (\mathbf{v} \wedge \mathbf{B})] + \ldots. \tag{8.30}$$

Putting $\tau = \varepsilon\tau_0$ and $\mathbf{v} = \boldsymbol{\eta}/\tau_0$, we obtain

$$\tilde{\mathbf{B}}(\mathbf{x}) = \mathbf{B}(\mathbf{x}) + \varepsilon\nabla \wedge (\boldsymbol{\eta} \wedge \mathbf{B}) + \tfrac{1}{2}\varepsilon^2\nabla \wedge [\boldsymbol{\eta} \wedge \nabla \wedge (\boldsymbol{\eta} \wedge \mathbf{B})] + \ldots. \tag{8.31}$$

This is the Eulerian equivalent of the Lagrangian statement

$$\tilde{\mathbf{B}}_i(\tilde{\mathbf{x}}) = B_k(\mathbf{x})\left(\delta_{ik} + \varepsilon\frac{\partial\eta_i}{\partial x_k} + \tfrac{1}{2}\varepsilon^2\frac{\partial}{\partial x_k}(\boldsymbol{\eta}\,.\,\nabla)\eta_i + \ldots\right). \tag{8.32}$$

Since these relationships are instantaneous, they remain valid when $\boldsymbol{\eta}$, $\tilde{\mathbf{B}}$ and $\mathbf{B}$ depend explicitly on t as well as on $\mathbf{x}$.

By virtue of (8.25), the corresponding relationship between $\tilde{\mathbf{U}}(\mathbf{x})$ and $\mathbf{U}(\mathbf{x})$ is, in compact notation,

$$\tilde{\mathbf{U}}(\mathbf{x}) = \sum_{n=0}^{\infty} \frac{\varepsilon^n}{n!} (\nabla \wedge \boldsymbol{\eta} \wedge)^n \left(\mathbf{U}(\mathbf{x}) - \frac{\partial \mathbf{x}}{\partial t} \right). \tag{8.33}$$

Here, as in (8.25), $\partial \mathbf{x}/\partial t$ is to be evaluated 'keeping $\tilde{\mathbf{x}}$ constant'; hence, from (8.27),

$$\frac{\partial x_i}{\partial t} = -\varepsilon \frac{\partial \eta_i}{\partial t} + \frac{1}{2}\varepsilon^2 \left(\frac{\partial \eta_j}{\partial t} \frac{\partial \eta_i}{\partial x_j} - \eta_j \frac{\partial^2 \eta_i}{\partial x_j \, \partial t} \right) + O(\varepsilon^3). \tag{8.34}$$

8.3. Effective variables in a Cartesian geometry

To simplify the discussion, let us for the moment use Cartesian coordinates $(x, y, z)(\equiv(x_1, x_2, x_3))$ instead of cylindrical polars (s, φ, z). Following the discussion of § 8.1 we consider velocity and magnetic fields of the form

$$\left.\begin{aligned} \tilde{\mathbf{U}}(\mathbf{x}, t) &= U(x, z)\mathbf{i}_y + \varepsilon \mathbf{u}'(\mathbf{x}, t) + \varepsilon^2 \mathbf{u}_P(x, z), \\ \tilde{\mathbf{B}}(\mathbf{x}, t) &= B(x, z)\mathbf{i}_y + \varepsilon \mathbf{b}'(\mathbf{x}, t) + \varepsilon^2 \mathbf{b}_P(x, z). \end{aligned}\right\} \tag{8.35}$$

Here the fields $U\mathbf{i}_y$, $B\mathbf{i}_y$ are the dominant toroidal fields, $\mathbf{u}_P$ and $\mathbf{b}_P$ are poloidal (so that $\mathbf{i}_y \, . \, \mathbf{b}_P = \mathbf{i}_y \, . \, \mathbf{u}_P = 0$), and $\langle \mathbf{u}' \rangle = \langle \mathbf{b}' \rangle = 0$, the brackets $\langle \cdots \rangle$ now indicating an average over the coordinate y. The essence of Soward's (1972) approach is now to 'accommodate' the $O(\varepsilon)$ terms in (8.35) through choice of $\boldsymbol{\eta}(\mathbf{x}, t)$, in such a way that the transformed fields $\mathbf{U}$, $\mathbf{B}$ take the form

$$\left.\begin{aligned} \mathbf{U}(\mathbf{x}, t) &= U(x, z)(1 + O(\varepsilon^2))\mathbf{i}_y + \varepsilon^2 \mathbf{u}_{eP}(x, z) + \varepsilon^2 \mathbf{u}''(\mathbf{x}, t), \\ \mathbf{B}(\mathbf{x}, t) &= B(x, z)(1 + O(\varepsilon^2))\mathbf{i}_y + \varepsilon^2 \mathbf{b}_{eP}(x, z) + \varepsilon^2 \mathbf{b}''(\mathbf{x}, t), \end{aligned}\right\} \tag{8.36}$$

where $\langle \mathbf{u}'' \rangle = \langle \mathbf{b}'' \rangle = 0$, and where the effective fields (or 'effective variables' as Braginskii, 1964*a*, called them) $\mathbf{u}_{eP}$, $\mathbf{b}_{eP}$ are to be determined. Substitution of (8.35*b*) and (8.36*b*) in (8.31) and equating terms of order ε gives immediately

$$\mathbf{b}'(\mathbf{x}, t) = \nabla \wedge (\boldsymbol{\eta}(\mathbf{x}, t) \wedge B(x, z)\mathbf{i}_y) = B \frac{\partial \boldsymbol{\eta}}{\partial y} - (\boldsymbol{\eta} \, . \, \nabla) B \mathbf{i}_y. \tag{8.37}$$

Similarly

$$\mathbf{u}'(\mathbf{x}, t) = \left(\frac{\partial}{\partial t} + U\frac{\partial}{\partial y}\right)\boldsymbol{\eta} - (\boldsymbol{\eta}\,.\,\nabla)U\mathbf{i}_y, \tag{8.38}$$

an equation which in principle serves to determine $\boldsymbol{\eta}$ if $\mathbf{u}'$ and U are given. It is in fact simpler now to regard $\boldsymbol{\eta}(\mathbf{x}, t)$ as given, and the fluctuating part of the velocity field as then given by (8.38).

Similarly at the $O(\varepsilon^2)$ level, the mean poloidal ingredient of (8.31) gives

$$\mathbf{b}_P(x, z) = \mathbf{b}_{eP}(x, z) + \frac{1}{2}\nabla \wedge \left\{\left\langle \boldsymbol{\eta} \wedge \frac{\partial \boldsymbol{\eta}}{\partial y}\right\rangle_y B\mathbf{i}_y\right\}. \tag{8.39}$$

Alternatively, defining vector potentials $a\mathbf{i}_y$ and $a_e\mathbf{i}_y$ by

$$\mathbf{b}_P = \nabla \wedge (a\mathbf{i}_y), \qquad \mathbf{b}_{eP} = \nabla \wedge (a_e\mathbf{i}_y), \tag{8.40}$$

the relation between a and a_e is evidently

$$a_e = a + \varpi B \quad \text{where } \varpi = -\tfrac{1}{2}\langle \boldsymbol{\eta} \wedge \partial\boldsymbol{\eta}/\partial y\rangle_y. \tag{8.41}$$

The quantity ϖ is a pseudo-scalar (having the dimensions of a length). It is therefore non-zero only if the statistical (i.e. y-averaged) properties of the function $\boldsymbol{\eta}(\mathbf{x}, t)$ lack reflexional symmetry.

The relationship between $\mathbf{u}_{eP}$ and $\mathbf{u}_P$ is analogous to (8.39), but with the difference again that $B\,\partial/\partial y$ is replaced by $\partial/\partial t + U\,\partial/\partial y$, i.e.

$$\mathbf{u}_{eP} = \mathbf{u}_P - \frac{1}{2}\nabla \wedge \left\{\left\langle \boldsymbol{\eta} \wedge \left(\frac{\partial}{\partial t} + U\frac{\partial}{\partial y}\right)\boldsymbol{\eta}\right\rangle_y \mathbf{i}_y\right\}. \tag{8.42}$$

If the displacement function $\boldsymbol{\eta}$ depends only on $\mathbf{x}$ (as is appropriate if the perturbation field is steady) then the analogy between (8.39) and (8.42) becomes precise.

The fact that the perturbation fields in (8.36) are relegated to $O(\varepsilon^2)$ means that when $\lambda = 0$ the relevant equations for the evolution of $\mathbf{B}(\mathbf{x}, t)$ at leading order are precisely the two-dimensional equations derived in § 3.6, but now expressed in terms of the effective poloidal fields, viz.

$$\left.\begin{aligned} \partial B/\partial t + \mathbf{U}_{eP}\,.\,\nabla B &= \mathbf{B}_{eP}\,.\,\nabla U, \\ \partial A_e/\partial t + \mathbf{U}_{eP}\,.\,\nabla A_e &= 0, \end{aligned}\right\} \tag{8.43}$$

where $\mathbf{U}_{eP} = \varepsilon^2 \mathbf{u}_{eP}$, $A_e = \varepsilon^2 a_e$. This holds because of the basic invariance of the induction equation (8.17) under the frozen-field transformation (8.20). It is evident from (8.43) that the distinction between $\mathbf{B}_P$ and $\mathbf{B}_{eP}$ will be significant only if $\nabla U \neq 0$, i.e. only if the mean flow exhibits shear. Similarly it is evident from (8.42) that $\mathbf{u}_{eP}$ differs from $\mathbf{u}_P$ only either when $\nabla U \neq 0$ or when $\boldsymbol{\eta}$ is statistically inhomogeneous in the (x, z) plane. It is therefore spatial inhomogeneity in the y-averaged (or equivalently azimuthally averaged) flow properties that leads to the natural appearance of effective variables; this aspect of Braginskii's theory has no perceptible counterpart in the mean-field electrodynamics described in chapter 7.

8.4. Lagrangian transformation including weak diffusion effects

Suppose now that we subject the full induction equation in the form

$$(\partial \tilde{\mathbf{B}}/\partial t - \tilde{\nabla} \wedge (\tilde{\mathbf{U}} \wedge \tilde{\mathbf{B}}))_i = -\lambda (\tilde{\nabla} \wedge (\hat{\nabla} \wedge \tilde{\mathbf{B}}))_i \tag{8.44}$$

to the transformation (8.20), (8.23) and (8.25). Multiplying by $\partial x_k/\partial \tilde{x}_i$, the left-hand side transforms by the result of § 8.2 to

$$(\partial \mathbf{B}/\partial t - \nabla \wedge (\mathbf{U} \wedge \mathbf{B}))_k. \tag{8.45}$$

The right-hand side transforms to

$$-\lambda \frac{\partial x_k}{\partial \tilde{x}_i} \varepsilon_{ijm} \frac{\partial x_p}{\partial \tilde{x}_j} \frac{\partial}{\partial x_p} (\tilde{\nabla} \wedge \tilde{\mathbf{B}})_m = -\lambda \varepsilon_{kpq} \frac{\partial \tilde{x}_m}{\partial x_q} \frac{\partial}{\partial x_p} (\tilde{\nabla} \wedge \tilde{\mathbf{B}})_m, \tag{8.46}$$

using (8.21). Now

$$(\tilde{\nabla} \wedge \tilde{\mathbf{B}})_m = \varepsilon_{mrs} \frac{\partial \tilde{B}_s}{\partial \tilde{x}_r} = \varepsilon_{mrs} \frac{\partial}{\partial x_i} \left(B_j \frac{\partial \tilde{x}_s}{\partial x_j} \right) \frac{\partial x_i}{\partial \tilde{x}_r}. \tag{8.47}$$

Hence, using the identities

$$\varepsilon_{kpq} \frac{\partial^2 \tilde{x}_m}{\partial x_p \, \partial x_q} = 0, \qquad \frac{\partial}{\partial x_j} \left(\frac{\partial \tilde{x}_s}{\partial x_i} \frac{\partial x_i}{\partial \tilde{x}_r} \right) = \frac{\partial}{\partial x_j} \delta_{rs} = 0, \tag{8.48}$$

the expression (8.46) reduces to the form

$$(\nabla \wedge \mathscr{E}')_k + \lambda (\nabla^2 \mathbf{B})_k, \tag{8.49}$$

where

$$\mathscr{E}'_i(\mathbf{x}, t) = \alpha'_{ij} B_j + \beta'_{ijk}\, \partial B_j/\partial x_k, \tag{8.50}$$

and where

$$\begin{aligned} \alpha'_{ij} &= \lambda \varepsilon_{ikl} \frac{\partial x_k}{\partial \tilde{x}_p} \frac{\partial}{\partial x_j}\left(\frac{\partial x_l}{\partial \tilde{x}_p}\right), \\ \beta'_{ijk} &= \lambda \varepsilon_{ijp}\left(\frac{\partial x_p}{\partial \tilde{x}_r} \frac{\partial x_k}{\partial \tilde{x}_r} - \delta_{pk}\right), \end{aligned} \tag{8.51}$$

and (8.44) becomes

$$\partial \mathbf{B}/\partial t - \nabla \wedge (\mathbf{U} \wedge \mathbf{B}) = \nabla \wedge \mathscr{E}' + \lambda \nabla^2 \mathbf{B}. \tag{8.52}$$

The structure of (8.52) and the expression (8.50) for $\mathscr{E}'$ are now strongly reminiscent of equations for mean fields encountered in chapter 7. Note however that in the present context the term $\nabla \wedge \mathscr{E}'$ is *wholly diffusive in origin*. As in the case of the random wave situation of § 7.8, α'_{ij} and β'_{ijk} are $O(\lambda)$ as $\lambda \to 0$. We have noted previously that although diffusion is responsible for the natural tendency of the field to decay, it is also of crucial importance in making field regeneration a possibility. In Braginskii's theory it is diffusion that, as in § 7.7, shifts the phase of field perturbations relative to velocity perturbations leading to the appearance of a mean toroidal electromotive force (proportional to λ) which is sufficient to provide closure of the dynamo cycle.

8.5. Dynamo equations for nearly rectilinear flow

As in § 8.3, we again specialise the argument to the nearly two-dimensional situation. The y-average of $\mathscr{E}'$ is then

$$\mathscr{E}_i(x, z) = \alpha_{ij} B_j + \beta_{ijk}\, \partial B_j/\partial x_k, \tag{8.53}$$

where, to leading order in ε,

$$\alpha_{ij} = \langle \alpha'_{ij} \rangle = \lambda \varepsilon^2 \varepsilon_{ikl} \langle \eta_{k,m} \eta_{l,mj} \rangle, \tag{8.54}$$

and

$$\beta_{ijk} = \langle \beta'_{ijk} \rangle = \lambda \varepsilon^2 \varepsilon_{ijp} \langle \eta_{p,r} \eta_{k,r} \rangle. \tag{8.55}$$

With $\lambda = O(\varepsilon^2)$, both of these pseudo-tensors are $O(\varepsilon^4)$. With $\mathbf{B}$ given by (8.36*b*), we have, at leading order,

$$\mathscr{E}_y \sim \alpha_{22} B = O(\varepsilon^4). \tag{8.56}$$

Note that the β-term in (8.53) makes no contribution at the $O(\varepsilon^4)$ level. The counterpart of (8.43*b*) incorporating diffusion effects now therefore takes the form (with $\alpha_{22} = \alpha$)

$$\partial A_e/\partial t + \mathbf{U}_{eP} \,.\, \nabla A_e = \alpha B + \lambda \nabla^2 A_e. \tag{8.57}$$

Note that all three terms contributing to $\partial A_e/\partial t$ are $O(\varepsilon^4)$.

Similarly $(\nabla \wedge \boldsymbol{\mathscr{E}})_y = O(\varepsilon^4)$ in general, and this is negligible compared with the natural diffusive term $\lambda \nabla^2 B$ in the mean y-component of (8.52). The counterpart of (8.43*a*) is therefore simply

$$\partial B/\partial t + \mathbf{U}_{eP} \,.\, \nabla B = \mathbf{B}_{eP} \,.\, \nabla U + \lambda \nabla^2 B. \tag{8.58}$$

In this equation, all three contributions to $\partial B/\partial t$ are $O(\varepsilon^2)$.

Apart from the appearance of effective variables, equations (8.57) and (8.58) are just what would be obtained on the basis of the mean-field electrodynamics of chapter 7, provided that in the 'toroidal' equation (8.58) the production of toroidal field by the term $\mathbf{B}_{eP} \,.\, \nabla U$ dominates over possible production by the α-effect. By contrast, the α-effect term αB in (8.57) is the only source term for the field A_e and hence for the poloidal field $\mathbf{B}_{eP}$, and it is therefore of central importance.

The expression (8.54) for α_{ij} is closely related to the expression (7.77) obtained on the basis of first-order smoothing theory. To see this, let us suppose for the moment that all y-averaged properties of the velocity field are independent of x and z (or so weakly dependent that they may be treated as locally uniform). In this situation, from (8.38),

$$\mathbf{u}' = \left(\frac{\partial}{\partial t} + U\frac{\partial}{\partial y}\right)\boldsymbol{\eta}, \tag{8.59}$$

and the corresponding Fourier transforms in the notation of § 7.8 are related by

$$\tilde{\mathbf{u}}' = (-\mathrm{i}\omega + Uk_2)\tilde{\boldsymbol{\eta}}. \tag{8.60}$$

The Fourier transform of $\eta_{k,m}$ is $\mathrm{i}k_m\tilde{\eta}_k$ and of $\eta_{l,mj}$ is $-k_mk_j\tilde{\eta}_l$, and so the expression (8.54) may be translated into spectral terms as

$$\alpha_{ij} = \mathrm{i}\lambda\varepsilon_{ikl}\iint \frac{k^2 k_j \Phi_{kl}(\mathbf{k}, \omega)\, \mathrm{d}\mathbf{k}\, \mathrm{d}\omega}{\omega^2 + U^2k_2^2}, \tag{8.61}$$

where $\Phi_{kl}(\mathbf{k}, \omega)$ is the spectrum tensor of the perturbation velocity field $\varepsilon\mathbf{u}'$. This expression agrees with (7.77) in the weak diffusion limit $\lambda k^2 \ll \omega$ (for all relevant $\mathbf{k}$, ω) and under the natural replacement of ω^2 by $\omega^2 + U^2k_2^2$ to take account of mean convection in the y-direction with velocity U. Note that in this important domain of overlap of Braginskii's theory and mean-field electrodynamics, the former theory undoubtedly gives the result $\alpha_{ij} \to 0$ as $\lambda \to 0$. We have already noted that this result holds (see 7.83) when the velocity spectrum contains no zero frequency ingredients (or, in the notation of the present section, ingredients with wave speed ω/k_2 equal to the mean velocity component U). If the velocity spectrum *does* contain such ingredients, then the displacement $\boldsymbol{\eta}(\mathbf{x}, t)$ will not remain bounded for all t, and the Braginskii approach, based implicitly on the assumption $|\boldsymbol{\eta}| = O(1)$, is no longer valid.

In general, of course, if U does depend significantly on x and z (as it must do if the production term $\mathbf{B}_P \,.\, \nabla U$ is to be important) then mean quantities such as $\langle \eta_{k,m}\eta_{l,mj}\rangle$ will also vary with x and z because of the dynamical interaction of mean and perturbation fields; hence in general α_{ij} as given by (8.54) is also a function of x and z.

8.6. Corresponding results for nearly axisymmetric flows

Let us now return to the notation of § 8.1, with cylindrical polar coordinates (s, φ, z). The relation between $\mathbf{b}'$ and $\boldsymbol{\eta}$ is now (cf. (8.37))

$$\mathbf{b}' = \nabla \wedge (\boldsymbol{\eta} \wedge \mathbf{i}_\varphi B). \tag{8.62}$$

The meridional projection of this equation (see (8.5)) is

$$\mathbf{b}'_M = \frac{B}{s} \frac{\partial_1}{\partial\varphi} \boldsymbol{\eta}_M, \tag{8.63}$$

where the operator $\partial_1/\partial\varphi$ is defined for any vector $\mathbf{f}$ by

$$\frac{\partial_1 \mathbf{f}}{\partial\varphi}=\mathbf{i}_s\frac{\partial f_s}{\partial\varphi}+\mathbf{i}_\varphi\frac{\partial f_\varphi}{\partial\varphi}+\mathbf{i}_z\frac{\partial f_z}{\partial\varphi}=\frac{\partial \mathbf{f}}{\partial\varphi}-\mathbf{i}_z\wedge\mathbf{f}. \tag{8.64}$$

Similarly the relation between $\mathbf{u}'_M$ and $\boldsymbol{\eta}_M$ is

$$\mathbf{u}'_M=\left(\frac{\partial}{\partial t}+\frac{U}{s}\frac{\partial_1}{\partial\varphi}\right)\boldsymbol{\eta}_M, \tag{8.65}$$

a relation which in principle determines $\boldsymbol{\eta}_M$ if $\mathbf{u}'_M$ and U are given.

Effective variables are now given by formulae closely analogous to (8.39) and (8.42), viz.

$$\mathbf{b}_{eP}=\mathbf{b}_P+\nabla\wedge(\varpi B\mathbf{i}_\varphi), \qquad \mathbf{u}_{eP}=\mathbf{u}_P+\nabla\wedge(\varpi_1 U\mathbf{i}_\varphi), \tag{8.66}$$

where now

$$\begin{aligned}\varpi&=-\frac{1}{2}\left\langle\boldsymbol{\eta}\wedge\frac{1}{s}\frac{\partial_1\boldsymbol{\eta}}{\partial\varphi}\right\rangle_{\mathrm{az}},\\ \varpi_1&=-\frac{1}{2U}\left\langle\boldsymbol{\eta}\wedge\left(\frac{\partial}{\partial t}+\frac{U}{s}\frac{\partial_1}{\partial\varphi}\right)\boldsymbol{\eta}\right\rangle_{\mathrm{az}}.\end{aligned} \tag{8.67}$$

The evolution equations for $B(s,z,t)$ and $A_e(s,z,t)$ analogous to (8.58) and (8.57) are

$$\frac{\partial B}{\partial t}+s(\mathbf{U}_{eP}\,.\,\nabla)\frac{B}{s}=s(\mathbf{B}_{eP}\,.\,\nabla)\frac{U}{s}+\lambda\left(\nabla^2-\frac{1}{s^2}\right)B, \tag{8.68}$$

and

$$\frac{\partial A_e}{\partial t}+\frac{1}{s}(\mathbf{U}_{eP}\,.\,\nabla)sA_e=\alpha B+\lambda\left(\nabla^2-\frac{1}{s^2}\right)A_e, \tag{8.69}$$

and the expression for α analogous to (8.54) (Soward, 1972) is given by

$$\begin{aligned}\frac{\alpha}{\lambda\varepsilon^2}=\frac{2}{s}&\left\langle\eta_{z,s}\eta_{s,s\varphi}+\frac{1}{s^2}\eta_{z,\varphi}\eta_{s,\varphi\varphi}+\eta_{z,z}\eta_{s,z\varphi}\right\rangle\\ &+\frac{2}{s^2}\langle s^{-1}\eta_s\eta_{z,\varphi}+\eta_{z,z}\eta_{z,\varphi}\rangle.\end{aligned} \tag{8.70}$$

Here the first group of terms is precisely analogous to the group appearing in (8.54); the second group arises from the curvature of the metric in cylindrical polar coordinates, and can be derived only by first converting the expression $\alpha_{ij}B_j$ in (8.53) to this coordinate system.

The equations (8.68) and (8.69) describe the evolution of the fields $\varepsilon^{-2}A_e$ and B correctly to order ε^2. The equations were obtained in this form by Braginskii (1964*a*, *b*) by a direct expansion procedure which made no appeal to the Lagrangian invariance property. The labour involved in this procedure is very considerable, and the pseudo-Lagrangian approach of Soward (1972) can now be seen as providing an important and illuminating simplification. The Braginskii expansion was continued to the next level ($O(\varepsilon^3)$) by Tough (1967) (see also Tough & Gibson, 1969) who showed that the structure of the equations is unaltered, but that ϖ, ϖ_1 and α as given by (8.67) and (8.70) need small corrections; the corrections to ϖ and ϖ_1 can be obtained by including the ε^3 terms in the expansions (8.31) and (8.33). Similarly the correction to $\alpha(=\alpha_{22})$ can be obtained by retaining terms up to $O(\varepsilon^3)$ in (8.51*a*); it is evident that this procedure leads to an expression of the form

$$\alpha/\lambda\varepsilon^2 = \Gamma_0 + \varepsilon\Gamma_1, \tag{8.71}$$

where Γ_0 is the right-hand side of (8.70) and Γ_1 involves the mean of an expression cubic in $\boldsymbol{\eta}$.[1]

If terms of order ε^4 are retained (Soward, 1972) then the structure of (8.68) and (8.69) *is* modified, firstly through the effects of components of α_{ij} other than α_{22} and secondly through the β_{ijk} terms of (8.53).

8.7. A limitation of the pseudo-Lagrangian approach

The arguments of Soward (1972) are expressed in general terms that do not for the most part invoke the limitation of small displacement functions. It is assumed that the magnetic Reynolds number

[1] An analogous correction to the first-order smoothing result (7.77) can be obtained by systematic expansion in powers of the amplitude of the velocity fluctuations; this correction likewise involves a cubic functional of the velocity field (a weighted integral in **k**-space of the Fourier transform of a triple velocity correlation).

R_m is large, and velocity fields[2] $\tilde{\mathbf{U}}(\tilde{\mathbf{x}})$ are considered having the property that there exists a continuous (1–1) mapping $\tilde{\mathbf{x}}\to\mathbf{x}$ such that the related velocity field $\mathbf{U}(\mathbf{x})$ given by (8.25) has the form

$$\mathbf{U}(\mathbf{x}) = U(s, z)\mathbf{i}_\varphi + R_m^{-1}\mathbf{u}_M(s, z) + R_m^{-1}\mathbf{u}''(\mathbf{x}), \qquad (8.72)$$

where $\langle \mathbf{u}'' \rangle_{az} = 0$. The wide choice of mappings implies a correspondingly wide family of velocity fields $\tilde{\mathbf{U}}(\tilde{\mathbf{x}})$ that can be subjected to this treatment. Even so, such velocity fields are rather special for the following reason.

Under the instantaneous transformation

$$U_k(\mathbf{x}) = \tilde{U}_i(\tilde{\mathbf{x}})\, \partial x_k/\partial \tilde{x}_i, \qquad (8.73)$$

it may readily be ascertained (with the help of (8.21)) that

$$(\mathrm{d}\mathbf{x} \wedge \mathbf{U}(\mathbf{x}))_i = (\mathrm{d}\tilde{\mathbf{x}} \wedge \tilde{\mathbf{U}}(\tilde{\mathbf{x}}))_j\, \partial \tilde{x}_j/\partial x_i, \qquad (8.74)$$

and hence it follows that the streamlines of the $\tilde{\mathbf{U}}$-field map onto streamlines of the $\mathbf{U}$-field. Let $\mathbf{A}$ and $\tilde{\mathbf{A}}$ be vector potentials of $\mathbf{U}$ and $\tilde{\mathbf{U}}$, and let

$$\mathscr{H} = \int \mathbf{U} \cdot \mathbf{A}\, \mathrm{d}V = \int \tilde{\mathbf{U}} \cdot \tilde{\mathbf{A}}\, \mathrm{d}V, \qquad (8.75)$$

the integral being throughout a volume V on whose (fixed) surface S it is assumed that $\mathbf{n} \cdot \mathbf{U} = \mathbf{n} \cdot \tilde{\mathbf{U}} = 0$. Equality of the two integrals in (8.75) follows from the interpretation of $\mathscr{H}$ as a topological measure for the $\mathbf{U}$ or $\tilde{\mathbf{U}}$-field (cf. the discussion of § 2.1), a quantity that is evidently invariant under continuous mappings $\mathbf{x}\to\tilde{\mathbf{x}}$ which conserve the identity of streamlines. Now the vector potential of $U(s, z)\mathbf{i}_\varphi$ is evidently poloidal, and so, from (8.72),

$$\int \mathbf{U} \cdot \mathbf{A}\, \mathrm{d}V = O(R_m^{-1}). \qquad (8.76)$$

Hence $\int \tilde{\mathbf{U}} \cdot \tilde{\mathbf{A}}\, \mathrm{d}V = O(R_m^{-1})$ also. Conversely, if we consider a velocity field $\tilde{\mathbf{U}}(\mathbf{x})$ for which $\int \tilde{\mathbf{U}} \cdot \tilde{\mathbf{A}}\, \mathrm{d}V = O(1)$ as $R_m \to \infty$, then there exists *no* continuous transformation $\mathbf{x} = \mathbf{x}(\tilde{\mathbf{x}})$ for which the related velocity $\mathbf{U}(\mathbf{x})$ is given by (8.72).

[2] For simplicity the discussion is here restricted to steady velocity fields, although Soward's general theory covers unsteady fields also.

In simpler terms, and rather loosely speaking, the velocity field (8.72) exhibits a vanishingly small degree of streamline linkage in the limit $R_m \to \infty$, and only velocity fields exhibiting the same small degree of linkage are amenable to Soward's Lagrangian analysis.

If

$$\tilde{\mathbf{U}}(\mathbf{x}) = U(s, z)\mathbf{i}_\varphi + \varepsilon\,\mathbf{u}'(\mathbf{x}) + \varepsilon^2\mathbf{u}_P(s, z) \tag{8.77}$$

where $\langle \mathbf{u}' \rangle_{az} = 0$ and $\varepsilon = O(R_m^{-1/2})$, as effectively assumed in the foregoing sections, then

$$\int \tilde{\mathbf{U}} \,.\, \tilde{\mathbf{A}}\, \mathrm{d}V = O(\varepsilon^2) = O(R_m^{-1}), \tag{8.78}$$

and the existence of the transformation function $\mathbf{x}(\tilde{\mathbf{x}})$ is not therefore excluded.

8.8. Matching conditions and the external field

In the context of the Earth's magnetic field, let us assume that the core–mantle boundary S is the sphere $r = R_C$, and that the medium in the external region $\hat{V}(r > R_C)$ is insulating (effects of weak mantle conductivity can be incorporated in an improved theory). The field $\hat{\mathbf{B}}(\mathbf{x}, t)$ in $\hat{V}$ is then a potential field, which matches continuously to all orders in ε to the total field $\tilde{\mathbf{B}}(\tilde{\mathbf{x}}, t)$ in the interior V. At leading order, the interior field in Braginskii's theory is the purely toroidal field $B(s, z)\mathbf{i}_\varphi$; since the external field is purely poloidal (cf. § 6.11), B must satisfy

$$B(s, z) = 0 \quad \text{on} \quad r^2 = s^2 + z^2 = R_C^2. \tag{8.79}$$

The mean field in the exterior region, $\hat{\mathbf{B}}_P(s, z)$ say, is given by

$$\hat{\mathbf{B}}_P(s, z) = \nabla\Psi(s, z), \qquad \nabla^2\Psi = 0, \tag{8.80}$$

and satisfies the boundary condition

$$\hat{\mathbf{B}}_P = \varepsilon^2\mathbf{b}_P(s, z) = \varepsilon^2\nabla \wedge (a\,\mathbf{i}_\varphi) \quad \text{on } r = R_C. \tag{8.81}$$

Now the displacement function $\boldsymbol{\eta}(\mathbf{x}, t)$ satisfies $\mathbf{n} \,.\, \boldsymbol{\eta} = 0$ on S (and $\mathbf{n} \wedge \boldsymbol{\eta} = 0$ also if viscous effects are taken into consideration) and so the pseudo-scalar ϖ defined by (8.67a) vanishes on S. Also

$$(\mathbf{n} \,.\, \nabla)\varpi B = \varpi(\mathbf{n} \,.\, \nabla)B + B(\mathbf{n} \,.\, \nabla)\varpi = 0 \quad \text{on } S, \tag{8.82}$$

and it follows that

$$a_e = a \quad \text{and} \quad (\mathbf{n} \,.\, \nabla)a_e = (\mathbf{n} \,.\, \nabla)a \quad \text{on } S, \tag{8.83}$$

where $a_e = a + \varpi B$, and so (8.81) becomes

$$\hat{\mathbf{B}}_P = \varepsilon^2 \nabla \wedge (a_e \mathbf{i}_\varphi) \quad \text{on } r = R_C. \tag{8.84}$$

Hence the external field matches continuously to the internal *effective* field, and is $O(\varepsilon^2)$ relative to the internal toroidal field.

The fluctuating ingredient of the external field, $\hat{\mathbf{B}}'$ say, is also of interest, since it is this ingredient that provides the observed secular variations and the slow drift of the dipole moment $\boldsymbol{\mu}^{(1)}(t)$ relative to the rotation axis. The fact that the magnetic and rotation axes are nearly coincident (see § 4.3) provides evidence that Coriolis forces arising from the rotation are of dominant importance in controlling the structure of core motions; the fact that they are not exactly coincident provides evidence that systematic deviations from exact axisymmetry may be an essential ingredient in the Earth's dynamo process. The internal fluctuating field is $O(\varepsilon)$ and this can penetrate to the external region only through the influence of diffusion; with $\lambda = O(\varepsilon^2)$, this gives a contribution to $\hat{\mathbf{B}}'$ of order ε^3. There is a second contribution also $O(\varepsilon^3)$ due to distortion by $\varepsilon \mathbf{u}'$ of the mean poloidal field $\varepsilon^2 \mathbf{b}_P$. Braginskii (1964*a*) has shown that if the velocity field $\varepsilon \mathbf{u}'$ is steady (or at any rate steady over the time-scale R_C/U_0 characterising the mean toroidal flow) then the relevant boundary condition for the determination of $\hat{\mathbf{B}}'$ is, from the superposition of these two effects[3],

$$\hat{B}'_r = 2\varepsilon \frac{\lambda}{U} \frac{\partial B}{\partial r} \frac{\partial \eta_r}{\partial r} + \varepsilon^3 \left(b_r \frac{\partial \eta_r}{\partial r} - \frac{\eta_\theta}{r} \frac{\partial b_r}{\partial \theta} \right) \quad \text{on } r = R_C, \tag{8.85}$$

where b_r is the radial component of $\mathbf{b}_P$. If $\boldsymbol{\eta}(\mathbf{x}, t)$ is unsteady admitting decomposition into 'azimuthal waves' proportional to $\exp m(\varphi - \omega t)$, then (Braginskii, 1964*b*) each such Fourier component makes a contribution to $\hat{B}'_r$ of the form (8.85) with U replaced by $U - \omega s$, the toroidal mean velocity relative to the frame of

[3] If the fluid is viscous then $U = 0$ on $r = R_C$ and an apparent singularity appears in the first term on the right of (8.85). However $\partial \eta_r / \partial r = 0$ on $r = R_C$ also under the no-slip condition. The appropriate modification of (8.85) requires close examination of the viscous boundary layer on $r = R_C$.

reference which rotates at the angular phase velocity ω of the component considered.

The boundary condition (8.85) is sufficient to determine the exterior fluctuating field $\hat{\mathbf{B}}' = \nabla\Psi'$ uniquely (of course under the additional condition $\Psi' = O(r^{-2})$ at infinity). The linearity of the relation between $\hat{\mathbf{B}}'$ and $\boldsymbol{\eta}$ implies that each Fourier component $\sim e^{im\varphi}$ in $\boldsymbol{\eta}$ generates a corresponding Fourier component in $\hat{\mathbf{B}}'$. In particular, if $m = 1$, a contribution to Ψ' of the form

$$A_1\varepsilon^3 r^{-2} P_1^1(\cos\theta)\cos(\varphi - \omega t) \tag{8.86}$$

is generated, representing a dipole whose moment rotates in the equatorial plane with angular velocity ω. In conjunction with the axisymmetric dipole whose potential is of the form

$$A_0\varepsilon^2 r^{-2} P_1(\cos\theta), \tag{8.87}$$

we have here the beginnings of a plausible explanation (in terms of core motions) for the tilt of the net vector dipole moment of the Earth, and the manner in which this drifts relative to the rotation axis. The observed angle of tilt ($\sim 11°$) is of course not infinitesimal, and corresponds to a value of the ratio $A_1\varepsilon^3/A_0\varepsilon^2 = A_1\varepsilon/A_0$ of order 0·2 (see the figures of table 4.2), i.e. $R_m \sim \varepsilon^{-2} \sim 25$. This value is perhaps (as recognised by Braginskii) uncomfortably low for the applicability of 'large R_m' expansions, although these perhaps provide a useful first step in the right direction.

CHAPTER 9

STRUCTURE AND SOLUTIONS OF THE DYNAMO EQUATIONS

9.1. Dynamo models of α^2- and $\alpha\omega$-type

In the previous two chapters it has been shown that in a wide variety of circumstances the mean field, which will now be denoted by $\mathbf{B}(\mathbf{x}, t)$ (the mean being over time or a Cartesian coordinate or the azimuth angle as appropriate), satisfies an equation of the form

$$\partial\mathbf{B}/\partial t = \nabla \wedge (\mathbf{U} \wedge \mathbf{B}) + \nabla \wedge \mathscr{E} + \lambda \nabla^2 \mathbf{B}, \tag{9.1}$$

where

$$\mathscr{E}_i = \alpha_{ij} B_j + \beta_{ijk}\, \partial B_j/\partial x_k, \tag{9.2}$$

and where α_{ij} and β_{ijk} are determined by the mean velocity $\mathbf{U}(\mathbf{x}, t)$, the statistical properties of the fluctuating field $\mathbf{u}(\mathbf{x}, t)$, and the parameter λ. In the mean-field approach of chapter 7, a useful idealisation is provided by the assumption of isotropy, under which

$$\alpha_{ij} = \alpha\delta_{ij}, \qquad \beta_{ijk} = \beta\varepsilon_{ijk}, \qquad \mathscr{E} = \alpha\mathbf{B} - \beta\nabla \wedge \mathbf{B}. \tag{9.3}$$

It must be recognised however that these expressions are unlikely to be realistic if the background turbulence (or random wave field) is severely anisotropic, as will clearly be the case if Coriolis forces are of dominant importance in controlling the statistical properties of the $\mathbf{u}$-field.

In Braginskii's model as presented in chapter 8, strong anisotropy is 'built in' at the outset through the assumption that the dominant ingredient of the velocity field is a strong toroidal flow $U(s, z)\mathbf{i}_\varphi$. In Braginskii's theory, a 'local' expression for $\mathscr{E}$ of the form (9.2) is obtained, not because the scale of the velocity fluctuations is small (for example, fluctuations proportional to $e^{i\varphi}$ have a scale of the same order as the global scale L), but rather because the departure of a fluid particle from its φ-averaged position is small compared with L. A substantial part of the $\mathscr{E}$ that is generated is 'absorbed' through the use of effective variables; the part of $\mathscr{E}$ that cannot be thus absorbed is given by

$$\mathscr{E} = \mathscr{E}_\varphi \mathbf{i}_\varphi, \qquad \mathscr{E}_\varphi = \alpha B_\varphi, \tag{9.4}$$

in cylindrical polars coordinates (s, φ, z). The *effective* expressions for α_{ij} and β_{ijk} in Braginskii's theory at leading order in the small parameter $\lambda^{1/2}$ are thus of the form

$$\alpha_{ij} = \begin{pmatrix} 0 & 0 & 0 \\ 0 & \alpha & 0 \\ 0 & 0 & 0 \end{pmatrix}, \qquad \beta_{ijk} = 0, \tag{9.5}$$

relative to local Cartesian coordinates (x, y, z) in the directions of increasing (s, φ, z) respectively.

When $\mathbf{U}$, $\mathbf{B}$ and $\mathscr{E}$ are axisymmetric, we have seen that (9.1) may be replaced by two scalar equations. Writing

$$\mathbf{U} = s\omega(s, z)\mathbf{i}_\varphi + \mathbf{U}_P, \qquad \mathbf{B} = B(s, z)\mathbf{i}_\varphi + \mathbf{B}_P, \qquad \mathscr{E} = \mathscr{E}_\varphi \mathbf{i}_\varphi + \mathscr{E}_P, \tag{9.6}$$

and with $\mathbf{B}_P = \nabla \wedge A(s, z)\mathbf{i}_\varphi$, these equations are

$$\partial B/\partial t + s(\mathbf{U}_P \,.\, \nabla)(s^{-1}B) = s(\mathbf{B}_P \,.\, \nabla)\omega + (\nabla \wedge \mathscr{E}_P)_\varphi + \lambda(\nabla^2 - s^{-2})B \tag{9.7}$$

and

$$\partial A/\partial t + s^{-1}(\mathbf{U}_P \,.\, \nabla)(sA) = \mathscr{E}_\varphi + \lambda(\nabla^2 - s^{-2})A. \tag{9.8}$$

When $\mathscr{E}$ is given by (9.3*c*), these equations become

$$\partial B/\partial t + s(\mathbf{U}_P \,.\, \nabla)(s^{-1}B) = s(\mathbf{B}_P \,.\, \nabla)\omega + \nabla \wedge (\alpha \mathbf{B}_P) + \lambda_e(\nabla^2 - s^{-2})B, \tag{9.9}$$

$$\partial A/\partial t + s^{-1}(\mathbf{U}_P \,.\, \nabla)(sA) = \alpha B + \lambda_e(\nabla^2 - s^{-2})A, \tag{9.10}$$

where $\lambda_e = \lambda + \beta$ is an effective diffusivity acting on the mean field; in (9.9) and (9.10) it is assumed for simplicity that β (and so λ_e) is uniform.

There are two source terms involving $\mathbf{B}_P$ on the right of (9.9), and the type of dynamo depends crucially on which of these dominates.

The ratio of these terms is in order of magnitude

$$\frac{|s(\mathbf{B}_P\,.\,\nabla)\omega|}{|\nabla\wedge(\alpha\mathbf{B}_P)|}=O\left(\frac{L^2\omega_0'}{\alpha_0}\right), \tag{9.11}$$

where α_0 is a typical value of α and ω_0' a typical value of $|\nabla\omega|$. If $|\alpha_0|\gg|L^2\omega_0'|$, then the differential rotation term in (9.9) is negligible and we have simply

$$\partial B/\partial t+s(\mathbf{U}_P\,.\,\nabla)(s^{-1}B)=\nabla\wedge(\alpha\mathbf{B}_P)+\lambda_e(\nabla^2-s^{-2})B. \tag{9.12}$$

The α-effect here acts both as the source of poloidal field (via the term αB in (9.10)) and as the source of toroidal field (via the term $\nabla\wedge(\alpha\mathbf{B}_P)$ in (9.12)). Dynamos that depend on this reciprocal process are described as 'α^2-dynamos'.

If on the other hand $|\alpha_0|\ll|L^2\omega_0'|$, then the differential rotation term in (9.9) dominates, so that

$$\partial B/\partial t+s(\mathbf{U}_P\,.\,\nabla)s^{-1}B=s(\mathbf{B}_P\,.\,\nabla)\omega+\lambda_e(\nabla^2-s^{-2})B. \tag{9.13}$$

Now, toroidal field is generated by differential rotation, and poloidal field is generated by the α-effect; dynamos that function in this way are described as '$\alpha\omega$-dynamos'. It will be noticed that, if the distinction between actual and effective variables is ignored, then (9.10) and (9.13) are precisely the equations (8.68) and (8.69) obtained by Braginskii. The assumption $|\alpha_0|\ll|L^2\omega_0'|$ is implicit in Braginskii's analysis; in fact from (8.61), $|\alpha_0|\sim\lambda u_0^2/LU_0^2\sim R_m^{-2}U_0$ so that, with $L^2|\omega_0'|\sim U_0$, in Braginskii's model with $R_m\gg1$,

$$|\alpha_0/L^2\omega_0'|\sim R_m^{-2}\ll 1. \tag{9.14}$$

9.2. Free modes of the α^2-dynamo

To exhibit dynamo action in its simplest form, suppose that the fluid domain V is of infinite extent, that $U\equiv0$, and that $\mathscr{E}$ is given by (9.3c) with α and β uniform and constant. Equation (9.1) then becomes

$$\partial\mathbf{B}/\partial t=\alpha\nabla\wedge\mathbf{B}+\lambda_e\nabla^2\mathbf{B}. \tag{9.15}$$

Now let $\hat{\mathbf{B}}(\mathbf{x})$ be any field satisfying the 'force-free' condition

$$\nabla \wedge \hat{\mathbf{B}}(\mathbf{x}) = K\hat{\mathbf{B}}(\mathbf{x}), \tag{9.16}$$

where K is constant. Examples of such fields have been given in § 2.4. For such a field,

$$\nabla^2\hat{\mathbf{B}} = -\nabla \wedge \nabla \wedge \hat{\mathbf{B}} = -K^2\mathbf{B}, \tag{9.17}$$

and it is evident from (9.15) that if $\mathbf{B}(\mathbf{x}, 0) = \hat{\mathbf{B}}(\mathbf{x})$, then

$$\mathbf{B}(\mathbf{x}, t) = \hat{\mathbf{B}}(\mathbf{x})\, e^{pt} \tag{9.18}$$

where

$$p = \alpha K - \lambda_e K^2. \tag{9.19}$$

Hence the field grows exponentially in strength (its force-free structure being preserved) provided

$$0 < \alpha K > \lambda_e K^2, \tag{9.20}$$

i.e. provided the initial scale of variation of the field $L = |K|^{-1}$ is sufficiently large.

To be specific, let us suppose that $\alpha > 0$, so that the growth condition is simply

$$0 < K < K_c = \alpha/\lambda_e. \tag{9.21}$$

The maximum growth rate p_m occurs for $K = \frac{1}{2}K_c$, and is

$$p_m = \alpha^2/4\lambda_e. \tag{9.22}$$

For self-consistency of the two-scale approach of chapter 7, we require that $K_c l (= l/L)$ should be small. In the case of turbulence with $R_m \ll 1$, this condition is certainly satisfied; for in this case, from (7.90) and (7.99),

$$\alpha \sim l u_0^2/\lambda, \qquad \beta \sim (l u_0)^2/\lambda \qquad (\ll \lambda), \tag{9.23}$$

and so

$$K_c l \sim (l u_0/\lambda)^2 = R_m^2 \ll 1. \tag{9.24}$$

Similarly in the case of a random wave field with no zero-frequency ingredients, and with spectral peak at frequency $\omega_0(=t_0^{-1})$ and wave-number $k_0(=l^{-1})$ satisfying

$$\omega_0 \gg \lambda k_0^2, \tag{9.25}$$

the expressions (7.78) and (7.98) lead to the estimates

$$\alpha \sim \lambda t_0^2 l^{-3} u_0^2, \qquad \beta \sim \lambda (u_0 t_0/l)^2 \ll \lambda, \tag{9.26}$$

and so in this case (under the condition (7.29))

$$K_c l \sim (u_0 t_0/l)^2 \ll 1. \tag{9.27}$$

However there is a potential inconsistency when there *are* zero frequency ingredients in the wave spectrum if (as is by no means certain) α and β do tend to finite limits as $\lambda \to 0$. If this is the case, then on dimensional grounds,

$$\alpha \sim u_0, \qquad \beta \sim u_0 l \qquad (\gg \lambda), \tag{9.28}$$

and so

$$K_c l = \alpha l/\lambda_e = O(1). \tag{9.29}$$

In this situation, the medium would in fact be most unstable to magnetic modes whose length-scale is of the same order as the scale of the background **u**-field; this conclusion is incompatible with the two-scale approach leading to equation (9.15), and indicates again that conclusions that lean heavily on the estimates (9.28) must be treated with caution.

Of course the estimate $\alpha \sim u_0$ is valid only if the **u**-field is *strongly helical* in the sense that

$$|\langle \mathbf{u} \cdot \boldsymbol{\omega} \rangle| \sim l^{-1} \langle \mathbf{u}^2 \rangle. \tag{9.30}$$

In a *weakly helical* situation, with

$$|\langle \mathbf{u} \cdot \boldsymbol{\omega} \rangle| = \varepsilon l^{-1} \langle \mathbf{u}^2 \rangle, \qquad \varepsilon \ll 1, \tag{9.31}$$

the estimate for α must be modified, while that for β remains unchanged:

$$\alpha \sim \varepsilon u_0, \qquad \beta \sim u_0 l. \tag{9.32}$$

Now we have

$$K_c l = O(\varepsilon) \ll 1, \tag{9.33}$$

and the conclusions are once again compatible with the underlying assumptions.

At this point, it may be useful to examine briefly the influence of subsequent terms in the expansion (7.10) of $\mathscr{E}$ in terms of derivatives of $\mathbf{B}$. In the isotropic situation, this expansion can only take the form

$$\mathscr{E} = \alpha \mathbf{B} - \beta \nabla \wedge \mathbf{B} + \gamma \nabla \wedge (\nabla \wedge \mathbf{B}) - \ldots, \tag{9.34}$$

where γ (like α) is a pseudo-scalar, and (9.15) is replaced by

$$\partial \mathbf{B}/\partial t = \alpha \nabla \wedge \mathbf{B} - \lambda_e \nabla \wedge (\nabla \wedge \mathbf{B}) + \gamma \nabla \wedge \nabla \wedge (\nabla \wedge \mathbf{B}) - \ldots. \tag{9.35}$$

The eigenfunctions of this equation are still the force-free modes and (9.19) is replaced by

$$p = \alpha K - \lambda_e K^2 + \gamma K^3 - \ldots. \tag{9.36}$$

If $\gamma > 0$, the last term can be destabilising if

$$K > \lambda_e/|\gamma|; \tag{9.37}$$

however, since λ_e and γ are both determined by the statistical properties of the $\mathbf{u}$-field, it is to be expected that $\lambda_e/|\gamma|$ is of order l (at least) on dimensional grounds. The condition (9.37) is then incompatible with the condition $Kl \ll 1$, and the conclusion is that dynamo instabilities associated with the third (and subsequent) terms of (9.34) are unlikely to arise within the framework of a double-length-scale theory.

9.3. Free modes when α_{ij} is anisotropic

Suppose now that α_{ij} is no longer isotropic, but still uniform and symmetric with principal values $\alpha^{(1)}, \alpha^{(2)}, \alpha^{(3)}$. We restrict attention here to the situation in which the β_{ijk} contribution to $\mathscr{E}$ is negligible (cf. (9.23*b*) or (9.26*b*)). Then, with $\mathbf{U} = 0$, (9.1) becomes

$$\frac{\partial B_i}{\partial t} = \varepsilon_{ijk} \alpha_{km} \frac{\partial B_m}{\partial x_j} + \lambda \nabla^2 B_i. \tag{9.38}$$

This equation admits plane wave solutions of the form

$$\mathbf{B} = \hat{\mathbf{B}}\, e^{pt}\, e^{i\mathbf{K}.\mathbf{x}}, \qquad \mathbf{K}\,.\,\hat{\mathbf{B}} = 0, \tag{9.39}$$

substitution in (9.38) giving

$$(p + \lambda K^2)\hat{B}_i = i\varepsilon_{ijk}\alpha_{km}K_j\hat{B}_m. \tag{9.40}$$

If we refer to the principal axes of α_{km}, the first component of (9.40) becomes

$$(p + \lambda K^2)\hat{B}_1 = i(\alpha^{(3)}K_2\hat{B}_3 - \alpha^{(2)}K_3\hat{B}_2), \tag{9.41}$$

and the two other components are given by cyclic permutation of suffixes. For a non-trivial solution $(\hat{B}_1, \hat{B}_2, \hat{B}_3)$, the determinant of the coefficients must vanish. This gives a cubic equation for p with roots

$$p_0 = -\lambda K^2, \qquad p_1 = -\lambda K^2 + Q, \qquad p_2 = -\lambda K^2 - Q, \tag{9.42}$$

where

$$Q^2 = \alpha^{(2)}\alpha^{(3)}K_1^2 + \alpha^{(3)}\alpha^{(1)}K_2^2 + \alpha^{(1)}\alpha^{(2)}K_3^2. \tag{9.43}$$

We are here only interested in the possibility of exponential growth of **B**, i.e. $\operatorname{Re} p > 0$. The condition for this to occur is evidently

$$Q^2 > \lambda^2 K^4. \tag{9.44}$$

The surface $Q^2 = \lambda^2 K^4$ is sketched in fig. 9.1 in the axisymmetric situation $\alpha^{(1)} = \alpha^{(2)}$ (in which case it is a surface of revolution about the K_3-axis) and for the two essentially distinct possibilities $\alpha^{(1)}\alpha^{(3)} >$ or <0. Wave amplification now depends, as might be expected, on the direction as well as the magnitude of the wave-vector **K**. When $\alpha^{(1)}\alpha^{(3)} > 0$, amplification occurs for all directions provided $|\mathbf{K}|$ is sufficiently small (as in the isotropic case). When $\alpha^{(1)}\alpha^{(3)} < 0$, however, amplification can only occur for wave-vectors within the cone $Q^2 > 0$ (and then only for sufficiently small $|\mathbf{K}|$).

The possibility $\alpha^{(1)}\alpha^{(3)} < 0$ is perhaps a little pathological in the context of turbulence as normally conceived; it would presumably

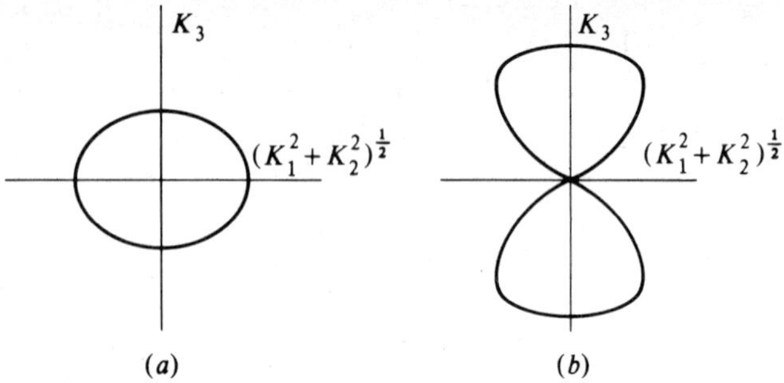

Fig. 9.1 Sketch of the surface $Q^2=\lambda^2K^4$ where Q^2 is given by (9.43) and $\alpha^{(1)}=\alpha^{(2)}$. Field amplification occurs if the vector $\mathbf{K}$ is inside this surface. (*a*) $\alpha^{(1)}\alpha^{(3)}>0$; (*b*) $\alpha^{(1)}\alpha^{(3)}<0$.

arise in a situation in which $\langle u_1\omega_1\rangle$ and $\langle u_3\omega_3\rangle$ have opposite signs where $\mathbf{u}$ and $\boldsymbol{\omega}$ are as usual the random velocity and vorticity distributions. Although it is possible to conceive of artificial methods of generating such turbulence (cf. the discussion of § 7.6), it is difficult to see how such a situation could arise without artificial helicity injection.

Anisotropic field amplification as discussed in this section has been encountered in the closely related context of dynamo action due to velocity fields that are steady and strictly space-periodic (Childress, 1970; G. O. Roberts, 1970, 1972); the methods of mean-field electrodynamics of course apply equally to this rather special situation, with the difference that the velocity spectrum tensor (which may be defined via the operation of spatial averaging) is discrete rather than continuous. In general this spectrum tensor (and similarly α_{ij}) will be anisotropic due to preferred directions that may be apparent in the velocity field. For example if

$$\mathbf{u}=u_0(\sin kz, \cos kz, 0), \tag{9.45}$$

then Oz is clearly a preferred direction; and we have seen in § 7.7 that in this situation

$$\alpha_{ij}=\alpha\delta_{i3}\delta_{j3} \quad \text{where } \alpha=-u_0^2/\lambda k. \tag{9.46}$$

If however we choose a space-periodic velocity field that exhibits cubic symmetry (invariance under the group of rotations of the cube), e.g.

$$\mathbf{u} = u_0(\sin kz + \cos ky, \sin kx + \cos kz, \sin ky + \cos kx) \tag{9.47}$$

(Childress, 1970), then

$$\langle u_1\omega_1\rangle = \langle u_2\omega_2\rangle = \langle u_3\omega_3\rangle = ku_0^2, \tag{9.48}$$

and, as may be easily shown by the method of § 7.7, provided $u_0/k\lambda \ll 1$,

$$\alpha_{ij} = \alpha\delta_{ij} \quad \text{where } \alpha = -u_0^2/\lambda k. \tag{9.49}$$

Hence although the velocity field (9.47) exhibits *three* preferred directions, the pseudo-tensor α_{ij} is nevertheless isotropic[1]. More generally, the velocity field

$$\mathbf{u} = u_0(\sin k_3 z + \cos k_2 y, \sin k_1 x + \cos k_3 z, \sin k_2 y + \cos k_1 x), \tag{9.50}$$

for which

$$\frac{\langle u_1\omega_1\rangle}{k_2+k_3} = \frac{\langle u_2\omega_2\rangle}{k_3+k_1} = \frac{\langle u_3\omega_3\rangle}{k_1+k_2} = \tfrac{1}{2}u_0^2, \tag{9.51}$$

yields the non-isotropic form

$$\alpha_{ij} = \alpha^{(1)}\delta_{i1}\delta_{j1} + \alpha^{(2)}\delta_{i2}\delta_{j2} + \alpha^{(3)}\delta_{i3}\delta_{j3}, \tag{9.52}$$

where

$$\alpha^{(1)}k_1 = \alpha^{(2)}k_2 = \alpha^{(3)}k_3 = -u_0^2/\lambda. \tag{9.53}$$

Here k_1, k_2 and k_3 may be positive or negative, and all possible sign combinations of $\alpha^{(1)}$, $\alpha^{(2)}$ and $\alpha^{(3)}$ can therefore in principle occur.

9.4. The α^2-dynamo in a spherical geometry

Suppose now that α is uniform and constant within the sphere $r<R$, the external region $r>R$ being insulating. With $\mathbf{U}=0$, equation (9.15) is still satisfied in the sphere and $\mathbf{B}$ must match to a potential

[1] The situation may be compared with other familiar situations – e.g. cubic symmetry of a mass distribution is sufficient to ensure isotropy of its inertia tensor.

field across $r=R$. This problem was first considered by Krause & Steenbeck (1967), who determined the possible steady field structures. The treatment that follows is a little more general.

Instead of **B**, let us consider the (mean) current distribution $\mathbf{J}(\mathbf{x}, t)=\mu_0^{-1}\nabla\wedge\mathbf{B}$, which from (9.15) satisfies (for $r<R$)

$$\partial\mathbf{J}/\partial t=\alpha\nabla\wedge\mathbf{J}-\lambda_e\nabla\wedge(\nabla\wedge\mathbf{J}), \tag{9.54}$$

together with the boundary condition $\mathbf{n}\,.\,\mathbf{J}=0$ on $r=R$. As suggested by the result of § 9.2, it is illuminating here to consider the evolution of current structures[2] for which

$$\nabla\wedge\mathbf{J}=K\mathbf{J}, \qquad \mathbf{J}\,.\,\mathbf{n}=0 \quad \text{on } r=R. \tag{9.55}$$

From (9.54), such a current distribution satisfies

$$\partial\mathbf{J}/\partial t=(\alpha K-\lambda_e K^2)\mathbf{J} \quad \text{for } r<R, \tag{9.56}$$

the structure (9.55) being conserved. As in § 9.2, we have exponential growth of **J** (and so of **B**) for any possible values of K satisfying

$$\alpha K>\lambda_e K^2 \tag{9.57}$$

Moreover the mode of maximum growth rate is that for which $(\alpha K-\lambda_e K^2)$ takes the largest positive value.

Field structures satisfying (9.55) may be easily determined by the procedure of § 2.4 (applied now to **J** rather than to **B**). Letting

$$\mathbf{J}=\nabla\wedge\nabla\wedge\mathbf{x}S+K\nabla\wedge\mathbf{x}S, \tag{9.58}$$

(9.55) is satisfied provided

$$(\nabla^2+K^2)S=0, \qquad S=0 \quad \text{on } r=R. \tag{9.59}$$

Solutions have the form

$$S=A_n r^{-1/2}J_{n+\frac{1}{2}}(|K|r)S_n(\theta,\varphi), \qquad (n=1,2,\ldots), \tag{9.60}$$

where possible values of K are determined by

$$J_{n+\frac{1}{2}}(|K|R)=0. \tag{9.61}$$

[2] Note that the corresponding fields are *not* force-free; we know from § 2.4 that force-free fields continuous everywhere and $O(r^{-3})$ at infinity do not exist. The non-existence theorem does not however apply to **J** which has a tangential discontinuity across $r=R$.

If any of the roots of this equation satisfy (9.57), then the corresponding current structure given by (9.55) and (9.60) grows exponentially in time. Note that if α is negative, then K must be chosen negative also in (9.58) to give a growing current mode.

The field **B** corresponding to the current (9.58) may be easily determined. By uncurling (9.55) we obtain, for $r<R$,

$$\nabla \wedge \mathbf{B} = K\mathbf{B} + K\nabla\psi \tag{9.62}$$

for some function ψ satisfying

$$\nabla^2\psi = 0. \tag{9.63}$$

The toroidal part of **B** (cf. 2.34) is simply

$$\mathbf{B}_T = \nabla \wedge (\mathbf{x}S) = -\mathbf{x} \wedge \nabla S, \tag{9.64}$$

and, from (9.62), the poloidal part is given by

$$\mathbf{B}_P = K^{-1}\nabla \wedge \mathbf{B}_T - \nabla\psi. \tag{9.65}$$

For $r>R$, $\mathbf{B}_P = -\nabla\hat{\psi}$, say, where $\nabla^2\hat{\psi}=0$. The harmonic functions ψ and $\hat{\psi}$ must be chosen so that $\mathbf{B}_P$ is continuous across $r=R$.

When S is given by (9.60), we must have evidently

$$\psi = C_n(r/R)^n S_n(\theta, \varphi), \qquad \hat{\psi} = D_n(R/r)^{n+1}S_n(\theta,\varphi), \tag{9.66}$$

the constants C_n and D_n being chosen so that $\mathbf{n}\,.\,\mathbf{B}_P$ and $\mathbf{n}\wedge\mathbf{B}_P$ are continuous across $r=R$; these conditions give

$$C_n = D_n = -\frac{A_n}{2n+1}|K|R\frac{\mathrm{d}}{\mathrm{d}R}(R^{-1/2}J_{n+\frac{1}{2}}(|K|R)), \tag{9.67}$$

and this completes the determination of the field structures. Evidently if $n=1$ we have a dipole field outside the sphere, if $n=2$ we have a quadrupole, and so on.

The roots of (9.61) are given by $|K|R = x_{nq}$ where x_{nq} are as tabulated on p. 39. Denoting the corresponding current structures by $\mathbf{J}^{(nq)}$, we have from (9.56)

$$\partial\mathbf{J}^{(nq)}/\partial t = -\lambda_e R^{-2}x_{nq}(x_{nq}-R_\alpha)\mathbf{J}^{(nq)}, \tag{9.68}$$

where

$$R_\alpha = |\alpha| R/\lambda_e. \tag{9.69}$$

R_α may be regarded as a magnetic Reynolds number based on the intensity of the α-effect. The smallest value of R_α for which dynamo action is possible is evidently

$$R_\alpha = x_{11} \approx 4{\cdot}49, \tag{9.70}$$

the corresponding mode being a dipole field for $r > R$ (fig. 9.2). If $x_{11} < R_\alpha < x_{21}$, then the mode $\mathbf{J}^{(11)}$ is the only mode that can be excited. If $x_{21} < R < x_{31}$, then the quadrupole mode ($n = 2, q = 1$) will be excited also. Note also that if $R_\alpha > x_{11} + x_{21}$, then the quadrupole mode is excited more intensely than the dipole mode. Similarly in general if $R_\alpha > x_{nq} + x_{n'q'}$ and $x_{n'q'} > x_{nq}$ then it is the *more structured* mode (n', q') which has the higher growth rate. This means that if $R_\alpha \gg 1$ (as would for example be the case if $|\alpha| \sim u_0$, $|\lambda_e| \sim u_o l$, and $l \ll R$) then the most rapidly excited modes will be

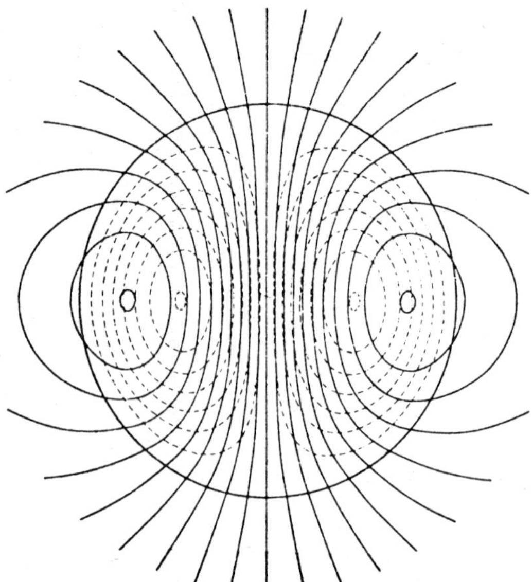

Fig. 9.2 Lines of force of the poloidal field (solid) and lines of constant toroidal field (dotted), for the α^2-dynamo with $\alpha = \text{cst.}$ in $r < R$ and with $n = q = 1$. (Krause & Steenbeck, 1967.)

highly structured modes corresponding to large values of n and/or q, having characteristic length-scale L *small* compared with R; as in the discussion of § 9.2, the results are meaningful only if this scale L remains large compared with the micro-scale l of the background random motions.

9.5. The α^2-dynamo with antisymmetric α

If attention is focussed on the possibility of *steady* dynamo action in a sphere due to an α-effect and still with zero mean velocity, then we are faced with the eigenvalue problem

$$\begin{aligned} &\nabla^2\mathbf{B}+\nabla\wedge(\alpha/\lambda_e)\mathbf{B}=0 \quad (r<R),\\ &\nabla\wedge\mathbf{B}=0 \quad (r>R),\\ &[\mathbf{B}]=0 \quad \text{across } r=R,\\ &\mathbf{B}=O(r^{-3}) \quad \text{at } \infty, \end{aligned} \tag{9.71}$$

and the analysis of the preceding section in effect provides the solution of this problem in the particular situation when α is uniform. However, in a rotating body, in which the value of α is controlled in some way by Coriolis forces, a more realistic theory must allow for variation of α on scales large compared with the underlying scale l of the turbulence. In particular, it is to be expected that $\alpha(\mathbf{x})$ will have the same antisymmetry about the equatorial plane as the vertical component $\Omega\cos\theta$ of the rotation vector $\mathbf{\Omega}$ where θ is the colatitude. A possible mechanism in the terrestrial context that makes the process physically explicit (Parker, 1955*b*) is the following: suppose that a typical 'event' in the northern hemisphere consists of the rising of a blob of fluid (as a result of a density defect relative to its surroundings); fluid must be entrained from the sides and conservation of angular momentum (in the body of fluid which rotates as a whole) implies that the blob acquires positive helicity. Now suppose that these events occur at random throughout the body of fluid, the upward motion of the blobs being compensated by a downward flow between blobs (note the need for a topological asymmetry between upward and downward flow – cf. the discussion of § 3.12); then a mean helicity distribution is generated, positive in the northern hemisphere, and (by the same argument) negative in the southern hemisphere. In the strong diffusion (or weak random wave) limit (§ 7.5) the corresponding value of α would be generally negative in the northern hemisphere and positive in the southern hemisphere; a simple assumption incorporating this antisymmetry is

$$\alpha(\mathbf{x}) = -\alpha_0 \hat{\alpha}(\hat{r}) \cos\theta, \tag{9.72}$$

where $\hat{r} = r/R$, and $\hat{\alpha}$ is dimensionless[3].

Unfortunately, under any assumption of the form (9.72), the eigenvalue problem (9.71) is no longer amenable to simple analysis and recourse must be had to numerical methods. The problem was studied by Steenbeck & Krause (1966, 1969b) and Roberts (1972*a*) using series expansions for **B** and truncating after a few terms; Roberts found that six terms were sufficient to give eigenvalues to within 0·1% accuracy.

As in § 9.4, the α-effect is here responsible for generating toroidal from poloidal field, *and* poloidal from toroidal field; the problem is essentially the determination of eigenvalues of the dimensionless parameter $R_\alpha = |\alpha_0| R/\lambda_e$. The eigenvalues corresponding to fields of dipole and quadrupole symmetries about the equatorial plane as obtained by Roberts (1972*a*) for various choices of the function $\hat{\alpha}(\hat{r})$ are shown in table 9.1. The most striking feature here is that the eigenvalues for fields of dipole and quadrupole symmetry are almost indistinguishable. This has been interpreted by Steenbeck & Krause (1969*b*) with reference to the field structures that emerge from the computed eigenfunctions. Fig. 9.3 shows the poloidal field lines and the isotors (i.e. curves of constant toroidal fields) for the case $\hat{\alpha}(\hat{r}) = \hat{r}(3\hat{r} - 2)$. The toroidal current (giving rise to poloidal field) is concentrated in high latitudes ($|\theta|, |\pi - \theta| \lesssim \pi/10$), and the mutual inductance (or coupling) between these toroidal current

[3] It may be argued that whereas a rising blob will converge near the bottom of a convection layer (thus acquiring positive helicity in the northern hemisphere) it will diverge near the top of this layer, thus acquiring negative helicity (again in the northern hemisphere). There is therefore good reason to consider models in which $\hat{\alpha}(\hat{r})$ in (9.72) changes sign for an intermediate value of $\hat{r}$, as has been done by Yoshimura (1975).

loops is very small. Consequently the toroidal current in one hemisphere can be reversed without greatly affecting conditions in the other. This operation transforms a field of dipole symmetry into one of quadrupole symmetry and vice versa.

Table 9.1. *Eigenvalues of* R_α *obtained by Roberts* (1972*a*), *following the* α^2*-dynamo model of Steenbeck & Krause* (1966, 1969*b*). *The numerical factors in* $\hat{\alpha}(\hat{r})$ *are chosen so that* $\hat{\alpha}_{max} = 1$. *The figures given correspond to truncation of the spherical harmonic expansion of* **B** *at the level* $n = 5$ *for the first three cases, and* $n = 4$ *for the last case, and to radial discretisation of the governing differential equations into* 30 *segments.*

$\hat{\alpha}(\hat{r})$	R_α (Dipole symmetry)	R_α (Quadrupole symmetry)
1	7·64	7·81
$\hat{r}(3\hat{r}-2)$	24·95	24·93
$7{\cdot}37\hat{r}^2(1-\hat{r})(5\hat{r}-3)$	14·10	14·10
$45{\cdot}56\hat{r}^8(1-\hat{r}^2)^2$	13·04	13·11

Note that the only O-type neutral points of the poloidal fields depicted in fig. 9.3 are situated in regions where $\alpha \neq 0$, and the field in the neighbourhood of these points can therefore be maintained by the α-effect. In each case (*a*) and (*b*), there is also a neutral point on the equatorial plane $\theta = \pi/2$ where $\alpha = 0$; this is however an X-type neutral point (i.e. a saddle point of the flux-function $\chi(s, z)$), and Cowling's neutral point argument (§ 6.5) does not therefore apply (Weiss, 1971).

Steenbeck & Krause (1969*b*) also consider choices of $\hat{\alpha}(\hat{r})$ vanishing for values of $\hat{r}$ less than some value $\hat{r}_I$ between 0 and 1 in order to estimate the effect of the solid inner body of the Earth (for which $\hat{r}_I \approx 0{\cdot}19$). The poloidal field lines and isotors were plotted for $\hat{r}_I = 0{\cdot}5$ and are reproduced by Roberts & Stix (1971). The change in the eigenvalues is no more than would be expected from the reduced volume of fluid in which the α-effect is operative.

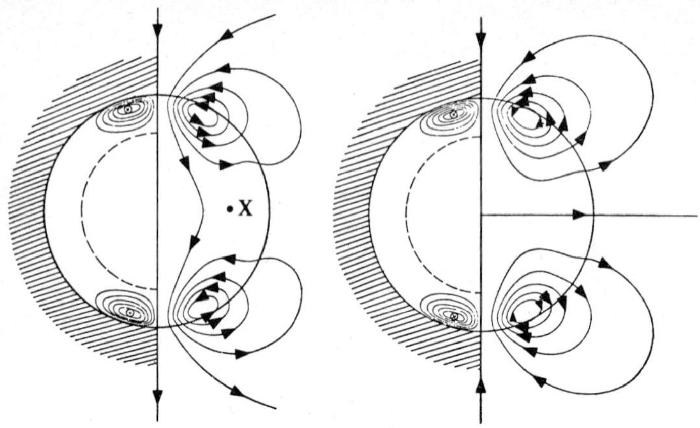

Fig. 9.3 Fields excited by α^2-dynamo action when $\alpha = \alpha_0 R^{-2} r \times (3r - 2R)\cos\theta$ (Steenbeck & Krause, 1969*b*): (*a*) dipole mode; (*b*) quadrupole mode. The poloidal field lines are shown on the right of each figure and the isotors on the left. Note that in case (*a*) the neutral point of $\mathbf{B}_P$ on the equatorial plane is X-type (rather than O-type), so that although $\alpha = 0$ at this point, Cowling's antidynamo theorem does not apply (Weiss, 1971).

9.6. Free modes of the $\alpha\omega$-dynamo

The $\alpha\omega$-dynamo, as discussed in § 9.1, is described by equations (9.10) and (9.13). Let us first consider the Cartesian analogue of these equations[4], viz.

$$\partial A/\partial t + \mathbf{U}_P . \nabla A = \alpha B + \lambda \nabla^2 A, \tag{9.73}$$

$$\partial B/\partial t + \mathbf{U}_P . \nabla B = \mathbf{B}_P . \nabla U + \lambda \nabla^2 B, \tag{9.74}$$

the mean velocity and mean magnetic field being $U\mathbf{i}_y + \mathbf{U}_P$ and $B\mathbf{i}_y + \mathbf{B}_P$ respectively, and with $\mathbf{B}_P = \nabla \wedge (A\mathbf{i}_y)$. These equations admit local solutions (Parker, 1955*b*) of the form

$$(A, B) = (\hat{A}, \hat{B}) \exp(pt + i\mathbf{K} . \mathbf{x}), \qquad \mathbf{K} = (K_x, 0, K_z), \tag{9.75}$$

[4] Here, and subsequently, λ will be understood as including turbulent diffusivity effects when these are present.

over regions of limited extent in which $\mathbf{U}_P$, α and ∇U may all be treated as uniform. Substitution gives

$$\tilde{p}\hat{A} = \alpha\hat{B}, \qquad \tilde{p}\hat{B} = -\mathrm{i}(\mathbf{K}\wedge\nabla U)_y\hat{A}, \tag{9.76}$$

where

$$\tilde{p} = p + \lambda K^2 + \mathrm{i}\mathbf{U}_p\,.\,\mathbf{K}.$$

Eliminating $\hat{A}$, $\hat{B}$ we obtain the dispersion relation

$$\tilde{p}^2 = 2\mathrm{i}\gamma \quad \text{where } \gamma = -\tfrac{1}{2}\alpha(\mathbf{K}\wedge\nabla U)_y. \tag{9.77}$$

The character of the solutions is largely determined by the sign of γ.

(i) *Case* $\gamma > 0$. In this case, $\tilde{p} = \pm(1+\mathrm{i})\gamma^{1/2}$, and so

$$p = -\lambda K^2 \pm \gamma^{1/2} + \mathrm{i}(\pm\gamma^{1/2} - \mathbf{U}_P\,.\,\mathbf{K}). \tag{9.78}$$

The solutions (9.75) do not decay if $\mathrm{Re}\,p \geqslant 0$, and this is satisfied by (9.78) (with the upper choice of sign) when $\gamma \geqslant \lambda^2 K^4$, i.e. when

$$-\alpha(\mathbf{K}\wedge\nabla U)_y \geqslant 2\lambda^2 K^4. \tag{9.79}$$

The phase factor for this wave of growing (or at least non-decaying) amplitude is

$$\exp\{\mathrm{i}\mathbf{K}\,.\,\mathbf{x} + \mathrm{i}(\gamma^{1/2} - \mathbf{U}_P\,.\,\mathbf{K})t\}, \tag{9.80}$$

and it propagates in the direction $\pm\mathbf{K}$ according as

$$\gamma^{1/2} - \mathbf{U}_P\,.\,\mathbf{K} < \text{or} > 0. \tag{9.81}$$

When $\mathbf{U}_P = 0$, the field necessarily has an oscillatory character (contrast the steady free modes of the α^2-dynamo discussed in § 9.2) and the wave propagates in the direction of the vector $-\mathbf{K}$.

(ii) *Case* $\gamma < 0$. In this case, $\tilde{p} = \pm(1-\mathrm{i})|\gamma|^{1/2}$, and so

$$p = -\lambda K^2 \pm |\gamma|^{1/2} + \mathrm{i}(\mp|\gamma|^{1/2} - \mathbf{U}_P\,.\,\mathbf{K}), \tag{9.82}$$

and the growing field, when $\mathbf{U}_P = 0$, propagates in the $+\mathbf{K}$ direction.

As recognised by Parker (1955*b*), these results are strongly suggestive in the solar context in explaining the migration of sunspots towards the equatorial plane (as described in § 5.3). In the outer convective layer of the Sun and in the northern hemisphere let $Oxyz$ be locally Cartesian coordinates with Ox south, Oy east and Oz vertically upwards (fig. 9.4) and suppose that vertical shear dominates in the toroidal flow so that

$$\gamma = \tfrac{1}{2}K_x\alpha\ \partial U/\partial z. \tag{9.83}$$

If magnetic disturbances can be represented in terms of simple migratory waves of the above kind, then the migration is towards or away from the equatorial plane according as $\alpha\ \partial U/\partial z <$ or >0. (The signs are reversed in the southern hemisphere.) The sign of the product $\alpha\ \partial U/\partial z$ (or of $\alpha\ \partial\omega/\partial r$ when we return to the spherical geometry) is of crucial importance. If sunspots are formed by distortion due to buoyant upwelling (see § 10.7) of any underlying toroidal field, then equatorial migration of sunspots reflects equatorial migration of the toroidal field pattern. This picture of equatorial migration as a result of the '$\alpha\omega$-effect' – i.e. the joint action of an α-effect and differential rotation (of appropriate sign) – is confirmed by numerical solutions which take due account of the

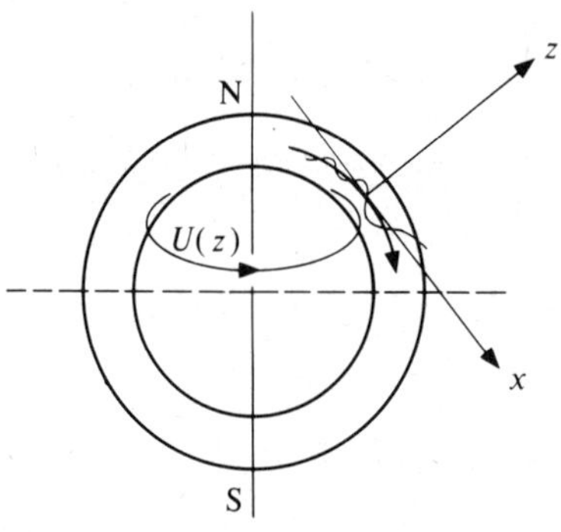

Fig. 9.4 Parker's (1955*b*) dynamo wave, which propagates with increasing amplitude towards the equatorial plane when $\alpha\ \partial U/\partial z < 0$.

spherical geometry and of spatial variation of α and $\nabla\omega$ (see § 9.12 below).

If $\mathbf{U}_P \neq 0$ in (9.78) or (9.82), the phase velocity of the 'dynamo-wave' modes (9.75) is modified. If $|\gamma|^{1/2} = \mathbf{U}_P\,.\,\mathbf{K}$, then the wave (9.78) (with upper sign) is stationary, while if $|\gamma|^{1/2} = -\mathbf{U}_P\,.\,\mathbf{K}$, the wave (9.82) (with upper sign) is stationary. The inference is that an appropriate poloidal mean velocity may transform a situation in which an oscillating field $\hat{\mathbf{B}}(\mathbf{x})\,e^{i\omega t}$ is maintained by the $\alpha\omega$-effect into one in which the field $\hat{\mathbf{B}}(\mathbf{x})$ may be maintained as a steady dynamo. The importance of meridional circulation $\mathbf{U}_P$ in determining whether the preferred mode of magnetic excitation has a steady or an oscillating character was recognised by Braginskii (1964*b*).

As in the case of the free modes discussed in § 9.2, it is desirable for consistency that the length-scale of the most unstable mode should be large compared with the scale l of the background random motions. Writing $(K \wedge \nabla U)_y = KG$, where G is a representative measure of the mean shear rate, the critical wave-number given by (9.79) has magnitude $K_c = (|\alpha G|/2\lambda^2)^{1/3}$, and the maximum growth rate (i.e. maximum value of $\operatorname{Re} p$ as given by (9.78)) occurs for $K = 2^{-4/3}K_c$. The scale of the most unstable modes is therefore given (in order of magnitude) by

$$L \sim K_c^{-1} \sim (\lambda^2/|\alpha G|)^{1/3}, \tag{9.84}$$

and the consistency condition $L \gg l$ becomes

$$\lambda^2/l^3|\alpha G| \gg 1. \tag{9.85}$$

If, in the weak diffusion limit, we adopt the estimates

$$|\alpha| \sim u_0, \qquad \lambda \sim u_0 l, \tag{9.86}$$

then (9.85) becomes simply

$$u_0/l \gg |G|, \tag{9.87}$$

so that the treatment is consistent if the 'random shear' (of order u_0/l) is large compared with the mean shear, a condition that is likely to be satisfied in the normal turbulence context. Note however that the $\alpha\omega$-model is appropriate only if $|\alpha| \ll L|G|$ (from the discussion of § 9.1) so that, with the estimate $|\alpha| \sim u_0$, we also require

$$u_0/l \ll |G|L/l. \tag{9.88}$$

9.7. Concentrated generation and shear

In order to shed some further light on the structure of equations (9.73) and (9.74), consider now the situation $\mathbf{U}_P = 0$, and $\alpha = \alpha(z)$, $U = U(z)$ where

$$\alpha(z) = \Delta_0 \delta(z), \qquad dU/dz = U_0 \delta(z - z_0), \tag{9.89}$$

so that the α-effect and the shear effect are concentrated in two parallel layers distance z_0 apart (fig. 9.5). This is of course an

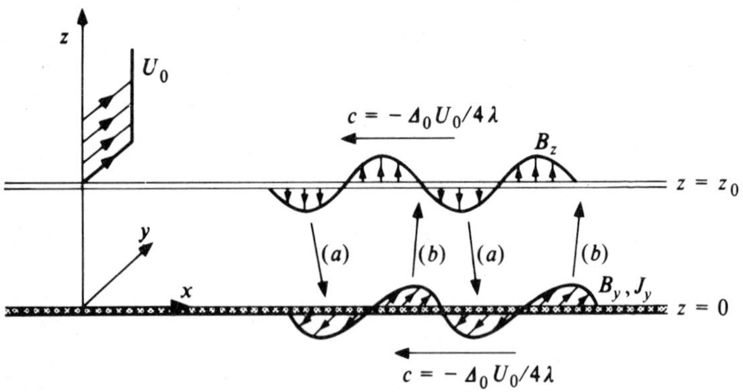

Fig. 9.5 Oscillatory dynamo action with concentrated α-effect ($z = 0$) and shear ($z = z_0$): (a) shearing of the field B_z generates B_y which gives rise to J_y by the α-effect; (b) this current is the source of a poloidal field $(B_x, 0, B_z)$. The complete field pattern propagates in the x-direction with phase velocity $c = -\Delta_0 U_0/4\lambda$, and the wave amplitude grows exponentially if (9.101) is satisfied.

idealisation, with the sole merit that it permits simple mathematical analysis in a situation in which α and ∇U are non-uniform. More realistic distributions of α and U in terrestrial and solar contexts generally require recourse to the computer. It is useful to have some firm analytical results if only to provide points of comparison with computational studies. In the same spirit of idealisation, let us for the moment ignore the influence of fluid boundaries (as in § 9.6), or equivalently assume that the fluid fills all space.

Under the assumptions (9.89), equations (9.73) and (9.74) become

$$\partial A/\partial t = \lambda \nabla^2 A, \qquad \partial B/\partial t = \lambda \nabla^2 B \qquad (z \neq 0, z_0). \tag{9.90}$$

The singularities (9.89) evidently induce discontinuities in $\partial A/\partial z$ across $z=0$ and in $\partial B/\partial z$ across $z=z_0$, but A and B remain continuous across both layers. Integration of (9.73) and (9.74) across the layers gives the jump conditions

$$[\partial B/\partial z]=0, \qquad \lambda[\partial A/\partial z]=-\Delta_0 B, \qquad \text{on } z=0, \tag{9.91}$$

$$[\partial A/\partial z]=0, \qquad \lambda[\partial B/\partial z]=-U_0\,\partial A/\partial x, \qquad \text{on } z=z_0. \tag{9.92}$$

We suppose further that

$$A, B\to 0 \quad \text{as } |z|\to\infty. \tag{9.93}$$

We look for solutions of the form

$$(A, B)=(\hat{A}(z), \hat{B}(z))\,\mathrm{e}^{pt+\mathrm{i}Kx}, \tag{9.94}$$

where K is real, but may be positive or negative. From (9.90), $\hat{A}$ and $\hat{B}$ can depend on z only through the factors $\mathrm{e}^{\pm mz}$ where

$$\lambda m^2=p+\lambda K^2, \tag{9.95}$$

and where we may suppose that $\operatorname{Re} m>0$. The conditions of continuity and the conditions at infinity then imply that

$$\hat{A}(z)=\begin{cases} A_1\,\mathrm{e}^{mz} & (z<0),\\ A_1\,\mathrm{e}^{-mz} & (0<z<z_0),\\ A_1\,\mathrm{e}^{-mz} & (z>z_0).\end{cases}$$
$$\hat{B}(z)=\begin{cases} B_1\,\mathrm{e}^{m(z-z_0)} & (z<0),\\ B_1\,\mathrm{e}^{m(z-z_0)} & (0<z<z_0),\\ B_1\,\mathrm{e}^{-m(z-z_0)} & (z>z_0).\end{cases} \tag{9.96}$$

The conditions (9.91*b*) and (9.92*b*) then give

$$2\lambda m A_1=\Delta_0 B_1\,\mathrm{e}^{-mz_0}, \qquad 2\lambda m B_1=U_0\mathrm{i}KA_1\,\mathrm{e}^{-mz_0}, \tag{9.97}$$

and elimination of $A_1:B_1$ gives

$$\mathrm{i}\Delta_0 U_0 K\,\mathrm{e}^{-2mz_0}=4\lambda^2 m^2. \tag{9.98}$$

In conjunction with (9.95), this determines the possible values of p for given K, z_0, Δ_0, U_0. In dimensionless form, we have a relation of the form

$$p/\lambda K^2 = F(X, Kz_0), \qquad X = \Delta_0 U_0/4\lambda^2 K^2. \tag{9.99}$$

Note first that if $z_0 = 0$, i.e. if the two discontinuities coincide, then m^2 is pure imaginary, and so from (9.95) Re $p < 0$, i.e. dynamo action cannot occur. Suppose now that $|Kz_0| \ll 1$; then provided p is such that $|mz_0| \ll 1$ also, we may approximate e^{-2mz_0} in (9.98) by $1-2mz_0$, and solve the resulting quadratic for m. Neglecting terms of order $X^{1/2}(Kz_0)$, this yields two values for p:

$$p = -\lambda K^2(1 - iX \pm 2^{1/2} K z_0 X^{3/2}). \tag{9.100}$$

The condition $|mz_0| \gg 1$ is satisfied provided $X(Kz_0)^2 \ll 1$. Evidently we have dynamo action (Re $p > 0$) whenever

$$X > 0 \quad \text{and} \quad 2^{1/2} X^{3/2} > |Kz_0|^{-1} \gg 1. \tag{9.101}$$

The condition $X > 0$ means that the dynamo mechanism selects waves for which K has the same sign as $\Delta_0 U_0$. If $\Delta_0 U_0 > 0$ then the lower sign in (9.100) gives the dynamo mode when the conditions (9.101) are satisfied. If $\Delta_0 U_0 < 0$, then the upper sign in (9.100) is relevant. The conditions (9.101) imply that dynamo action will always occur when $z_0 \neq 0$, $\Delta_0 U_0 \neq 0$ and K is sufficiently small, i.e. when the disturbance wave-length in the x-direction is sufficiently large.

As in § 9.6, we have here solutions of the dynamo equations (9.73) and (9.74) that represents migrating dynamo waves, with phase speed

$$-\text{Im}\, p/K = -\lambda K X = -\Delta_0 U_0/4\lambda. \tag{9.102}$$

The physical nature of the dynamo process is illustrated in fig. 9.5. Shear distortion of the field component B_z at the layer $z = z_0$ generates the 'toroidal' field $B\mathbf{i}_y$ which diffuses (with a phase lag) to the neighbourhood of the layer $z = 0$; here the α-effect generates a toroidal current $J\mathbf{i}_y$ whose associated poloidal field diffuses (again with phase lag) back to the layer $z = z_0$, the net effect being maximally regenerative when the phase speed is given by (9.102). Of course a uniform superposed velocity $\Delta_0 U_0/4\lambda$ in the x-direction makes the field pattern stationary.

9.8. Symmetric $U(z)$ and antisymmetric $\alpha(z)$

A case of greater interest in terrestrial and astrophysical contexts is that in which $U(z)$ is symmetric and $\alpha(z)$ is antisymmetric about the equatorial plane $z = 0$. With $\mathbf{U}_P = 0$ and $U = U(z)$, $\alpha = \alpha(z)$, equations (9.73) and (9.74) admit solutions of the form (9.94) where

$$\begin{aligned}(p+\lambda K^2)\hat{A} &= \alpha(z)\hat{B} + \lambda\hat{A}''(z)\\ (p+\lambda K^2)\hat{B} &= \mathrm{i}KG(z)\hat{A} + \lambda\hat{B}''(z)\end{aligned} \qquad (9.103)$$

with $G = \mathrm{d}U/\mathrm{d}z$. If we suppose that

$$\alpha(z) = -\alpha(-z), \qquad G(z) = -G(-z), \qquad (9.104)$$

then the equations (9.103) are satisfied by solutions of dipole symmetry

$$\hat{A}(z) = \hat{A}(-z), \qquad \hat{B}(z) = -\hat{B}(-z), \qquad (9.105)$$

or solutions of quadrupole symmetry

$$\hat{A}(z) = -\hat{A}(-z), \qquad \hat{B}(z) = \hat{B}(-z). \qquad (9.106)$$

The linearity of the equations (9.73) and (9.74) of course permits superposition of solutions of the different symmetries, each with its appropriate value of p and K.

An example similar to that treated in § 9.7 is provided by the choice

$$\begin{aligned}\alpha(z) &= \Delta_0\delta(z - z_0) - \Delta_0\delta(z + z_0),\\ G(z) &= U_0\delta(z - z_1) - U_0\delta(z + z_1).\end{aligned} \qquad (9.107)$$

Solutions of dipole symmetry then have the form (for $z>0$)

$$\hat{A}(z)=\begin{cases} A_1\dfrac{\cosh mz}{\cosh mz_0} & (z<z_0),\\ A_1\,\mathrm{e}^{-m(z-z_0)} & (z>z_0),\end{cases}$$
$$\hat{B}(z)=\begin{cases} B_1\dfrac{\sinh mz}{\sinh mz_1} & (z<z_1),\\ B_1\,\mathrm{e}^{-m(z-z_1)} & (z>z_1).\end{cases} \tag{9.108}$$

The conditions (9.91) and (9.92), applied now across the discontinuities at $z=z_0$ and $z=z_1$ respectively, give two linear relations between A_1 and B_1, from which we obtain the dispersion relation (valid for $z_1>z_0$ or $z_1<z_0$)

$$\Delta_0 \mathrm{U}_0 \mathrm{i}K(\mathrm{e}^{-2m|z_1-z_0|}-\mathrm{e}^{-2m(z_0+z_1)})=4\lambda^2 m^2. \tag{9.109}$$

This *same* dispersion relation may be derived for the quadrupole modes

$$\hat{A}(z)=\begin{cases} A_1\dfrac{\sinh mz}{\sinh mz_0} & (z<z_0),\\ A_1\,\mathrm{e}^{-m(z-z_0)} & (z>z_0),\end{cases}$$
$$\hat{B}(z)=\begin{cases} B_1\dfrac{\cosh mz}{\cosh mz_1} & (z<z_1),\\ B_1\,\mathrm{e}^{-m(z-z_1)} & (z>z_1),\end{cases} \tag{9.110}$$

indicating that for each unstable dipole mode there is a corresponding unstable quadrupole mode with the same (complex) growth rate[5].

If $z_0+z \gg |z_1-z_0|$, then (9.109) becomes

$$\Delta_0 U_0 \mathrm{i}K\,\mathrm{e}^{-2m|z_1-z_0|}\approx 4\lambda^2 m^2, \tag{9.111}$$

and we revert essentially to the situation studied in § 9.7, except that we now have two double layers at (z_0, z_1) and $(-z_0, -z_1)$ which interact negligibly with each other. If z_0 and z_1 are of the same order of magnitude, however, the situation appears to be quite different.

[5] A discussion of more general circumstances which permit this type of correspondence between modes of dipole and quadrupole symmetries has been given recently by Proctor (1977*b*).

For suppose that K is such that $|m|\,|z_1-z_0|\ll 1$ and $|m|\,|z_0+z_1|\ll 1$; then, expanding the left-hand side of (9.109) to order m^2, we obtain

$$m/K = \mathrm{i}XKz_m(1-4\mathrm{i}X\,.\,Kz_0\,.\,Kz_1+\ldots), \tag{9.112}$$

where z_m is the smaller of z_0 and z_1 and X is given by (9.99*b*); note this satisfies $\mathrm{Re}\, m>0$ as required. From (9.112), $\mathrm{Re}\,(m/K)^2<0$ and so $\mathrm{Re}\, p<0$ and dynamo action cannot occur. This argument of course does not exclude the possibility of dynamo action for modes for which $|m|\,|z_1-z_0|$ and $|m|\,|z_0+z_1|$ are of order unity.

If $z+z_0$ is decreased from large values, keeping $|z_1-z_0|$ and K fixed, with $K|z_1-z_0|X^{1/2}\gg 1$, the modes that are unstable when $z_1+z_0\gg|z_1-z_0|$ must disappear when z_1+z_0 becomes of the same order as $|z_1-z_0|$, due to destructive interference between the two double layers.

9.9. A model of the galactic dynamo

It has been suggested by Parker (1971*a*, *e*) that the galactic magnetic field may be maintained by a combination of toroidal shear (due to differential rotation in the galactic disc) and cyclonic turbulence. Although a model of this kind is by no means universally accepted (see for example Piddington, 1972*a*, *b*), the analysis nevertheless provides useful insight into the structure of the dynamo equations (9.73) and (9.74), and is particularly interesting in that it provides an example of dynamo action in which steady or oscillatory dynamo modes can be excited with almost equal ease. The model described below is more idealised than that of Parker, but permits a simpler analysis, while retaining the same essential physical features.

The origin O of a local Cartesian coordinate system is taken (fig. 9.6) at a point of the plane of symmetry of the galactic disc, with Oz normal to the disc, Ox radially outwards in the plane of the disc, and Oy in the azimuth direction. The toroidal velocity is taken to be a function of x only, with $G=\mathrm{d}U/\mathrm{d}x$ locally uniform, and equations (9.73) and (9.74) with $\mathbf{U}_P=0$ then become

$$\frac{\partial A}{\partial t}=\alpha B+\lambda\nabla^2 A, \tag{9.113}$$

$$\frac{\partial B}{\partial t}=-G\frac{\partial A}{\partial z}+\lambda\nabla^2 B. \tag{9.114}$$

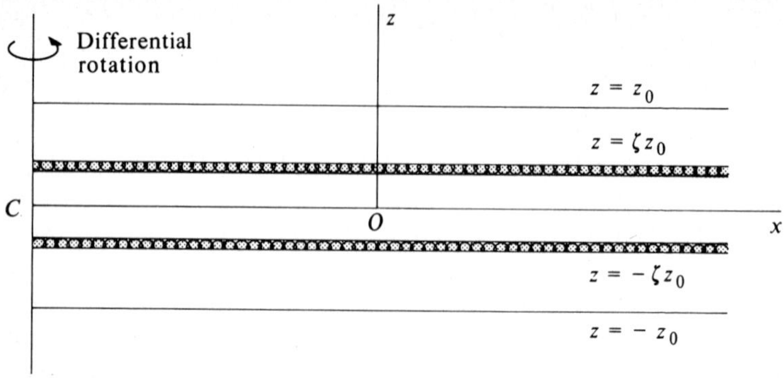

Fig. 9.6 Idealised model for study of the galactic dynamo mechanism; C is the galactic centre, and O is a point on the plane of symmetry of the galactic disc. Ox is the radial extension of CO, and Oz is normal to the galactic disc. The disc boundary is represented by the planes $z = \pm z_0$; the α-effect is supposed concentrated in neighbourhoods of the planes $z = \pm \zeta z_0$ $(\zeta < 1)$. Differential rotation of the galaxy provides a mean velocity field, which in the neighbourhood of 0 has the form $(0, U(x), 0)$, with $dU/dx = G$(cst.).

For the same reasons as discussed in § 9.5 in the solar and terrestrial contexts, it is appropriate to restrict attention to the situation in which $\alpha(z)$ is a (prescribed) *odd* function of z. Parker chose a step-function,

$$\alpha(z) = \begin{cases} \alpha_0 & (0 < z < z_0) \\ -\alpha_0 & (-z_0 < z < 0), \end{cases} \tag{9.115}$$

and he matched the resulting solutions of (9.113) and (9.114) to vacuum fields in the 'extra-galactic' region $|z| > z_0$, in order to obtain an equation for the (generally complex) growth rate p for field modes. Although the equations (9.113) and (9.114) appear simple enough, the dispersion relation turns out to be very complicated indeed and elaborate asymptotic procedures are required in the process of solution.

In order to avoid these difficulties, we shall simplify the problem by supposing that, instead of (9.115), $\alpha(z)$ is given (as in § 9.8) by

$$\alpha(z) = \alpha_0 \delta(\hat{z} - \zeta) - \alpha_0 \delta(\hat{z} + \zeta), \tag{9.116}$$

where $\hat{z} = z/z_0$ is a dimensionless coordinate, and $0 < \zeta < 1$, i.e. the α-effect is supposed concentrated in two layers $z = \pm \zeta z_0$, the 'mean value' of $\alpha(z)$ throughout either half of the galactic disc being $\pm \alpha_0$

(as for (9.115)). Of course the choice (9.116) is an idealisation (as is (9.115) also) but in retaining the essential antisymmetry of $\alpha(z)$, it may be expected that the qualitative behaviour of the system (9.113), (9.114) should not be dramatically affected. We shall find that this expectation is in fact only partially fulfilled!

As in the previous sections, we now investigate solutions of (9.113), (9.114) of the form

$$A = \hat{A}(\hat{z})\,e^{pt+iKx}, \qquad B = \frac{Gz_0}{\lambda}\hat{B}(\hat{z})\,e^{pt+iKx}, \tag{9.117}$$

the factor Gz_0/λ being introduced simply for convenience. For $|\hat{z}|<1$, $|\hat{z}|\neq\zeta$, (9.113) and (9.114) become

$$\frac{d^2\hat{A}}{d\hat{z}^2}-q^2\hat{A}=0, \tag{9.118}$$

$$\frac{d^2\hat{B}}{d\hat{z}^2}-q^2\hat{B}=\frac{d\hat{A}}{d\hat{z}}, \tag{9.119}$$

where

$$q^2=(p+\lambda K^2)z_0^2/\lambda, \tag{9.120}$$

and we have now (using (9.116)) the jump conditions

$$[d\hat{A}/d\hat{z}]=-X\hat{B}, \qquad [d\hat{B}/d\hat{z}]=0 \qquad \text{on } z=\zeta, \tag{9.121}$$

where

$$X=\alpha_0 G z_0^3/\lambda^2, \tag{9.123}$$

(and similarly across $\hat{z}=-\zeta$). The dimensionless number X (Parker's 'dynamo number') clearly provides a measure of the joint influence of the α-effect and the shear G in the layer, and we are chiefly concerned to find p as a function of X (and the other parameters of the problem) and in particular to determine whether $\operatorname{Re} p$ can be positive for a prescribed value of X.

There are of course further boundary conditions that must be satisfied. Firstly, in the vacuum region $\hat{z}>1$, the toroidal field B must vanish, and A must be a harmonic function, i.e. $\hat{A}(\hat{z})\propto e^{-|Kz_0|\hat{z}}$; continuity of $\hat{A}$, $\hat{B}$ and $d\hat{A}/d\hat{z}$ across $\hat{z}=1$ then gives

$$\hat{B}=0, \qquad d\hat{A}/d\hat{z}\pm|Kz_0|\hat{A}=0 \quad \text{on } \hat{z}=\pm 1. \tag{9.124}$$

When $|Kz_0| \ll 1$, these conditions become

$$\hat{B}(\hat{z})=0, \qquad d\hat{A}/d\hat{z}=0 \quad \text{on } \hat{z}=\pm 1. \tag{9.125}$$

We shall adopt these conditions, and provide restrospective justification (see discussion following (9.133) below).

It is clear that with $\alpha(z)=-\alpha(-z)$, equations (9.113) and (9.114) admit solutions of either dipole symmetry or quadrupole symmetry (cf. (9.105) and (9.106)). These modes satisfy the symmetry conditions

$$\hat{B}=d\hat{A}/d\hat{z}=0 \quad \text{on } \hat{z}=0 \tag{9.126}$$

for 'dipole' modes, and

$$\hat{A}=d\hat{B}/d\hat{z}=0 \quad \text{on } \hat{z}=0 \tag{9.127}$$

for 'quadrupole' modes. If these conditions are used (and with the implication that the symmetry conditions (9.105) and (9.106) are satisfied) then it is sufficient to restrict attention to the region $\hat{z}>0$.

Dipole modes

The solution of (9.118) and (9.119) satisfying (9.125) and (9.126) can readily be obtained in the form

$$\left.\begin{aligned} &\left.\begin{aligned} \hat{A}&=A_1\frac{\cosh q\hat{z}}{\cosh q\zeta} \\ \hat{B}&=\tfrac{1}{2}\hat{z}\hat{A}(\hat{z})-\tfrac{1}{2}A_1\zeta\frac{\sinh q\hat{z}}{\sinh q\zeta}+B_1\frac{\sinh q\hat{z}}{\sinh q\zeta} \end{aligned}\right\} \quad (z<\zeta), \\ &\left.\begin{aligned} \hat{A}&=A_1\frac{\cosh q(\hat{z}-1)}{\cosh q(\zeta-1)} \\ \hat{B}&=\tfrac{1}{2}(\hat{z}-1)\hat{A}(\hat{z})-\tfrac{1}{2}A_1(\zeta-1)\frac{\sinh q(\hat{z}-1)}{\sinh q(\zeta-1)} \\ &\quad +B_1\frac{\sinh q(\hat{z}-1)}{\sinh q(\zeta-1)} \end{aligned}\right\} \quad (\zeta<z<1), \end{aligned}\right. \tag{9.128}$$

where A_1 and B_1 are constants, and where we may clearly suppose that $\operatorname{Re} q \geqslant 0$. Application of the jump conditions (9.121) and elimination of A_1 and B_1 gives, after some simplification,

$$\frac{4q\sinh^2 q}{X} = 2\zeta \sinh q \cosh q(2\zeta-1) - \sinh 2q\zeta = f(q,\zeta), \text{ say,} \tag{9.129}$$

from which $q(X,\zeta)$ (and hence p from (9.120)) is in principle determinate. If $|q|>0$ and $|\arg q|<\pi/4$, then when $|Kz_0|$ is sufficiently small, Re p as given by (9.120) will be positive and we have dynamo action.

Consider first the possibility of non-oscillatory modes, for which q is real and positive. The function $f(q,\zeta)$ defined by (9.129) vanishes at $\zeta = 0, \frac{1}{2}, 1$ and satisfies

$$f(q,\zeta)\begin{cases} <0, & \text{for } 0<\zeta<\frac{1}{2}, \\ >0, & \text{for } \frac{1}{2}<\zeta<1. \end{cases} \tag{9.130}$$

It is moreover easily shown from (9.129) that for $\zeta \neq 0$

$$X^{-1} \sim \begin{cases} \zeta(2\zeta-1)/2q & \text{as } q\to 0, \\ (2\zeta-1)\,\mathrm{e}^{-2q(1-\zeta)}/4q & \text{as } q\to\infty. \end{cases} \tag{9.131}$$

Curves of X^{-1} as functions of q (for ζ less than and greater than $\frac{1}{2}$) are sketched in fig. 9.7(*a*). It is clear that for the model considered, modes with dipole symmetry can be excited only if

$$(\zeta-\tfrac{1}{2})X>0, \tag{9.132}$$

and that for each (X,ζ) satisfying this condition there is a unique real positive value of q determined by the graph of fig. 9.7(*a*).

The growth rate p is given in general by the formula

$$pz_0^2/\lambda = -(Kz_0)^2 + (q(X,\zeta,|Kz_0|))^2, \tag{9.133}$$

and the formula (9.129) in effect determines $q(X,\zeta,0)$. It is clear however that q must be a continuous function of $|Kz_0|$, and that if $q(X,\zeta,0)>0$, then there exists a range of (small) values of $|Kz_0|$ for which $pz_0^2/\lambda>0$ also; hence if (9.132) is satisfied, modes of dipole symmetry are excited for sufficiently small values of $|Kz_0|$, i.e. provided the scale K^{-1} of the magnetic perturbation in the x-direction is sufficiently large.

Consider now the possibility of oscillatory dynamo modes for which Re $q>0$, Im $q\neq 0$ and $|\arg q|<\pi/4$. When $|q|$ is small,

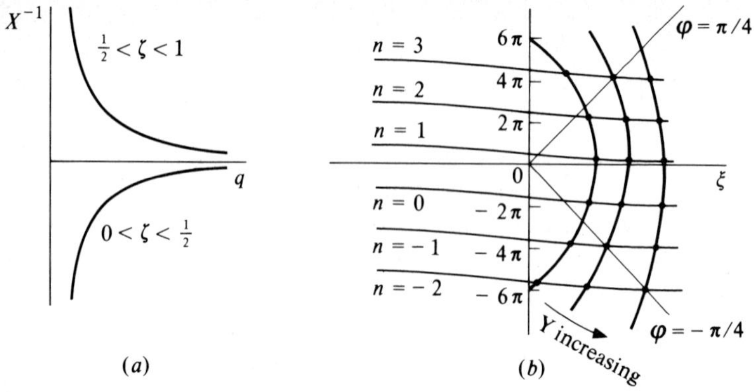

Fig. 9.7 (*a*) Sketch of the dependence of X^{-1} on q for values of ζ less than and greater than $\frac{1}{2}$. (*b*) Geometrical construction of solutions of $Q\,e^{Q} = Y$ with $Q = \xi + i\eta$; solutions are indicated by the intersections marked •; those for which $|\varphi| < \pi/4$ correspond to modes of the form (9.117) with exponentially increasing amplitude.

(9.129) takes the form (9.131*a*), and it is clear that in this limit (equivalently $X \to 0$), there is only the real solution

$$q \sim \tfrac{1}{2}X\zeta(2\zeta - 1). \tag{9.134}$$

When $|q|$ is large, however, (9.131*b*) gives, with $q = \frac{1}{2}Q/(1-\zeta)$,

$$Q\,e^{Q} = -\tfrac{1}{2}X(1-2\zeta)(1-\zeta) = Y \quad \text{say,} \tag{9.135}$$

and this equation does admit complex solutions $Q(Y)$. To see this, let

$$Q = \xi + i\eta = \rho\,e^{i\varphi}, \qquad 0 \leqslant \varphi < 2\pi, \tag{9.136}$$

so that (9.135) gives

$$\rho\,e^{\xi} = |Y|, \qquad \varphi + \eta = \begin{cases} 2n\pi & \text{if } Y > 0 \\ (2n+1)\pi & \text{if } Y < 0 \end{cases} \tag{9.137}$$

where n is an integer. The curves given by (9.137*a, b*) in the case $Y > 0$ are sketched in fig. 9.7(*b*), and it is clear that when $|Y|$ is sufficiently large, complex roots Q with $|\arg Q| < \pi/4$ do indeed exist; in fact, for $Y \to \infty$, these roots are given by

$$Q \sim Q_0 + 2n\pi i \qquad (n = 1, 2, \ldots, n_0), \tag{9.138}$$

where n_0 is the integral part of $Q_0/8$, and Q_0 is the unique real solution of $Q_0\,e^{Q_0}=Y$ satisfying

$$Q_0 \sim \log Y \quad \text{as } Y\to\infty. \tag{9.139}$$

The growth rates of such oscillatory modes are all slightly less than the growth rate of the non-oscillatory mode (for which $\eta=0$), the difference becoming less and less significant as Y increases.

Quadrupole modes

The solution of (9.118) and (9.119) satisfying (9.125) and (9.127) is similarly of the form

$$\left.\begin{aligned} \hat{A} &= A_1\frac{\sinh q\hat{z}}{\sinh q\zeta} \\ \hat{B} &= \tfrac{1}{2}\hat{z}\hat{A}(\hat{z})-\tfrac{1}{2}A_1\zeta\frac{\cosh q\hat{z}}{\cosh q\zeta} \\ &\quad +B_1\frac{\cosh q\hat{z}}{\cosh q\zeta}\end{aligned}\right\}\quad(\hat{z}<\zeta),$$

$$\left.\begin{aligned} \hat{A} &= A_1\frac{\cosh q(\hat{z}-1)}{\cosh q(\zeta-1)} \\ \hat{B} &= \tfrac{1}{2}(\hat{z}-1)\hat{A}(\hat{z})-\tfrac{1}{2}A_1(\zeta-1)\frac{\sinh q(\hat{z}-1)}{\sinh q(\zeta-1)} \\ &\quad +B_1\frac{\sinh q(\hat{z}-1)}{\sinh q(\zeta-1)}\end{aligned}\right\}\quad(\zeta<\hat{z}<1), \tag{9.140}$$

and the relation between q, X and ζ corresponding to (9.129) takes the form

$$\frac{4q\cosh^2 q}{X}=2\zeta\cosh q\sinh q(2\zeta-1)-\sinh 2q\zeta=g(q,\zeta),\text{ say.} \tag{9.141}$$

The function $g(q,\zeta)$ vanishes for $\zeta=0$, 1 and satisfies

$$g(q,\zeta)<0 \quad \text{for } 0<\zeta<1. \tag{9.142}$$

The asymptotic behaviour of X^{-1}, from (9.141), is given by

$$X^{-1} \sim -\zeta(1-\zeta)(1-\tfrac{2}{3}q^2\zeta(1+\zeta-\zeta^2)+O(q^4)) \quad \text{as } q \to 0, \tag{9.143}$$

$$\left.\begin{aligned} X^{-1} &\sim -\tfrac{1}{2}\zeta q^{-1}\,\mathrm{e}^{-2q\zeta} \quad (0<\zeta\leq\tfrac{1}{2}) \\ & -\tfrac{1}{4}(2\zeta-1)q^{-1}\,\mathrm{e}^{-2q(1-\zeta)} \quad (\tfrac{1}{2}<\zeta<1) \end{aligned}\right\} \text{as } q \to \infty. \tag{9.144}$$

Typical curves of X^{-1} as functions of q are sketched in fig. 9.8. It is clear that a necessary and sufficient condition for dynamo excitation of modes of quadrupole symmetry is (in the limit $|Kz_0| \to 0$)

$$X < -\zeta^{-1}(1-\zeta)^{-1}, \tag{9.145}$$

and that such modes are most readily excited if $\zeta \approx \frac{1}{2}$.

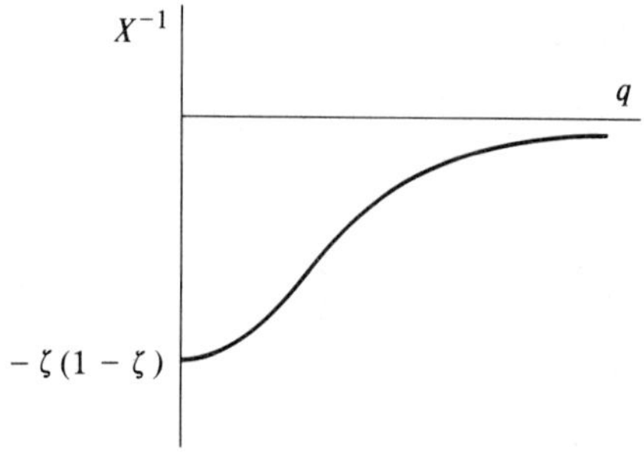

Fig. 9.8 A typical curve of X^{-1} vs. q as given by (9.141).

The regions of the (X, ζ) plane for which non-oscillatory dipole and quadrupole excitation are possible according to the criteria (9.132) and (9.145) are indicated in fig. 9.9(a), in which a small allowance is made for the effect of the small terms involving $|Kz_0|$ neglected in the above approximate analysis.

The manner in which the behaviour of the system changes as ζ changes (i.e. as the distribution of α as a function of z changes) is particularly striking, and it may be expected that other choices for $\alpha(z)$ may give equally varied behaviour. With the choice (9.115), Parker (1971a) found[6] that the system exhibited excitation of

[6] The results (9.146) and (9.147) may be extracted from Parker's equations (55) and (71).

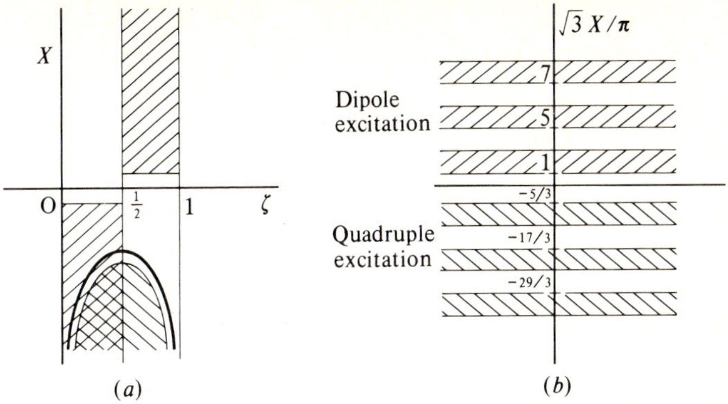

Fig. 9.9 (*a*) Region of X–ζ plane in which non-oscillatory dynamo excitation occurs for modes of dipole symmetry (///) and quadrupole symmetry (\\\\\\), as given by (9.132) and (9.145); (*b*) excitation bands for Parker's (1971*a*) model, as given by (9.146) and (9.147).

non-oscillatory quadrupole modes if X is positive and within any of the bands defined by

$$(4n-3)\pi/\sqrt{3} \leqslant X \leqslant (4n-1)\pi/\sqrt{3}, \tag{9.146}$$

and non-oscillatory dipole modes if X is negative and within any of the bands

$$-(4n+1)\pi/\sqrt{3} \leqslant X + 2\pi/3\sqrt{3} \leqslant -(4n-1)\pi/\sqrt{3}, \tag{9.147}$$

while outside these bands, all non-oscillatory modes necessarily decay. In (9.146) and (9.147), n is (strictly) a large positive integer (to justify the asymptotic methods used); however the results remain approximately correct even when n takes *small* integral values 1, 2, 3, ... The contrast between this behaviour (fig. 9.9(*b*)) and that of fig. 9.9(*a*) is remarkable, and should perhaps serve as a warning that the behaviour of an $\alpha\omega$-system (i.e. one operating under the joint action of an α-effect and a mean shear) may depend qualititatively as well as quantitatively on the precise distribution of α and of the shear that is presupposed.

As regards the possibility of oscillatory modes ($\text{Re}\,p > 0$, $\text{Im}\,p \neq 0$) with quadrupole symmetry, note that the change of variables

$$Q = 2q\zeta, \qquad Y = -X\zeta^2, \qquad (0 < \zeta \leq \tfrac{1}{2}),$$

or

$$Q = 2q(1-\zeta), \qquad Y = -\tfrac{1}{2}(2\zeta - 1)(1-\zeta)X, \quad (\tfrac{1}{2} < \zeta < 1),$$

reduces the asymptotic from (9.144) in either case to

$$Q\,\mathrm{e}^{Q} \sim Y \qquad (|Q| \gg 1),$$

with complex solutions $Q(Y)$ as discussed below (9.135). When $\frac{1}{2} < \zeta < 1$, replacement of X by $-X$ clearly makes (9.131*b*) and (9.144*b*) identical, so that if a dipole oscillatory mode exists for $X = X_0$ say, then a quadrupole oscillatory mode with the same (complex) growth rate p exists for $X = -X_0$. A similar property was noted by Parker (1971*a*) for the case when $\alpha(z)$ is given by (9.115). It is noteworthy however that the property does not extend to the case $0 < \zeta \leq \frac{1}{2}$ described (in the quadrupole case) by the asymptotic form (9.144*a*). Symmetry properties of this kind have been studied in a more general context by Proctor (1977*c*).

Finally, we should note that an improved model of galactic dynamo action has been analysed by Stix (1975) who expressed the dynamo equations (9.9) and (9.10) in spheroidal coordinates and determined (numerically) the critical value of X (defined analogously to the definition (9.123)) for which $\text{Re}\,p = 0$ for a galaxy whose 'boundary' is supposed to be a severely flattened spheroid[7]. In this treatment, Stix found that $\text{Im}\,p = 0$ when $\text{Re}\,p = 0$, and he inferred that the non-oscillatory modes are more easily excited, consistent with the discussion above (which indicates that when $|X|$ is small, oscillatory modes certainly cannot be excited, whereas non-oscillatory modes can, provided $|Kz_0|$ is sufficiently small).

9.10. Generation of poloidal fields by the α-effect

In a spherical geometry, and still restricting attention to the situation $\mathbf{U}_P = 0$, the equations governing $\alpha\omega$-dynamo models are (from

[7] The dynamo numbers computed by Stix are incorrect and should all be multiplied by a/b where a and b are the major and minor axes of the spheroid (Stix, private communication). This correction does not affect the qualitative conclusions mentioned in the text above.

(9.10) and (9.13))

$$\partial A/\partial t = \alpha B + \lambda(\nabla^2 - s^{-2})A, \tag{9.148}$$

$$\partial B/\partial t = s(\mathbf{B}_P \,.\, \nabla)\omega + \lambda(\nabla^2 - s^{-2})B, \tag{9.149}$$

(where any eddy diffusivity effect is still assumed incorporated in the value of the parameter λ). We have in § 3.11 considered the process of generation of toroidal field $B\mathbf{i}_\varphi$ when the poloidal field $\mathbf{B}_P$ is supposed given. It is natural now to consider the complementary process of generation of poloidal field $\mathbf{B}_P = \nabla \wedge (A\mathbf{i}_\varphi)$ by the α-effect when the toroidal field $B\mathbf{i}_\varphi$ is supposed given. The similar structure of (9.148) and (9.149) makes this a straightforward task, in the light of the results of § 3.11.

Suppose that $B(r, \theta)\mathbf{i}_\varphi$ is a given steady toroidal field such that B/s is everywhere bounded (a natural condition to impose if the poloidal current associated with $B\mathbf{i}_\varphi$ is finite). Suppose further that $\alpha(r, \theta)$ is given as a function of r and θ and that $A = 0$ at time $t = 0$. Then for $t \ll L^2/\lambda$, where L is the characteristic scale of the source term αB in (9.148), we have simply

$$A(r, \theta, t) = \alpha(r, \theta)B(r, \theta)t. \tag{9.150}$$

When t is of order L^2/λ and greater, diffusion of the poloidal field thus generated becomes important, and for $t \gg L^2/\lambda$, a steady state is approached in which

$$(\nabla^2 - s^{-2})A = -\lambda^{-1}\alpha B \tag{9.151}$$

within the conducting region. Let us as usual suppose that this is the region $r < R$, and that the region $r > R$ is insulating so that

$$(\nabla^2 - s^{-2})A = 0, \quad (r > R). \tag{9.152}$$

As in § 3.3, it is apparent that if $\alpha B/\lambda$ has the expansion

$$\alpha B/\lambda = \sum_1^\infty F_n(r)\, \mathrm{d}P_n(\cos\theta)/\mathrm{d}\theta, \tag{9.153}$$

then the solution of (9.151), (9.152) is

$$A(r, \theta) = \sum_1^\infty G_n(r)\, \mathrm{d}P_n(\cos\theta)/\mathrm{d}\theta, \tag{9.154}$$

where

$$r^{-2}(r^2G_n')'-r^{-2}n(n+1)G_n=F_n(r), \tag{9.155}$$

and where we adopt the convention $F_n(r)\equiv 0$ for $r>R$. Continuity of $\mathbf{B}_P$ across $r=R$ requires that

$$[G_n]=[G_n']=0 \quad \text{on } r=R. \tag{9.156}$$

The solution (cf. (3.105)) is given by

$$G_n(r)=-\frac{1}{2n+1}\left\{\frac{1}{r^{n+1}}\int_0^r x^{n+2}F_n(x)\,\mathrm{d}x+r^n\int_r^\infty \frac{F_n(x)}{x^{n-1}}\,\mathrm{d}x\right\}. \tag{9.157}$$

Suppose for example that, for $r<R$,

$$B(r,\theta)=B_0(r/R)\cos\theta\sin\theta, \qquad \alpha(r,\theta)=\alpha_0\cos\theta. \tag{9.158}$$

Then

$$\frac{\alpha B}{\lambda}=\frac{\alpha_0B_0r}{\lambda R}\cos^2\theta\sin\theta=-\frac{\alpha_0B_0r}{\lambda R}\frac{\mathrm{d}}{\mathrm{d}\theta}\left[\frac{2}{15}P_3(\cos\theta)+\frac{1}{5}P_1(\cos\theta)\right]. \tag{9.159}$$

Hence, for $r<R$,

$$F_1(r)=-\alpha_0B_0/5\lambda R, \qquad F_3(r)=-2\alpha_0B_0r/15\lambda R, \tag{9.160}$$

and $F_n(r)\equiv 0$ ($n\neq 1$ or 3). For $r>R$, (9.144) then gives

$$G_1(r)=\alpha_0B_0R^4/75r^2\lambda, \qquad G_3(r)=2\alpha_0B_0R^6/735r^4\lambda, \tag{9.161}$$

and, from (9.154), we then have for $r>R$

$$A(r,\theta)=-\frac{\alpha_0B_0R^2}{15\lambda}\left[\frac{1}{5}\left(\frac{R}{r}\right)^2\sin\theta+\frac{3}{49}\left(\frac{R}{r}\right)^4(5\cos^2\theta-1)\sin\theta\right]. \tag{9.162}$$

The associated poloidal field $\mathbf{B}_P=\nabla\wedge(A\,\mathbf{i}_\varphi)$ has dipole symmetry about the equatorial plane $\theta=\pi/2$, but contains both dipole and octupole ingredients. In particular, the radial component of $\mathbf{B}_P$ is given (for $r>R$) by

$$B_r = \frac{1}{r\sin\theta}\frac{\partial}{\partial\theta}(A\sin\theta)$$

$$= -\frac{\alpha_0 B_0 R}{15\lambda}\left[\frac{2}{5}\left(\frac{R}{r}\right)^3\cos\theta + \frac{3}{49}\left(\frac{R}{r}\right)^5(5\cos\theta\cos 2\theta - \sin 2\theta)\right]. \tag{9.163}$$

9.11. The $\alpha\omega$-dynamo with periods of stasis

In order to demonstrate that the α-effect in conjunction with differential rotation can act as a dynamo, it is useful to adopt the artifice of 'stasis' as described in § 6.12. Suppose that we start with a purely poloidal field $\mathbf{B}_{P_0} = \nabla \wedge (A_0\mathbf{i}_\varphi)$, with

$$A_0 = C_0 r^{-1/2} J_{3/2}(kr)\sin\theta \quad (r<R), \tag{9.164}$$

where kR is the first zero $(=\pi)$ of $J_{1/2}(kR)$; i.e. we start with the fundamental dipole mode of lowest natural decay rate (§ 2.7). The field for $r>R$ is of course harmonic and matches smoothly with the interior field. Let us now subject this field to a short period t_1 of intense differential rotation $\omega(r)$. Diffusion is negligible if this process is sufficiently rapid, and from equation (3.96) a toroidal field is generated in $r<R$ and is given at $t=t_1$ by

$$\mathbf{B}_T = B\mathbf{i}_\varphi = r\sin\theta B_r \omega'(r) t_1 \mathbf{i}_\varphi$$

$$= 2C_0 r^{-1/2} J_{3/2}(kr)\omega'(r) t_1 \sin\theta\cos\theta\,\mathbf{i}_\varphi. \tag{9.165}$$

We now stop the differential rotation and 'switch on' an intense α-effect with $\alpha = \alpha_0(r)\cos\theta$, which is maintained for a second short interval t_2 during which again diffusion is negligible. From (9.150), at time $t=t_1+t_2$, an additional poloidal field $\mathbf{B}_P = \nabla \wedge (A\mathbf{i}_\varphi)$ has been generated, where

$$A = 5f(r)\sin\theta\cos^2\theta = -f(r)\frac{\mathrm{d}}{\mathrm{d}\theta}\left(P_1(\cos\theta) + \tfrac{2}{3}P_3(\cos\theta)\right), \tag{9.166}$$

with

$$f(r) = \frac{2C_0}{5r^{1/2}} J_{3/2}(kr)\omega'(r)\alpha_0(r) t_1 t_2. \tag{9.167}$$

Finally we allow a period t_3 of stasis (i.e. pure ohmic decay) so that only the slowest decaying ingredient of (9.166) will survive after a long time. The function $f(r)$ may be expanded as an infinite sum of radial functions

$$f(r) = \sum_{q=1}^{\infty} C_{nq} f_{nq}(r), \tag{9.168}$$

with $n = 1$ or 3 as appropriate (n and q being the same labels as used in § 2.7). The slowest decaying mode is that for which $n = q = 1$, i.e. precisely the mode (9.164) with which we started. By waiting long enough, we can ensure that the contamination by higher modes is negligible; and by increasing $|\omega'(r)\alpha_0(r)|$ (keeping other things constant), we can further ensure that at time $t = t_1 + t_2 + t_3$ the field that survives is more intense than the initial field given by (9.164). The process may then be repeated indefinitely to give sustained dynamo action.

This type of dynamo may be oscillatory or non-oscillatory, according as $C_{11}/C_0 \lesseqgtr 0$. This depends on the precise radial dependence of the product $\omega'(r)\alpha_0(r)$. Reversing the sign of this product will clearly convert a non-oscillatory dynamo to an oscillatory dynamo, or vice versa.

9.12. Numerical investigations of $\alpha\omega$-dynamos

A number of numerical studies have been made of equations (9.10) and (9.13) in a spherical geometry and with a variety of prescribed axisymmetric forms for the functions $\alpha(\mathbf{x})$, $\omega(\mathbf{x})$ and $\mathbf{U}_P(\mathbf{x})$. These studies are for the most part aimed at constructing plausible models for possible dynamo processes within either the liquid core of the Earth or the convective envelope of the Sun. A few general conclusions have emerged from these studies. In describing these, we may limit attention to the situation in which $\alpha(\mathbf{x})$ and $U_z(\mathbf{x})$ are odd functions of the coordinate z normal to the equatorial plane, and $\omega(\mathbf{x})$ and $U_s(\mathbf{x})$ are even functions of z. In this case equations (9.10) and (9.13) admit solutions of dipole symmetry (A even, B odd in z) or quadrupole symmetry (A odd, B even). In all cases of course the field must be matched to an irrotational field across the spherical boundary $r = R$.

The behaviour of the system is characterised (as in § 9.9) by a dimensionless dynamo number (cf. (9.99*b*), (9.123))

$$X = \alpha_0 \omega_0' R^3/\lambda^2, \qquad (9.169)$$

where α_0 and ω_0' are typical values of α and $\partial\omega/\partial r$. To be specific, following the convention adopted by Roberts (1972*a*), let α_0 be the value of α at the point in the northern hemisphere where $|\alpha|$ has its maximum value (so that α_0 may be positive or negative) and let ω_0' be the value of $\partial\omega/\partial r$ where $|\partial\omega/\partial r|$ is maximal. X may then likewise be positive or negative.

Equation (9.10) and (9.13) admit solutions proportional to e^{pt}, and (as for the laminar dynamo theories discussed in § 6.11) the problem that presents itself is essentially the determination of possible values of p as functions of X (and any other dimensionless parameter that may appear in the specification of the velocity field). Interest centres on that value of p for which $\mathrm{Re}\, p$ first becomes positive as $|X|$ increases continuously from zero. The corresponding field structure may have dipole or quadrupole symmetry – and we then say that the dipolc (or quadrupole) mode is the *preferred* mode of excitation. Moreover if $\mathrm{Im}\, p = 0$ when $\mathrm{Re}\, p = 0$, this preferred mode is non-oscillatory, while if $\mathrm{Im}\, p \neq 0$ when $\mathrm{Re}\, p = 0$ it is oscillatory.

Roberts (1972*b*) studied a number of models, both with and without meridional circulation $\mathbf{U}_P$, for various smooth choices of $\alpha(\mathbf{x})$ and $\omega(\mathbf{x})$ (satisfying the symmetry conditions specified above) with the following conclusions:

(i) When $X > 0$, and $\mathbf{U}_P = 0$, the mode that is preferred is quadrupole and oscillatory; critical values of X ranging between 76 and 212 were obtained depending on the particular distribution of α and ω adopted. Introduction of a small amount of meridional circulation can however yield a preferred mode that is dipole and non-oscillatory, the critical value of X being reduced by a factor of order $\frac{1}{2}$ or less in the process. This rather dramatic effect of meridional velocity was first recognized by Braginskii (1964*b*) and presumably admits interpretation in terms of the convective effect of $\mathbf{U}_P$ on dynamo waves in a sense contrary to their natural phase velocity (cf. the discussion at the end of § 9.6). Unfortunately however it appears (Robert, 1972) that the *sense* of the meridional circulation (i.e. whether from poles to equator or equator to poles on the surface $r = R$) that will lead to a reduction in the critical value of X depends on the model (i.e. on the particular choice of $\alpha(\mathbf{x})$ and $\boldsymbol{\omega}(\mathbf{x})$), and a simple physical interpretation of the effect of $\mathbf{U}_P$ therefore seems unlikely.

(ii) When $X < 0$, the conclusions are reversed; i.e. when $\mathbf{U}_P = 0$, the preferred mode is dipole and oscillatory, critical values of X ranging between -74 and -206 for the models studied; and introduction of suitable meridional velocity $\mathbf{U}_P$ (the sense being again model dependent) substantially reduces the critical value of $|X|$, the preferred mode becoming quadrupole and non-oscillatory.

The oscillatory character of the modes when $\mathbf{U}_P = 0$ has been confirmed by Jepps (1975) who carried out a direct numerical integration of equations (9.10) and (9.13) again with specified simple forms for $\alpha(\mathbf{x})$ and $\omega(\mathbf{x})$ with $|X|$ marginally greater than its critical value X_c as determined by the eigenvalue approach. For initial condition at time $t = 0$, Jepps assumed the field to be purely poloidal and in the fundamental decay mode for a sphere (§ 2.7); in the subsequent evolution, the field rapidly settled down to a time-periodic behaviour (modulated by the slow amplification expected as a result of the supercritical choice of X).

In the case $X < 0$, and when $\mathbf{U}_P = 0$, it is a characteristic feature of the periodic solutions that both poloidal and toroidal field ingredients appear to originate in the polar regions and then to amplify during a process of propagation towards the equatorial plane; there, diffusion eliminates toroidal field of opposite signs from the two hemispheres. This type of behaviour can be seen clearly in fig. 9.10 (from Roberts, 1972*a*) which shows a half-cycle of both poloidal and toroidal fields for a model in which

$$\alpha = \frac{729}{16}\alpha_0\hat{r}^8(1-\hat{r}^2)^2\cos\theta, \qquad \omega = -\frac{19683}{40960}\omega_0'(1-\hat{r}^2)^5, \tag{9.170}$$

where $\hat{r} = r/R$, the coefficients being chosen so that the maximum values of $|\alpha|$ and $|\partial\omega/\partial r|$ are respectively $|\alpha_0|$ and $|\omega_0'|$. This apparent propagation from poles to equator is of course consistent with the behaviour of the plane wave solutions discussed in § 9.6: in that case

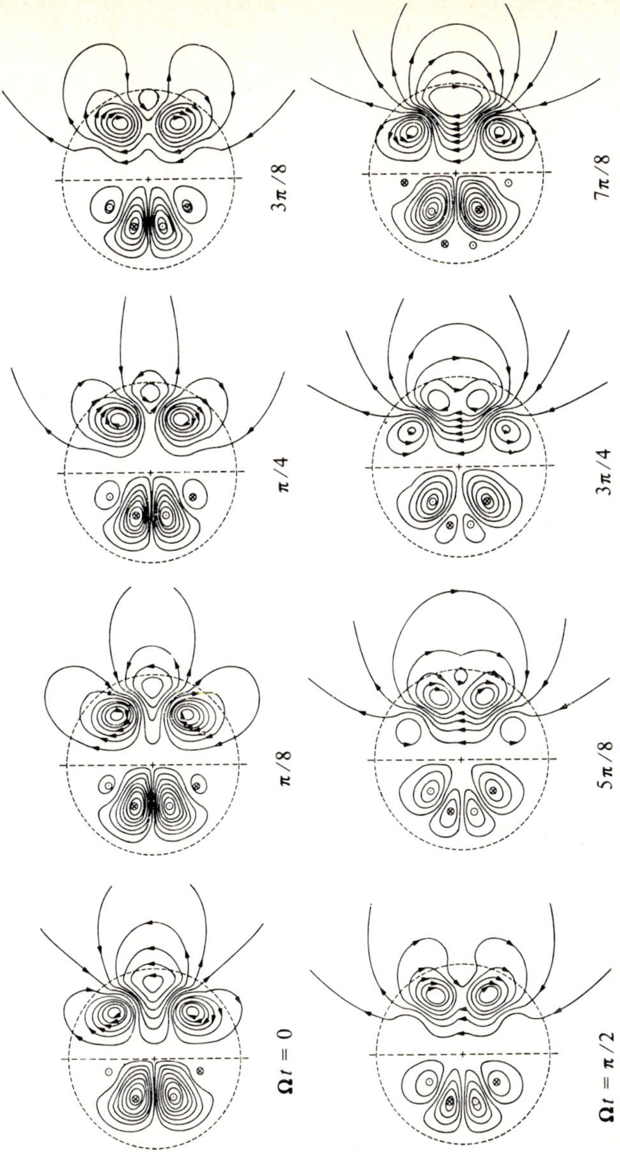

Fig. 9.10 Evolution of the marginal dipole oscillation of the $\alpha\omega$-dynamo defined by (9.170), with $X = -206{\cdot}1$, $\Omega = \lambda^{-1}R^2 \operatorname{Im} p = 47{\cdot}44$. Meridian sections are shown, the symmetry axis being dotted. The $\mathbf{B}_P$-lines (χ = cst.) are shown on the right of this axis, and the lines of constant toroidal field (B_φ = cst.) on the left, at equal intervals of χ and B_φ respectively; B_φ is positive in regions marked $\odot$, and negative in regions marked $\otimes$. The progression of the pattern from poles to equator, as the half-cycle proceeds, is apparent. (From Roberts, 1972.)

also, the necessary condition for propagation towards the equatorial plane was $\alpha\ \partial\omega/\partial r<0$.

On the assumption that sunspots form by a process of eruption when the subsurface toroidal field exceeds some critical value (see § 10.7), a number of authors have, on the basis of time-periodic solutions of the dynamo equations such as that depicted in fig. 9.10, constructed butterfly diagrams (§ 5.3) defined in this context as the family of curves

$$B(r,\theta,t)=kB_{\max}(r) \tag{9.171}$$

in the plane of the variables θ and t, for a fixed representative value of r; here $B_{\max}$ is simply the maximum value of $B(r,\theta,t)$ for $0\leqslant\theta\leqslant\pi$, $0\leqslant t\leqslant 2\pi/\operatorname{Im} p$, and k is a constant between 0 and 1. With the expectation that sunspots may be expected to form in any region where $|B|>k|B_{\max}|$ (for some k), these diagrams are directly comparable with Maunder's butterfly diagram (fig. 5.3) depicting observed occurrences of sunspots (again in the θ, t plane). Fig. 9.11 shows a diagram obtained by Steenbeck & Krause (1969*a*), with[8]

$$\alpha=\frac{1}{2}\alpha_0\left(1+\operatorname{erf}\frac{\hat{r}-0{\cdot}9}{0{\cdot}075}\right)\cos\theta,\qquad \omega=\frac{1}{2}\omega_0'\left(1-\operatorname{erf}\frac{\hat{r}-0{\cdot}7}{0{\cdot}075}\right). \tag{9.172}$$

The *qualitative* resemblance with fig. 5.3 is impressive. Quantitative comparison of course requires that the period $2\pi/\operatorname{Im} p$ of the theoretical solution be comparable with the period of the sunspot cycle, i.e. about 22 years. The various computed solutions (e.g. Roberts, 1972) give

$$\Omega=\operatorname{Im} p\approx 100\lambda/R^2,\qquad 2\pi/\Omega\approx 2\pi R^2/100\lambda,$$

and with $R\approx 7\times 10^5$ km, and a turbulent diffusivity[9] $\lambda\approx u_0 l\approx 10^2\ \text{km}^2\ \text{s}^{-1}$, we get an estimated period $2\pi/\Omega\approx 10$ yr, which is of the right order of magnitude.

8 These distributions of α and ω were chosen to provide a model in which the two types of inductive activity are separated in space; the choice is clearly arbitrary.

9 If instead we choose $\lambda\approx 10^3\ \text{km}^2\ \text{s}^{-1}$ as suggested by the granulation scales (see chapter 5), then we get $2\pi/\Omega\approx 1$ yr, an order of magnitude smaller than the observed period.

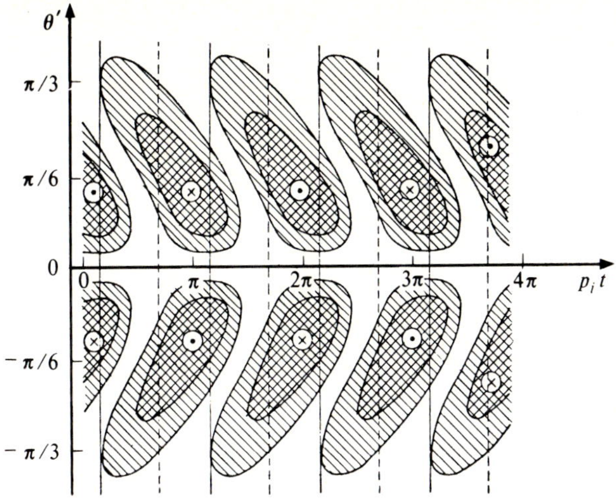

Fig. 9.11 Butterfly diagram corresponding to the $\alpha\omega$-dynamo given by (9.172); the area $|B_\varphi|>\frac{1}{3}B_{\max}$ is hatched, and the area $|B_\varphi|>\frac{2}{3}B_{\max}$ is cross-hatched. The unbroken lines mark the phase at which the polarity at the poles changes; the dashed lines mark the phase at which the sign of the dipole moment changes. The phase $(\operatorname{Im} p)t=0$ corresponds to the maximum value of the toroidal field. (From Steenbeck & Krause, 1969*a*.)

The condition $\alpha_0\omega_0'<0$ is necessary to give propagation of field patterns towards the equatorial plane (and hence butterfly diagrams with the right qualitative characteristics). This condition would appear to leave open the two possibilities

$$\alpha_0>0, \omega_0'<0 \quad \text{or} \quad \alpha_0<0, \omega_0'>0. \tag{9.173}$$

(Recollect here that α_0 represents the extreme value of $\alpha(\mathbf{x})$ in the *northern* hemisphere.) It has however been pointed out by Stix (1976) that the linear relation between the fields A and B (and hence between $\mathbf{B}_P$ and B) involves α and $\partial\omega/\partial r$ separately and not merely via their product $\alpha\ \partial\omega/\partial r$ (cf. equation (9.76)) and that the observed phase relation between the radial component of the Sun's general field and the toroidal component (as revealed by the sunspot pattern) is in fact incompatible with the second possibility in (9.173). Hence if the Sun does act as an $\alpha\omega$-dynamo, then the indications are that α is predominantly positive in the northern hemisphere (and negative in the southern) and that $\omega(r, \theta)$ increases with increasing depth.

There are independent arguments (Steenbeck, Krause & Rädler, 1966) for the conclusion that $\alpha > 0$ in the northern hemisphere, based on simple dynamical considerations. We have already commented (§ 9.5) on one physical mechanism (viscous entrainment) that may generate positive helicity and so *negative* α in the northern hemisphere when blobs of hot fluid rise due to buoyancy forces. In the convective envelope of the Sun, there is however a second mechanism which has the contrary effect: compressibility. A blob of fluid rising through several scale heights will expand and so will tend to rotate in a sense opposite to the mean solar rotation (to conserve its absolute angular momentum); this leads to generation of negative helicity, and so positive α in the northern hemisphere, and it seems at least plausible that this is the dominant effect in the solar context. The detailed dynamical calculations of § 10.7 below, in which the value of α associated with buoyancy instabilities is calculated, suggests that the true picture may be rather more complicated than suggested by these simple arguments.

Likewise, increase of ω with increasing depth is what one would naïvely expect as a result of conservation of angular momentum; in conjunction with meridional circulation, this might be expected to lead to a situation in which $\omega \propto r^{-2}$ in the convection zone. The same argument would suggest however that on the Sun's surface the rotation rate should be greater in the polar regions than in the equatorial zone, whereas the observed situation is quite the opposite. The dynamical theory of differential rotation of the Sun is a very large subject in its own right (see e.g. Durney, 1976), and lies outside the scope of this book. Clearly however dynamo models can serve to eliminate those distributions $\omega(r, \theta)$ that are totally incompatible with observed solar magnetic activity. For example, Stix (1976) has commented that models in which $\omega = \omega(s)$ with $s = r \sin\theta$ are constrained by the observed distribution of ω on the solar surface $r = R$, and therefore satisfy $\partial\omega/\partial r > 0$; since, as mentioned above, this is incompatible with observed phase relation between A and B on $r = R$, one may reasonably conclude that ω is *not* constant on cylindrical surfaces $s = s_0$ through the convection zone.

The models discussed so far (particularly those of Roberts, 1972*a*) have the property that, when $\mathbf{U}_P = 0$, the preferred modes are oscillatory. Examples are known however for which this is not the case. Deinzer *et al.* (1974) have studied the effect of concentrated layers of inductive activity, i.e.

$$\partial\omega/\partial r = \omega_0'\delta(\hat{r}-\hat{r}_1), \qquad \alpha = \alpha_0\delta(\hat{r}-\hat{r}_2)\cos\theta, \qquad (9.175)$$

where $\hat{r} = r/R$, and $0 < \hat{r}_1, \hat{r}_2 < 1$. It appears that when $\hat{r}_1$ and $\hat{r}_2$ are sufficiently separated, the preferred modes are non-oscillatory. Fig. 9.12 indicates the character of the preferred mode for the possible

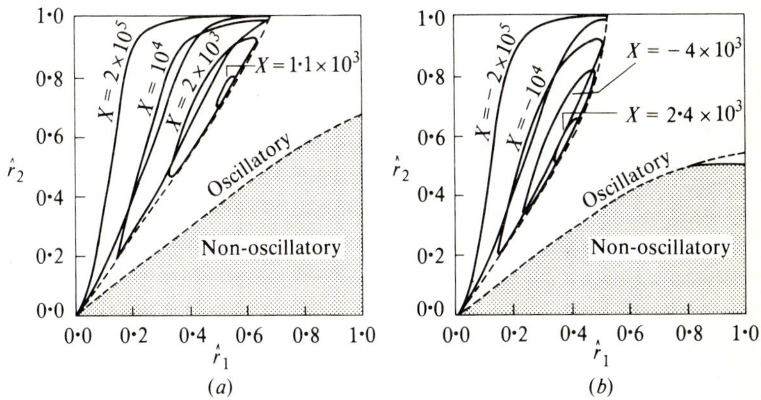

Fig. 9.12 Regions of the $\hat{r}_1-\hat{r}_2$ space for which oscillatory or non-oscillatory modes are preferred for the $\alpha\omega$-dynamo given by (9.175); both (*a*) (dipole symmetry) and (*b*) (quadrupole symmetry) are symmetric about the line $\hat{r}_1 = \hat{r}_2$. Variation of the critical dynamo number X for different values of $\hat{r}_1$ and $\hat{r}_2$ is indicated by the solid curves. (From Deinzer *et al.*, 1974.)

values of $\hat{r}_1$ and $\hat{r}_2$. Here again (as in Herzenberg's two sphere dynamo discussed in § 6.9) spatial separation of the regions of inductive activity appears to encourage the possibility of non-oscillatory dynamos through the filtering out of 'unwanted' harmonics of the main field. This effect would be most easily analysed for the distribution (9.175) of α and ω in the extreme case $\hat{r}_1 \ll \hat{r}_2 \ll 1$ in which only the leading field harmonics generated at either layer have any significant influence at the other layer.

A related situation has been studied by Levy (1972), viz.

$$\begin{aligned} \partial\omega/\partial r &= \omega_0'\delta(\hat{r}-\hat{r}_1), \\ \alpha &= \alpha_0\delta(\hat{r}-\hat{r}_2)(\delta(\cos\theta-\cos\alpha)-\delta(\cos\theta+\cos\alpha)). \end{aligned} \tag{9.176}$$

The α-effect is here supposed concentrated in two 'rings of cyclonic activity' at $\hat{r}=\hat{r}_2$, $\theta=\alpha$, $\pi-\alpha$. The same situation was further analysed by Stix (1973). The conclusion of these studies is that when $X>0$, the preferred mode is non-oscillatory and dipole, while if $X<0$, the preferred mode is non-oscillatory and quadrupole if $|90°-\alpha|\leqslant 54°$, otherwise oscillatory and dipole. It is perhaps not surprising that, as in the case of the δ-function model analysed in § 9.9, the preferred mode can depend in quite a sensitive manner on just where the α-effect is concentrated. The true distribution of $\alpha(\mathbf{x})$ (and the related distribution of helicity density) is largely controlled by the *dynamics* of the background random motions, a topic treated in the following chapters.

Finally, it is worth noting that the transition between oscillatory $\alpha\omega$-dynamo behaviour and non-oscillatory α^2-dynamo behaviour has been examined numerically by Roberts & Stix (1972). It is to be expected that if $|\alpha_0/\omega_0'R^2|$ is increased from small values, the $\alpha\omega$-behaviour will give way to the α^2-behaviour when this parameter reaches a value of order unity. For the particular model considered, Roberts & Stix found that this transition in behaviour in fact occurred when $|\alpha_0/\omega_0'R^2|\approx 0.1$. This treatment was based on an isotropic α-effect; it is perhaps worth remembering that in the presence of strong differential rotation, the assumption of isotropy in the background turbulence (or random wave motion) is really untenable; a severely non-isotropic α-effect of the form

$$\mathcal{E}=\alpha\mathbf{B}_T+\alpha'\mathbf{B}_P, \quad |\alpha'|\ll|\alpha|, \tag{9.177}$$

seems more plausible, and is moreover indicated by the approach of Braginskii as elaborated in chapter 8.

Although results of the models discussed above, and in particular their ability to reproduce butterfly diagrams with the right qualitative structure, are suggestive in the solar context, nevertheless they must as yet be viewed with caution. To get the right time-scale a turbulent diffusivity of order 100 km^2 s^{-1} is necessary; as noted in chapter 5, turbulence of the required intensity corresponds to a magnetic Reynolds number of order 10^4, and the turbulent magnetic fluctuations may then be expected to be an order of magnitude stronger than the mean field itself. Although the surface field of the Sun does exhibit intermittency on observable scales down to $\sim$100 km, the associated observed fluctuations do not have the entirely random character that the theory strictly requires, and there is no direct evidence for field or velocity fluctuations on scales smaller than $\sim$100 km. The mean field is heavily disguised by its fine-structure, and the analysis of Altschuler *et al.* (1974) suggests that in fact equatorial dipole and quadrupole ingredients are at least as strong, if not stronger, than the axial ingredients which appear most naturally in the theory of the $\alpha\omega$-dynamo in so far as it can explain the solar cycle. As emphasised recently by Cowling (1975*b*), although mean-field electrodynamics is alluring in its relative simplicity, certain inconsistencies need to be resolved before it can be definitely accepted as providing the correct basic description of solar field generation.

CHAPTER 10

WAVES OF HELICAL STRUCTURE INFLUENCED BY CORIOLIS, LORENTZ AND BUOYANCY FORCES

10.1. The momentum equation and some elementary consequences

So far we have regarded the velocity field $\mathbf{u}(\mathbf{x}, t)$ as given. In this chapter, we turn to the study of dynamic effects in which the evolution of $\mathbf{u}$ is determined by the Navier–Stokes equation

$$\rho(\partial\mathbf{u}/\partial t + \mathbf{u}\,.\,\nabla\mathbf{u}) = -\nabla p + \mathbf{J}\wedge\mathbf{B} + \rho\nu\nabla^2\mathbf{u} + \rho\mathbf{F}. \qquad (10.1)$$

Here $\mathbf{J}\wedge\mathbf{B}$ is the Lorentz force distribution, ν is the kinematic viscosity of the fluid (assumed uniform) and $\mathbf{F}$ represents any further force distribution that may be present. We limit attention to incompressible flows for which

$$\mathrm{D}\rho/\mathrm{D}t = 0 \quad \text{and} \quad \nabla\,.\,\mathbf{u} = 0, \qquad (10.2)$$

so that bulk viscosity effects do not appear in (10.1). Although we shall be primarily concerned with motion in a rotating system of reference, for the moment we adopt an inertial frame in which acceleration is represented adequately by the left-hand side of (10.1).

The integral equivalent of (10.1) is

$$\frac{\mathrm{d}}{\mathrm{d}t}\int_V \rho u_i\,\mathrm{d}V = \int_S (\sigma_{ij} + T_{ij})n_j\,\mathrm{d}S + \int \rho F_i\,\mathrm{d}V, \qquad (10.3)$$

where S is a closed surface moving with the fluid, V its interior, σ_{ij} the stress tensor given by the Newtonian relation

$$\sigma_{ij} = -p\delta_{ij} + \rho\nu(\partial u_i/\partial x_j + \partial u_j/\partial x_i), \qquad (10.4)$$

and T_{ij} the Maxwell stress tensor given by

$$T_{ij} = \mu_0^{-1}(B_iB_j - \tfrac{1}{2}\mathbf{B}^2\,\delta_{ij}). \qquad (10.5)$$

With (10.3) we must clearly associate a jump condition

$$[(\sigma_{ij} + T_{ij})n_j] = 0, \qquad (10.6)$$

across any surface of discontinuity of physical properties; this may be either a fixed fluid boundary on which $\mathbf{n} \cdot \mathbf{u} = 0$, or an interior surface of discontinuity moving with the fluid.

If ρ is uniform, it is convenient to introduce the new variable

$$\mathbf{h} = (\mu_0 \rho)^{-1/2} \mathbf{B}, \tag{10.7}$$

in terms of which

$$\mathbf{J} \wedge \mathbf{B} = \mu_0^{-1} (\nabla \wedge \mathbf{B}) \wedge \mathbf{B} = \rho (\mathbf{h} \cdot \nabla \mathbf{h} - \tfrac{1}{2} \nabla \mathbf{h}^2), \tag{10.8}$$

and (10.1) becomes

$$\frac{\partial \mathbf{u}}{\partial t} + \mathbf{u} \cdot \nabla \mathbf{u} = -\nabla P + \mathbf{h} \cdot \nabla \mathbf{h} + \nu \nabla^2 \mathbf{u} + \mathbf{F}, \tag{10.9}$$

where $P = p/\rho + \frac{1}{2}\mathbf{h}^2$. In terms of $\mathbf{h}$, the induction equation becomes

$$\partial \mathbf{h}/\partial t + \mathbf{u} \cdot \nabla \mathbf{h} = \mathbf{h} \cdot \nabla \mathbf{u} + \lambda \nabla^2 \mathbf{h}. \tag{10.10}$$

$\mathbf{h}$ evidently has the dimensions of a velocity. There is a symmetry of structure in the non-linear terms of (10.9) and (10.10), which leads to some simple and important results when $\mathbf{F} = 0$ and in the limiting situation $\lambda = \nu = 0$. It is a trivial matter to verify that (10.9) and (10.10) are then both satisfied by solutions of the form

$$\mathbf{u} = \mathbf{f}(\mathbf{x} - \mathbf{h}_0 t), \qquad \mathbf{h} = \mathbf{h}_0 - \mathbf{f}(\mathbf{x} - \mathbf{h}_0 t), \tag{10.11}$$

or

$$\mathbf{u} = \mathbf{g}(\mathbf{u} + \mathbf{h}_0 t), \qquad \mathbf{h} = \mathbf{h}_0 + \mathbf{g}(\mathbf{x} + \mathbf{h}_0 t), \tag{10.12}$$

where $\mathbf{f}$ and $\mathbf{g}$ are arbitrary functions of their arguments and $\mathbf{h}_0$ is constant. When $\mathbf{f}$ and $\mathbf{g}$ are localised (e.g. square-integrable) functions, these solutions represent waves which propagate without change of shape in the directions $\pm \mathbf{h}_0$. These are known as Alfvén waves (Alfvén, 1942), and $\mathbf{h}_0 = (\mu_0 \rho)^{-1/2} \mathbf{B}_0$, where $\mathbf{B}_0$ is a uniform field, is known as the Alfvén velocity. In view of the non-linearity of (10.9) and (10.10), a linear superposition of (10.11) and (10.12) does *not* in general satisfy the equations. However, if $|\mathbf{f}| \ll h_0$, $|\mathbf{g}| \ll h_0$, and if squares and products of $\mathbf{f}$ and $\mathbf{g}$ are neglected, then

$$\left.\begin{aligned} \mathbf{u} &= \mathbf{f}(\mathbf{x} - \mathbf{h}_0 t) + \mathbf{g}(\mathbf{x} + \mathbf{h}_0 t) \\ \mathbf{h} &= \mathbf{h}_0 - \mathbf{f}(\mathbf{x} - \mathbf{h}_0 t) + \mathbf{g}(\mathbf{x} + \mathbf{h}_0 t) \end{aligned}\right\} \tag{10.13}$$

provides a solution of the linearised equations. In general, however, solutions of the form (10.11) and (10.12) will interact in a non-linear manner when the functions **f** and **g** are overlapping (fig. 10.1).

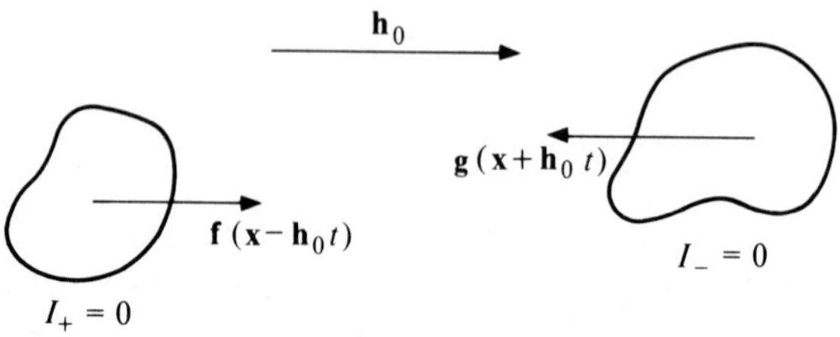

Fig. 10.1 Disturbances represented by the functions $\mathbf{f}(\mathbf{x}-\mathbf{h}_0t)$ and $\mathbf{g}(\mathbf{x}+\mathbf{h}_0t)$ interact only while they overlap in **x**-space. The time-scale characteristic of non-linear interaction of disturbances of length-scale L is evidently at most of order L/h_0.

Putting $\mathbf{h}=\mathbf{h}_0+\mathbf{h}_1$ (with $\mathbf{h}_1=-\mathbf{f}(\mathbf{x}-\mathbf{h}_0t)$ in (10.11) and $+\mathbf{g}(\mathbf{x}+\mathbf{h}_0t)$ in (10.12)) it is evident that the solutions (10.11) and (10.12) are characterised by the properties $\mathbf{u}\pm\mathbf{h}_1=0$ respectively. Let us define the integrals

$$I_\pm=\tfrac{1}{2}\int(\mathbf{u}\pm\mathbf{h}_1)^2\,\mathrm{d}V, \tag{10.14}$$

on the assumption that the disturbances are sufficiently localised for these integrals to exist; here we imagine the fluid to extend to infinity in all directions, and the integral is over all space. Then $I_+=0$ for the solution (10.11) and $I_-=0$ for the solution (10.12).

The integrals $I_\pm$ are in fact invariants of the pair of equations (10.9) and (10.10), when $\nu=\lambda=0$. For

$$\begin{aligned}\frac{\mathrm{d}I_+}{\mathrm{d}t}&=\int(\mathbf{u}+\mathbf{h}_1)\,.\,\frac{\mathrm{D}}{\mathrm{D}t}(\mathbf{u}+\mathbf{h}_1)\,\mathrm{d}V\\&=\int(\mathbf{u}+\mathbf{h}_1)\,.\,(-\nabla P+(\mathbf{h}_0+\mathbf{h}_1)\,.\,\nabla(\mathbf{u}+\mathbf{h}_1))\,\mathrm{d}V\\&=\int\nabla\,.\,\{-P(\mathbf{u}+\mathbf{h}_1)+\tfrac{1}{2}(\mathbf{h}_0+\mathbf{h}_1)(\mathbf{u}+\mathbf{h}_1)^2\}\,\mathrm{d}V,\end{aligned} \tag{10.15}$$

where we have used $\nabla\,.\,\mathbf{u}=0$, $\nabla\,.\,\mathbf{h}_1=0$. Using the divergence theorem and the vanishing of $\mathbf{u}+\mathbf{h}_1$ at infinity ($\mathbf{u}+\mathbf{h}_1=o(r^{-2})$ is clearly sufficient here) it follows that $dI_+/dt=0$, and hence that $I_+=\text{cst}$. Similarly $I_-=\text{cst}$. Note that

$$I_E=\tfrac{1}{2}(I_++I_-)=\tfrac{1}{2}\int(\mathbf{u}^2+\mathbf{h}^2)\,dV \tag{10.16}$$

is just the total energy of the disturbance (kinetic plus magnetic), and the invariance of this quantity is of course no surprise. We have also the related invariant (Woltjer, 1958)

$$I_C=\tfrac{1}{2}(I_+-I_-)=\int\mathbf{u}\,.\,\mathbf{h}_1\,dV. \tag{10.17}$$

The invariance of this integral admits topological interpretation (cf. the magnetic helicity integral (2.8)). In fact I_C provides a measure of the degree of linkage of the vortex lines of the $\mathbf{u}$-field with the lines of force of the $\mathbf{h}_1$-field. To see this, consider the particular situation in which $\mathbf{h}_1\equiv 0$ except in a single flux-tube of vanishing cross-section in the neighbourhood of the closed curve C, ad let Φ_1 be the flux of $\mathbf{h}_1$ in the tube. Then from (10.17),

$$I_C=\Phi_1\int_C\mathbf{u}\,.\,d\mathbf{x}=\Phi_1 K, \tag{10.18}$$

where K is the circulation round C, or equivalently the flux of vorticity across a surface S spanning C. Hence I_C is non-zero or zero according as the $\mathbf{h}_1$-lines do or do not enclose a net flux of vorticity; I_C is the *cross-helicity* of the fields $\mathbf{u}$ and $\mathbf{b}$.

The invariance of I_+ and I_- provides some indication of the nature of the interaction of two Alfvén waves of the form (10.11) and (10.12). If $\mathbf{f}$ and $\mathbf{g}$ are both localised functions with spatial extent of order L, then the duration of the interaction (which may be thought of as a collision of two 'blobs' represented by the functions $\mathbf{f}$ and $\mathbf{g}$) will be at most of order L/h_0. Choosing the origin of time to be during the interaction, for $t\ll -L/h_0$ (i.e. before the interaction) the solution has the form (10.13) while for $t\gg L/h_0$ (i.e. after the interaction) we must have

$$\left.\begin{aligned}\mathbf{u}&=\mathbf{f}_1(\mathbf{x}-\mathbf{h}_0t)+\mathbf{g}_1(\mathbf{x}+\mathbf{h}_0t)\\ \mathbf{h}&=\mathbf{h}_0-\mathbf{f}_1(\mathbf{x}-\mathbf{h}_0t)+\mathbf{g}_1(\mathbf{x}+\mathbf{h}_0t)\end{aligned}\right\} \tag{10.19}$$

where $\mathbf{f}_1$ and $\mathbf{g}_1$ are related in some way to $\mathbf{f}$ and $\mathbf{g}$. The invariance of I_+ and I_- then tells us that

$$\int\mathbf{f}_1^2\,dV=\int\mathbf{f}^2\,dV,\qquad\int\mathbf{g}_1^2\,dV=\int\mathbf{g}^2\,dV, \tag{10.20}$$

i.e. there can be no transfer of energy between the disturbances during the interaction. The spatial structure of each disturbance will however presumably be modified by the interaction. The nature of this modification presents an intriguing problem that does not appear yet to have been studied.

10.2. Waves influenced by Coriolis and Lorentz forces

In a rotating mass of fluid, such as the liquid core of the Earth, or the Sun, it is appropriate to refer both velocity and magnetic field to a frame of reference rotating with the fluid. If $\mathbf{\Omega}$ is the angular velocity of this frame, then when ρ is constant the momentum equation becomes

$$\partial \mathbf{u}/\partial t + \mathbf{u} \cdot \nabla \mathbf{u} + 2\mathbf{\Omega} \wedge \mathbf{u} = -\nabla P + \mathbf{h} \cdot \nabla \mathbf{h} + \nu \nabla^2 \mathbf{u} + \mathbf{F}, \tag{10.21}$$

where P now includes the centrifugal potential $-\frac{1}{2}(\mathbf{\Omega} \wedge \mathbf{x})^2$ as well as the magnetic pressure term $\frac{1}{2}\mathbf{h}^2$. The induction equation is however invariant, i.e. we still have

$$\partial \mathbf{h}/\partial t + \mathbf{u} \cdot \nabla \mathbf{h} = \mathbf{h} \cdot \nabla \mathbf{u} + \lambda \nabla^2 \mathbf{h}, \tag{10.22}$$

where $\mathbf{h}(\mathbf{x}, t)$, $\mathbf{u}(\mathbf{x}, t)$ are now measured relative to the rotating frame of reference. Physically, this is obvious: rigid body rotation simply rotates a magnetic field without distortion. Changes in $\mathbf{h}(\mathbf{x}, t)$ in the rotating frame are therefore caused only by motion relative to the rotating frame, and by the usual process of ohmic diffusion.[1]

Suppose now that

$$\mathbf{h} = \mathbf{h}_0 + \mathbf{h}_1, \qquad \mathbf{u} = \mathbf{u}_0 + \mathbf{u}_1, \tag{10.23}$$

where $\mathbf{h}_0$ and $\mathbf{u}_0$ are uniform, and $\mathbf{h}_1$ and $\mathbf{u}_1$ represent small perturbations. Neglecting squares and products of $\mathbf{u}_1$ and $\mathbf{h}_1$, and supposing $\mathbf{F} = 0$, the linearised forms of (10.21) and (10.22) are

$$\partial \mathbf{u}_1/\partial t + \mathbf{u}_0 \cdot \nabla \mathbf{u}_1 + 2\mathbf{\Omega} \wedge \mathbf{u}_1 = -\nabla P_1 + \mathbf{h}_0 \cdot \nabla \mathbf{h}_1 + \nu \nabla^2 \mathbf{u}_1, \tag{10.24}$$

[1] Note that this simple statement must break down at distances from the axis of rotation of order c/Ω, where c is the speed of light. At such distances, displacement currents cannot be ignored and the field inevitably lags behind the rotating frame of reference.

and

$$\partial \mathbf{h}_1/\partial t + \mathbf{u}_0 \,.\, \nabla \mathbf{h}_1 = \mathbf{h}_0 \,.\, \nabla \mathbf{u}_1 + \lambda \nabla^2 \mathbf{h}_1, \tag{10.25}$$

where P_1 is the associated perturbation in P. These equations admit wave-type solutions (Lehnert, 1954) of the form

$$(\mathbf{u}_1, \mathbf{h}_1, P_1) = \mathrm{Re}\,(\hat{\mathbf{u}}, \hat{\mathbf{h}}, \hat{P}) \exp \mathrm{i}(\mathbf{k} \,.\, \mathbf{x} - (\omega + \mathbf{u}_0 \,.\, \mathbf{k})t), \tag{10.26}$$

and the fact that $\mathbf{u}_1$ and $\mathbf{h}_1$ are solenoidal means that

$$\mathbf{k} \,.\, \hat{\mathbf{u}} = \mathbf{k} \,.\, \hat{\mathbf{h}} = 0, \tag{10.27}$$

i.e. these are transverse waves. Substitution first in (10.25) gives the relation between $\hat{\mathbf{h}}$ and $\hat{\mathbf{u}}$

$$\hat{\mathbf{h}} = -(\omega + \mathrm{i}\lambda k^2)^{-1}(\mathbf{k} \,.\, \mathbf{h}_0)\hat{\mathbf{u}} \tag{10.28}$$

that we have already encountered (equation (7.75)), and substitution in (10.24) and rearrangement of the terms gives

$$-\mathrm{i}\sigma\hat{\mathbf{u}} + 2\mathbf{\Omega} \wedge \hat{\mathbf{u}} = -\mathrm{i}\mathbf{k}\hat{P}, \tag{10.29}$$

where

$$\sigma = (\omega + \mathrm{i}\nu k^2) - (\omega + \mathrm{i}\lambda k^2)^{-1}(\mathbf{h}_0 \,.\, \mathbf{k})^2. \tag{10.30}$$

The effect of the magnetic field in (10.29) is entirely contained in the coefficient σ. This leads to an important modification of the dispersion relationship between ω and $\mathbf{k}$; but the spatial structure of the velocity field is unaffected by the presence of $\mathbf{h}_0$.

To get the dispersion relationship, we first take the cross-product of (10.29) with $\mathbf{k}$; since $\mathbf{k} \,.\, \hat{\mathbf{u}} = 0$, this gives

$$-\mathrm{i}\sigma\mathbf{k} \wedge \hat{\mathbf{u}} - 2(\mathbf{k} \,.\, \mathbf{\Omega})\hat{\mathbf{u}} = 0. \tag{10.31}$$

Taking the cross-product again with $\mathbf{k}$ gives

$$\mathrm{i}\sigma k^2\hat{\mathbf{u}} - 2(\mathbf{k} \,.\, \mathbf{\Omega})\mathbf{k} \wedge \hat{\mathbf{u}} = 0. \tag{10.32}$$

Elimination of $\hat{\mathbf{u}}$ and $\mathbf{k} \wedge \hat{\mathbf{u}}$ from (10.31) and (10.32) gives

$$\sigma^2 = 4(\mathbf{k} \,.\, \mathbf{\Omega})^2/k^2, \quad \text{or} \quad \sigma = \mp 2(\mathbf{k} \,.\, \mathbf{\Omega})/k, \tag{10.33}$$

and from either (10.31) or (10.32) we have the corresponding simple relation between $\hat{\mathbf{u}}$ and $\mathbf{k} \wedge \hat{\mathbf{u}}$,

$$\mathrm{i}\mathbf{k} \wedge \hat{\mathbf{u}} = \pm k\hat{\mathbf{u}}. \tag{10.34}$$

Since $\hat{\boldsymbol{\omega}} = \mathrm{i}\mathbf{k} \wedge \hat{\mathbf{u}}$ is the Fourier transform of the vorticity $\boldsymbol{\omega} = \nabla \wedge \mathbf{u}$ associated with the waves, it is evident from (10.34) that each constituent wave is of maximal helicity, i.e. for each wave

$$\langle \mathbf{u} \,.\, \boldsymbol{\omega} \rangle = \pm k |\hat{\mathbf{u}}|^2 \, \mathrm{e}^{-2|\omega_i|t}, \tag{10.35}$$

where $\omega = \omega_r + \mathrm{i}\omega_i$; the decay of the waves ($\omega_i \neq 0$) is of course associated with the processes of viscous and ohmic diffusion. Such motions are particularly effective in generating an α-effect; hence their particular interest in the dynamo context.

The structure of a motion satisfying (10.34) may be easily understood by choosing axes $OXYZ$ with OX parallel to $\mathbf{k}$, so that $\mathbf{k} = (k, 0, 0)$; then $\hat{\mathbf{u}} = (0, \hat{v}, \hat{w})$ and (10.34) gives $\hat{w} = \pm \mathrm{i}\hat{v}$; in conjunction with the factor $\mathrm{e}^{\mathrm{i}\mathbf{k}.\mathbf{x}} = \mathrm{e}^{\mathrm{i}kX}$ in (10.26), the motion then has a spatial structure given by

$$\mathbf{u}_1 \propto (0, \cos(kX + \psi), \pm \sin(kX + \psi)), \tag{10.36}$$

where the phase ψ is time-dependent. This is a circularly polarised wave-motion (fig. 10.2), the velocity vector $\mathbf{u}_1$ being constant in

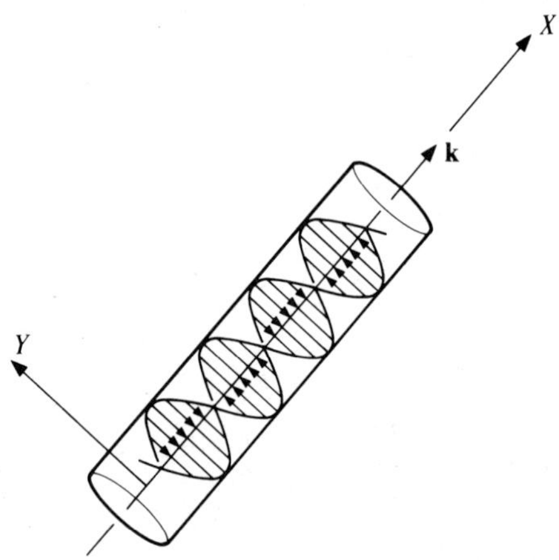

Fig. 10.2 Sketch of the structure of the wave-motion given by (10.36). The case illustrated, in which the velocity vector rotates in a right-handed sense as X increases, corresponds to the lower choice of sign in (10.36), with $k > 0$; the helicity of this motion as given by (10.35) is then negative.

magnitude, but rotating in direction as X increases; the sense of rotation is left-handed or right-handed according as the associated helicity $\mathbf{u}_1 \cdot \boldsymbol{\omega}$ is positive or negative.

When $\mathbf{h}_0 = 0$, and when viscous effects are negligible, (10.30) and (10.33) give

$$\omega = \mp 2(\mathbf{k} \cdot \boldsymbol{\Omega})/k, \tag{10.37}$$

the dispersion relation for pure inertial waves in a rotating fluid (see e.g. Greenspan, 1968). The group velocity for such waves is given by

$$\mathbf{c}_g = \nabla_{\mathbf{k}}\omega = \mp 2k^{-3}(k^2\boldsymbol{\Omega} - \mathbf{k}(\mathbf{k} \cdot \boldsymbol{\Omega})). \tag{10.38}$$

Since $\mathbf{c}_g \cdot \mathbf{k} = 0$, this is perpendicular to the phase velocity $\omega\mathbf{k}/k^2$. Moreover, since

$$\mathbf{c}_g \cdot \boldsymbol{\Omega} = \mp 2k^{-1}\Omega^2 \sin^2\theta, \tag{10.39}$$

where θ is the angle between $\mathbf{k}$ and $\boldsymbol{\Omega}$, the value of $\mathbf{c}_g \cdot \boldsymbol{\Omega}$ is negative or positive according as the helicity of the group is positive or negative. Loosely speaking, we may say that negative helicity is associated with upward propagation (relative to $\boldsymbol{\Omega}$) and positive helicity is associated with downward propagation. A random superposition of such waves in equal proportions would give zero net helicity; but if any mechanism is present which leads to preferential excitation of upward or downward propagating waves, then the net helicity will be negative or positive respectively.

Suppose now that $\mathbf{h}_0 \neq 0$, but that the Coriolis effect is still dominant in the sense that

$$|\sigma| = 2k^{-1}|\boldsymbol{\Omega} \cdot \mathbf{k}| \gg |\mathbf{h}_0 \cdot \mathbf{k}|. \tag{10.40}$$

(Of course if $\mathbf{h}_0$ is not parallel to $\boldsymbol{\Omega}$ there will always be some wave-vectors perpendicular or nearly perpendicular to $\boldsymbol{\Omega}$ for which (10.40) is *not* satisfied.) Then, provided the diffusion effects associated with ν and λ are weak, the two roots of (10.30) (regarded as a quadratic in ω) are given by

$$\omega + i\nu k^2 \approx \sigma + \frac{(\mathbf{h}_0 \cdot \mathbf{k})^2}{\sigma + i(\lambda - \nu)k^2}, \tag{10.41}$$

and

$$\omega + i\lambda k^2 \approx -\sigma^{-1}(\mathbf{h}_0 . \mathbf{k})^2, \tag{10.42}$$

where σ is still given by (10.33). Clearly (10.41) still represents an inertial wave whose frequency is weakly modified by the presence of the magnetic field and which is weakly damped by both viscous and ohmic effects; in fact from (10.41) we have $\omega = \omega_r + i\omega_i$ where (provided $|\lambda - \nu|k^2 \ll |\sigma|$),

$$\omega_r \approx \sigma + \sigma^{-1}(\mathbf{h}_0 . \mathbf{k})^2, \qquad \omega_i \approx -\nu k^2 - (\mathbf{h}_0 . \mathbf{k})^2(\lambda - \nu)k^2\sigma^{-2}. \tag{10.43}$$

Equation (10.42) on the other hand represents a wave which has no counterpart when $\mathbf{h}_0 = 0$. In this wave, the relation between $\hat{\mathbf{h}}$ and $\hat{\mathbf{u}}$ from (10.28) is given by

$$\hat{\mathbf{h}} = \sigma(\mathbf{k} . \mathbf{h}_0)^{-1}\hat{\mathbf{u}}, \tag{10.44}$$

so that, from (10.40), $|\hat{\mathbf{h}}| \gg |\hat{\mathbf{u}}|$, i.e. the magnetic perturbation dominates the velocity perturbation. The dispersion relation (10.42) may be obtained by neglecting the contribution $(\partial/\partial t - \nu\nabla^2)\mathbf{u}_1$ to (10.24) (or equivalently the contribution $\omega + i\nu k^2$ to (10.30)); setting aside the trivial effect of convection by the uniform velocity $\mathbf{u}_0$, the force balance in this wave is therefore given by

$$2\mathbf{\Omega} \wedge \mathbf{u}_1 \approx -\nabla P_1 + \mathbf{h}_0 . \nabla \mathbf{h}_1, \tag{10.45}$$

i.e. a balance between Coriolis, pressure and Lorentz forces. Such a force balance is described as *magnetostrophic* (by analogy with the term *geostrophic* used in meteorological contexts to describe balance between Coriolis and pressure forces alone). We shall describe waves of this second category with dispersion relation (10.42) as *magnetostrophic waves*. Other authors (e.g. Acheson & Hide, 1973) use the term 'hydromagnetic inertial waves'.

10.3. Modification of α-effect by the Lorentz force

We now consider a rather idealised model which gives some insight into the nature of the back-reaction of the Lorentz force on the α-effect which is (at least in part) responsible for the generation of the magnetic field. Let us include a body force $\mathbf{F}(\mathbf{x}, t)$ (random or periodic) on the right-hand side of (10.24), and let us suppose that this force is prescribed (statistically or in detail) and independent of the magnetic field $\mathbf{h}_0$. The response $(\mathbf{u}_1, \mathbf{h}_1, P_1)$ does depend on the field $\mathbf{h}_0$ as well as on $\mathbf{\Omega}$, λ and ν. We shall suppose that the dissipative effects represented by λ and ν are weak. The effect of $\mathbf{u}_0$ is trivial, and for simplicity we take $\mathbf{u}_0 = 0$.

The force field $\mathbf{F}$ must of course be sufficiently weak for the linearised equations to be applicable. We may suppose[2] that $\nabla \cdot \mathbf{F} = 0$, and we suppose further that $\mathbf{F}$ is reflexionally symmetric, in particular that $\langle \mathbf{F} \cdot \nabla \wedge \mathbf{F} \rangle = 0$. Any lack of reflexional symmetry in the $\mathbf{u}_1$-field then arises through the influence of rotation.

Consider first the effect of a single Fourier component of $\mathbf{F}$ of the form $\mathrm{Re}\,(\hat{\mathbf{f}}\,\mathrm{e}^{\mathrm{i}(\mathbf{k}.\mathbf{x}-\omega t)})$ where $\mathbf{k}$ and ω are real. The result (10.28) still holds, from which we deduce a contribution to $\langle \mathbf{u}_1 \wedge \mathbf{h}_1 \rangle$ of the form

$$\tfrac{1}{2}\,\mathrm{Re}\,\hat{\mathbf{u}} \wedge \hat{\mathbf{h}}^* = \frac{\lambda k^2}{2(\omega^2 + \lambda^2 k^4)} (\mathrm{i}\hat{\mathbf{u}} \wedge \hat{\mathbf{u}}^*)\mathbf{k} \cdot \mathbf{h}_0. \tag{10.46}$$

So far, this is just as in § 7.7. Now, however, we can no longer regard $\hat{\mathbf{u}}$ as 'given', but must express this in terms of $\hat{\mathbf{f}}$ by means of the equation

$$-\mathrm{i}\sigma\hat{\mathbf{u}} + 2\mathbf{\Omega} \wedge \hat{\mathbf{u}} = -\mathrm{i}\mathbf{k}\hat{P} + \hat{\mathbf{f}}, \tag{10.47}$$

where σ is still given by (10.30). When diffusion effects are weak, this has the form $\sigma = \sigma_r + \mathrm{i}\sigma_i$, where

$$\sigma_r \approx \omega - \omega^{-1}(\mathbf{h}_0 \cdot \mathbf{k})^2, \qquad \sigma_i \approx k^2(\nu + \lambda\omega^{-2}(\mathbf{h}_0 \cdot \mathbf{k})^2). \tag{10.48}$$

Taking the cross-product with $\mathbf{k}$ twice, and solving for $\hat{\mathbf{u}}$, we obtain

$$\hat{\mathbf{u}} = D^{-1}(2(\mathbf{k} \cdot \mathbf{\Omega})\mathbf{k} \wedge \hat{\mathbf{f}} + \mathrm{i}\sigma k^2 \hat{\mathbf{f}}), \tag{10.49}$$

where

$$D = \sigma^2 k^2 - 4(\mathbf{k} \cdot \mathbf{\Omega})^2. \tag{10.50}$$

Note that when ν and λ are small, there is a resonant response (i.e. $|D|$ has a sharp minimum) when

$$\sigma_r^2 k^2 = 4(\mathbf{k} \cdot \mathbf{\Omega})^2, \tag{10.51}$$

[2] The general $\mathbf{F}$ may be represented as the sum of solenoidal and irrotational ingredients; the latter leads only to a trivial modification of the pressure distribution (when $\nabla \cdot \mathbf{u} = 0$).

i.e. when the forcing wave excites at a natural frequency (or rather frequency–wave-vector combination) of the undamped system.

From (10.49), the value of $i\hat{\mathbf{u}} \wedge \hat{\mathbf{u}}^*$ can now be calculated. Using $i\hat{\mathbf{f}} \wedge \hat{\mathbf{f}}^* = 0$ (from the reflexional symmetry of $\mathbf{F}$), we obtain

$$i\hat{\mathbf{u}} \wedge \hat{\mathbf{u}}^* = -4|D|^{-2}(\mathbf{k}\,.\,\mathbf{\Omega})k\sigma_r k^2|\hat{\mathbf{f}}|^2. \qquad (10.52)$$

In conjunction with (10.46), we then have a contribution to the tensor α_{ij} of the form

$$\hat{\alpha}_{ij}(\mathbf{k}, \omega) = -2|D|^{-2}\sigma_r(\mathbf{k}\,.\,\mathbf{\Omega})\frac{\lambda k^4}{\omega^2 + \lambda^2 k^4}|\hat{f}|^2 k_i k_j, \qquad (10.53)$$

and in this expression, from (10.50), we have

$$|D|^2 \approx (\sigma_r^2 k^2 - 4(\mathbf{k}\,.\,\mathbf{\Omega})^2)^2 + 4\sigma_r^2 k^4(\nu + \lambda\omega^{-2}(\mathbf{h}_0\,.\,\mathbf{k})^2)^2. \quad (10.54)$$

The behaviour of $\hat{\alpha}_{ij}$ as $|\mathbf{h}_0\,.\,\mathbf{k}|$ increases from small values is of considerable interest. First when $\mathbf{h}_0\,.\,\mathbf{k} \approx 0$, we have $\sigma_r = \omega$ and the expression for $i\hat{\mathbf{u}} \wedge \hat{\mathbf{u}}^*$ (related to the helicity of the forced wave) is non-zero only when

$$\omega(\mathbf{k}\,.\,\mathbf{\Omega}) \neq 0.$$

This condition means simply that the forcing wave must either propagate 'upwards' or 'downwards' relative to the direction of $\mathbf{\Omega}$; as expected from the discussion of § 10.2, the velocity field will lack reflexional symmetry only if the forcing is such as to provide a net energy flux either upwards or downwards.

As $|\mathbf{h}_0\,.\,\mathbf{k}|$ increases from zero, the variation of $\hat{\alpha}_{ij}$ is determined by the behaviour of the scalar coefficient in (10.53), $|D|^{-2}\sigma_r$, σ_r being given by (10.48*a*). To fix ideas, suppose that $\omega > 0$ and $\mathbf{k}\,.\,\mathbf{\Omega} > 0$. The resonance condition (10.51) may be written in the form

$$|\mathbf{h}_0\,.\,\mathbf{k}| = \omega - 2(\mathbf{k}\,.\,\mathbf{\Omega})/k, \qquad (10.55)$$

and the behaviour therefore depends on whether the excitation frequency ω is greater or less than $2\mathbf{k}\,.\,\mathbf{\Omega}/k$. (i) If $\omega < 2\mathbf{k}\,.\,\mathbf{\Omega}/k$, then resonance does not occur for any value of $\mathbf{h}_0$; σ_r changes sign when $\mathbf{h}_0\,.\,\mathbf{k}$ passes through the value ω ($\mathbf{h}_0$ being then the Alfvén velocity for waves characterised by wave-vector $\mathbf{k}$ and frequency ω); and for large values of $|\mathbf{h}_0\,.\,\mathbf{k}|$, $\sigma_r|D|^{-2}$ is asymptotically proportional to

$|\mathbf{h}_0 . \mathbf{k}|^{-6}$. (ii) If $\omega > 2\mathbf{k} . \mathbf{\Omega}/k$, then resonance occurs when (10.55) is satisfied; for this value of $|\mathbf{h}_0 . \mathbf{k}|$,

$$\sigma_r|D|^{-2} = [4\sigma_r k^4(\nu + \lambda\omega^{-2}(\mathbf{h}_0 . \mathbf{k})^2)^2]^{-1}, \qquad (10.56)$$

indicating the nature of the resonance as λ, $\nu \to 0$. As $|\mathbf{h}_0 . \mathbf{k}|$ increases beyond the resonance value, the behaviour is again as in case (i). This behaviour is summarised in the qualitative sketches of $\hat{\alpha} = \frac{1}{3}\hat{\alpha}_{ii}$ against $|\mathbf{h}_0 . \mathbf{k}|$ in fig. 10.3.

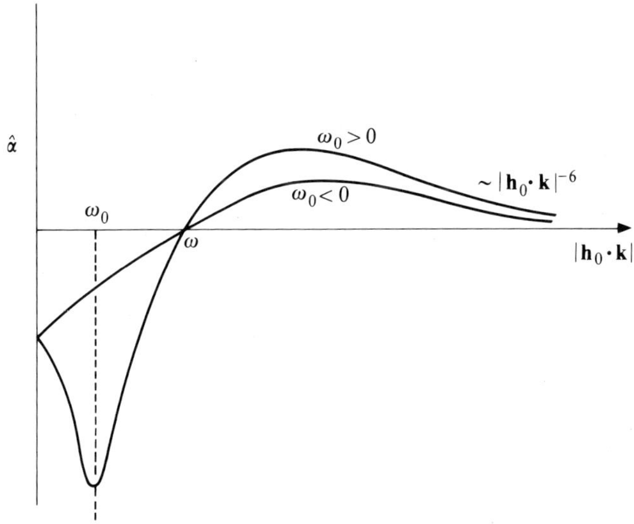

Fig. 10.3 Qualitative sketch of the behaviour of $\hat{\alpha} = \frac{1}{3}\hat{\alpha}_{ii}$ as a function of $|\mathbf{h}_0 . \mathbf{k}|$, as given by (10.53), with $\omega > 0$, $\mathbf{k} . \mathbf{\Omega} > 0$, for positive and negative values of the parameter $\omega_0 = \omega - 2\mathbf{k} . \mathbf{\Omega}/k$. In the former case, a resonant effect occurs at $|\mathbf{h}_0 . \mathbf{k}| = \omega$, the nature of the resonance (for small λ and ν) being given by (10.56).

A pseudo-tensor of the form (10.53) having only one non-zero principal value is not in itself adequate to provide a regenerative dynamo of α^2-type (see § 9.3); however it is easy to see that a superposition of forcing waves (at different values of $\mathbf{k}$ and ω) will give a pseudo-tensor α_{ij} which is just a sum (or integral) of contribution of the form (10.53):

$$\alpha_{ij} = \sum_{\mathbf{k},\omega} \quad \text{or} \quad \iint d^3\mathbf{k}\, d\omega\, \hat{\alpha}_{ij}(\mathbf{k}, \omega).$$

It is clear that this pseudo-tensor is in general non-zero if $|\hat{\mathbf{f}}(k_1, k_2, k_3, \omega)|^2 \neq |\hat{\mathbf{f}}(k_1, k_2, -k_3, \omega)|^2$ where $k_3 = (\mathbf{k} \,.\, \mathbf{\Omega})/k$, and this is just the condition that there should be a lack of symmetry in the average (or spectral) properties of the forcing field $\mathbf{F}(\mathbf{x}, t)$ with respect to any plane perpendicular to $\mathbf{\Omega}$.

Two broad conclusions can be drawn from the above analysis. First it is clear that if the condition

$$\omega > 2(\mathbf{k} \,.\, \mathbf{\Omega})/k \tag{10.57}$$

is satisfied for every $(\omega, \mathbf{k})$ in the spectrum of $\mathbf{F}(\mathbf{x}, t)$, and if a locally uniform field $\mathbf{h}_0$ grows from an initially weak level due to dynamo action, then the α-effect *will in general intensify* as the field grows towards the 'resonance level' (given by (10.55)) for any pair $(\omega, \mathbf{k})$. This intensification of the α-effect is associated with large amplitudes of the response $\hat{\mathbf{u}}(\mathbf{k}, \omega)$ at resonance. Rotation by itself keeps this response at a low level, and thus acts as a constraint on the motion. The growing magnetic field can, at an appropriate level, release this constraint and can, as it were, trigger large velocity fluctuations which make a correspondingly large contribution to $\hat{\alpha}_{ij}$. This may seem paradoxical. Analogous behaviour is however well-known in the context of the stability of hydrodynamic systems subject to the simultaneous action of Coriolis and Lorentz forces (see e.g. Chandrasekhar, 1961, chapter 5): whereas the effects of rotation and magnetic field are separately stabilising, the two effects can work against each other in such a way that a flow that is stable under the action of rotation alone becomes unstable when a magnetic field is also introduced. This type of behaviour has also been noted in the dynamo context by Busse (1976).

The second conclusion is perhaps less unexpected: no matter what the spectral properties of $\mathbf{F}(\mathbf{x}, t)$ may be, all ingredients of α_{ij} ultimately decrease as $|\mathbf{h}_0|$ increases and would tend to zero if $|\mathbf{h}_0|$ were (through some external agency) increased without limit[3]. Defining $\alpha = \frac{1}{3}\alpha_{ii}$, it is evident that $|\alpha|$ is certainly (asymptotically) a decreasing function of the local energy density $M = \frac{1}{2}h_0^2$ associated with the mean field. The precise functional dependence of $|\alpha|$ on M

[3] Wave-vectors $\mathbf{k}$ for which $\mathbf{h}_0 \,.\, \mathbf{k} \approx 0$ contribute singular behaviour in expressions such as (10.56) as $|\mathbf{h}_0| \to \infty$; such ingredients are however of little concern as they make a vanishingly small contribution to $\langle \mathbf{u}_1 \wedge \mathbf{h}_1 \rangle$ (see (10.46)).

depends in a complicated way on the spectral properties of $\mathbf{F}(\mathbf{x}, t)$: if $|\hat{\mathbf{f}}(\mathbf{k}, \omega)|^2$ is non-zero only when $\omega < 2\mathbf{k} \cdot \mathbf{\Omega}/k$, then the resonance phenomenon mentioned above does not occur, and evidently (from fig. 10.3)

$$|\alpha| \propto M^{-3} \quad \text{as} \quad M \to \infty. \tag{10.58}$$

If on the other hand a resonant response is possible, and if the Fourier amplitudes $\hat{\mathbf{f}}(\mathbf{k}, \omega)$ are isotropically distributed over the half-space $\mathbf{k} \cdot \mathbf{\Omega} > 0$, then in general a resonant response remains possible (for *some* $(\mathbf{k}, \omega)$) even when $\mathbf{h}_0$ becomes very strong (Moffatt, 1972); in this case $|\alpha|$ decreases less rapidly with M as $M \to \infty$; the detailed calculation in fact gives

$$|\alpha| \propto M^{-3/2} \quad \text{as} \quad M \to \infty. \tag{10.59}$$

What is important in either case is that if a field $\mathbf{h}_0$ grows on a large length-scale due to the α-effect then ultimately the intensity of the α-effect (as measured simply by $|\alpha|$) must decrease, and clearly an equilibrium situation (as in the case of the simple disc dynamo discussed in chapter 1) will be attained.

This same influence of a magnetic field in tending to reduce the α-effect has been discussed from different points of view by Vainshtein & Zel'dovich (1972) and by Pouquet, Frisch & Léorat (1976), in the turbulent (rather than the random wave) context. We shall defer consideration of this more complicated situation to the following chapter.

10.4. Dynamic equilibration due to reduction of α-effect

The α-effect associated with the type of waves discussed in §§ 10.2 and 10.3 will in general be anisotropic due to the influence of the preferred directions of both $\mathbf{\Omega}$ and $\mathbf{h}_0$. Nevertheless the qualitative nature of the process of equilibration (i.e. of limitation in the growth of magnetic energy) is most simply appreciated by supposing that $\alpha_{ij} = \alpha(M)\delta_{ij}$ where $M = \frac{1}{2}\mathbf{h}_0^2$ is the local energy density associated with the large-scale magnetic field. The arguments of the preceding paragraph guarantee that $|\alpha(M)| \to 0$ as $M \to \infty$; moreover $\alpha(0)$ is the value of α associated with pure inertial waves, unaffected by Lorentz forces.

Neglecting the influence of fluid boundaries, we know that the free modes of the α^2-dynamo (§ 9.2) have typical structure

$$\mathbf{h}_0 = h_{00}(t)(\sin Kz, \cos Kz, 0). \tag{10.60}$$

Such magnetic modes have uniform magnetic energy $M(t) = \frac{1}{2}h_{00}^2(t)$, and if attention is focussed on just one such mode (the mode of maximum growth rate on the kinematic theory being the natural one to choose), then α remains uniform in space as the mode intensifies. The equation for the evolution of $\mathbf{h}_0$ is then

$$\partial \mathbf{h}_0/\partial t = \alpha(M)\nabla \wedge \mathbf{h}_0 + \lambda \nabla^2 \mathbf{h}_0 = \alpha(M)K\mathbf{h}_0 - \lambda K^2 \mathbf{h}_0, \tag{10.61}$$

and so $M(t)$ evidently satisfies the equation

$$\mathrm{d}M/\mathrm{d}t = K\alpha(M)M - \lambda K^2 M. \tag{10.62}$$

Dynamo action occurs (as in § 9.2) provided

$$K\alpha(0) > \lambda K^2. \tag{10.63}$$

As M grows with time, $|\alpha(M)|$ may increase initially if resonant responses of the type discussed in § 10.3 play an important role; but ultimately $|\alpha(M)|$ must decrease and the magnetic energy must level off at a value M_c determined by

$$K\alpha(M_c) = \lambda K^2. \tag{10.64}$$

In the particular model of Moffatt (1972) (which takes account of the non-isotropy of α_{ij}) the equilibrium level M_c determined by the above type of argument is given by

$$M_c^{3/2} = \frac{al^2L}{\lambda}\left(\frac{\omega_0}{\Omega}\right)^{1/2}\langle \mathbf{F}^2 \rangle, \tag{10.65}$$

where l is a typical scale of the forcing field $\mathbf{F}(\mathbf{x}, t)$, ω_0 a typical frequency in its spectrum (assumed small compared with Ω), L is the scale of the large-scale field $\mathbf{h}_0$ (so that $L \gg l$), and a is a constant of order unity. The same model (with $\nu \gg \lambda$) gives an equilibrium kinetic energy density of the forced wave field

$$E_c = \frac{bl^3}{(\lambda\nu)^{1/2}M_c^{1/2}}\left(\frac{\omega_0}{\Omega}\right)\langle \mathbf{F}^2 \rangle, \tag{10.66}$$

where b is a second constant of order unity. Viscosity appears in this expression essentially because resonant forcing of waves for which $\mathbf{h}_0 \,.\, \mathbf{k} \approx 0$ generates waves whose amplitudes are controlled only by viscous dissipation – these waves make negligible contribution to the α-effect (see footnote on p. 256) but a large contribution to the kinetic energy density of the wave-field. The decrease of E_c with increasing M_c also deserves comment: this occurs essentially because of the general tendency of a strong magnetic field to inhibit response to forcing – in the conceptual limit $|\mathbf{h}_0| \to \infty$, the fluid acquires infinite rigidity and there is zero response to forcing of finite intensity.

The non-linear character of the relations (10.65) and (10.66) is also noteworthy. In an entirely linear theory, E_c would be proportional to $\langle \mathbf{F}^2 \rangle$; eliminating M_c from (10.65) and (10.66) gives, in contrast, $E_c \propto \langle \mathbf{F}^2 \rangle^{2/3}$.

The ratio of M_c to E_c is given by

$$\frac{M_c}{E_c} = C\left(\frac{\Omega}{\omega_0}\right)^{1/2}\left(\frac{\nu}{\lambda}\right)^{1/2}\frac{L}{l}, \tag{10.67}$$

where $C = a/b$. With $\omega_0 \ll \Omega$ and $\nu \ll \lambda$, the factors $(\Omega/\omega_0)^{1/2}$ and $(\nu/\lambda)^{1/2}$ tend to compensate each other. In any case it is evident that if L/l is sufficiently large (i.e. if the space available for the growth of magnetic modes is sufficient) then the magnetic energy density may be expected to rise to a level large compared with the kinetic energy density of any random wave motion. This conclusion is not model-dependent; indeed it is evident from (10.64) that M_c may generally be expected to be an increasing function of $L(\sim K^{-1})$, whereas E_c will generally be either independent of L or a decreasing function of L (due to the magnetic suppression effect as represented by a formula of type (10.66)).

An alternative means of injecting energy into the fluid system has been considered by Soward (1975). In Soward's model the fluid (supposed inviscid and highly conducting) is contained between two parallel boundaries $z = 0$, z_0, perpendicular to the rotation vector $\mathbf{\Omega}$; energy is injected by random mechanical excitation at $z = 0$, and is absorbed (without reflexion) at $z = z_0$. In the absence of a magnetic field, the mean helicity (averages being defined over planes $z = \text{cst.}$) is independent of z, and there is an associated non-isotropic α-effect, also independent of z. When z_0 is sufficiently large, this situation is unstable to the growth of magnetic field perturbations of the form (10.60). As soon as these reach a significant level, however, weak ohmic dissipation intervenes, leading to *spatial attenuation* of wave groups propagating from $z=0$ to $z=z_0$, and hence the α-effect decreases in intensity with increasing z. The equilibrium level of the magnetic field at any level is determined by the local value of the pseudo-tensor α_{ij}, and so the equilibrium magnetic energy density $M=\frac{1}{2}\langle \mathbf{h}_0^2\rangle$ is also a decreasing function of z. The scale characteristic of this spatial attenuation is found to be

$$L=lR_o^{-2}, \qquad R_o=U_0/l\omega_0, \tag{10.68}$$

where U_0 is the rms velocity on $z=0$, l is a length-scale characteristic of the excitation on $z=0$, and ω_0 a frequency characteristic of the waves excited (assumed small compared with Ω). The treatment is based on the assumption that the Rossby number R_o is small, so that, as in other two-scale approaches, $L \gg l$. Fig. 10.4 shows solutions obtained by Soward for the two horizontal mean field components in the equilibrium situation when $z_0/L=16.6$; a noteworthy feature of the solution is that the field rotates about the direction of $\mathbf{\Omega}$ with angular velocity approximately $0.2(R_o^4/Q)\Omega$ where $Q=\Omega l^2/\lambda\,(\gg 1)$.

The situation in the limit $z_0/L\to\infty$ (when fluid fills the half-space $z>0$) is rather curious. Soward argues that due to the decrease of the α-effect with height, the magnetic field must for $z \gtrsim L$ decay exponentially (the α-effect being inadequate to maintain it), and that a non-zero flux of wave energy must then propagate freely for $z \gg L$, 'while the strength of the α-effect remains constant'. It is difficult to accept this picture, because a non-zero α-effect always gives rise to magnetic instability when sufficient space is available, and in the situation envisaged an infinite half-space is available for the development of such instabilities. The only alternative is that the α-effect vanishes for large z, or equivalently that *all* the wave energy is dissipated in the layer $z=O(L)$ of field generation, and *none* survives to propagate to $z=\infty$.

Soward draws attention to a further complication that must in general be taken into account when slow evolution of a large-scale

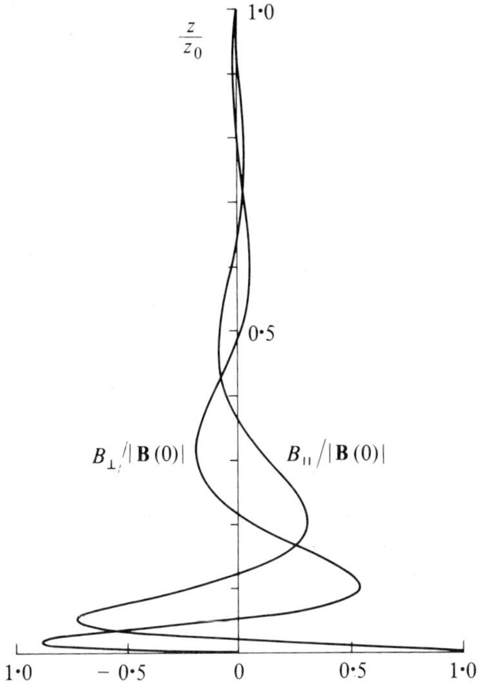

Fig. 10.4 Field components parallel and perpendicular to the field at the boundary $z = 0$ associated with random excitation on $z = 0$, the boundary $z = z_0$ being perfectly absorbing; the fluid rotates about the z-axis with angular velocity Ω, and the magnetic field rotates relative to the fluid with angular velocity $0{\cdot}2(R_o^4/Q)\Omega$ where R_o and Q are as defined in the text. (From Soward, 1975.)

mean magnetic field is considered. This is that, on the long time-scale associated with the evolution, non-linear interactions between the constituent background waves may also cause a systematic evolution of the wave spectra. This means that, no matter how small the perturbations $\mathbf{u}_1$ and $\mathbf{h}_1$ may be in equations (10.24) and (10.25), neglect of the non-linear terms may not be justified on the long time-scales associated with dynamo action caused by the mean electromotive force $\langle \mathbf{u}_1 \wedge \mathbf{h}_1 \rangle$, which is itself quadratic in small quantities.

An α that decreases with increasing $|\mathbf{B}|$ has been incorporated in a number of numerical studies of $\alpha\omega$-dynamos of the type discussed in § 9.12 (e.g. Braginskii, 1970; Stix, 1972; Jepps, 1975). Stix adopted a simple 'cut-off' formula

$$\alpha = \begin{cases} \alpha_0 & (|B| < B_c) \\ 0 & (|B| > B_c) \end{cases} \tag{10.69}$$

and studied the effect on the oscillatory dipole mode excited in a slab geometry (cf. § 9.9, but with $U = U(z)$). For the particular conditions adopted, the critical dynamo number for this mode was found to be $X_c = -89{\cdot}0$, and the period of its oscillation was $0{\cdot}993R^2/\lambda$, where R is the length-scale transverse to the slab (the same scale being used in the definition of X). When $X = aX_c$ with $a > 1$, linear kinematic theory gives an oscillatory mode of exponentially increasing amplitude. The cut-off effect (10.69) limits the amplitude to a value of order B_c, and also tends to lengthen the period of the oscillation: Stix found for example that when $a = 7$, the maximum field amplitude is between $2B_c$ and $3B_c$, and the period is about $1{\cdot}7R^2/\lambda$. Moreover the field variation with time, although periodic, is very far from sinusoidal: bursts of poloidal field are produced by the α-effect when $|B| < B_c$, and these are followed by less pronounced bursts of toroidal field (of intensity greater than B_c) due to the effect of shear. The resulting 'spikiness' of the field (as a function of time) was also noticed by Jepps (1975) who carried out similar computations (and for a range of cut-off functions) in a spherical geometry.

10.5. Helicity generation due to interaction of buoyancy and Coriolis forces

We have noted on several occasions that a lack of symmetry about planes perpendicular to the rotation vector $\mathbf{\Omega}$ is necessary to provide the essential lack of reflexional symmetry in random motions that can lead to an α-effect. As first pointed out by Steenbeck, Krause & Rädler (1966), this lack of symmetry is present when buoyancy forces $\rho'\mathbf{g}$ (with $\mathbf{g} \cdot \mathbf{\Omega} \neq 0$) act on fluid elements with density perturbation ρ' relative to the local undisturbed density ρ_0.

A simple and illuminating discussion of the mutual role of buoyancy and Coriolis forces in generating helicity has been recently given by Hide (1976). Suppose that conditions are geostrophic and that the Boussinesq approximation (in which account is taken of density fluctuations only in the buoyancy force term of the equation of motion, and not in the inertia term) is valid. Then the equation of motion reduces to

$$2\rho_0\mathbf{\Omega}\wedge \boldsymbol{u} = -\nabla p + \rho'\mathbf{g}, \tag{10.70}$$

wherein for the moment we neglect any Lorentz forces. The curl of this equation (with $\mathbf{g}\wedge\nabla\rho_0 = 0$) gives

$$2\rho_0\nabla\wedge(\mathbf{\Omega}\wedge\mathbf{u}) = -\mathbf{g}\wedge\nabla\rho', \tag{10.71}$$

and hence

$$\begin{aligned} 2\rho_0(\mathbf{\Omega}\wedge\mathbf{u}).\nabla\wedge(\mathbf{\Omega}\wedge\mathbf{u}) &= -(\mathbf{\Omega}\wedge\mathbf{u}).(\mathbf{g}\wedge\nabla\rho') \\ &= -\mathbf{\Omega}.\mathbf{u}\wedge(\mathbf{g}\wedge\nabla\rho'). \end{aligned} \tag{10.72}$$

Writing $\mathbf{u}_\perp = \mathbf{u} - (\mathbf{u}.\mathbf{\Omega})\mathbf{\Omega}/\Omega^2$ for the projection of $\mathbf{u}$ on planes perpendicular to $\mathbf{\Omega}$, it may readily be verified that (10.72) may be written in the form

$$2\rho_0\Omega^2(\mathbf{u}_\perp.\nabla\wedge\mathbf{u}_\perp) = -\mathbf{\Omega}.\mathbf{u}\wedge(\mathbf{g}\wedge\nabla\rho'), \tag{10.73}$$

or, taking the average over horizontal planes,

$$2\rho_0\Omega^2\langle\mathbf{u}_\perp.\nabla\wedge\mathbf{u}_\perp\rangle = -(\mathbf{\Omega}.\mathbf{g})\langle\mathbf{u}.\nabla\rho'\rangle + \mathbf{g}.\langle\mathbf{u}\nabla\rho'\rangle.\mathbf{\Omega}. \tag{10.74}$$

In the particular case where $\mathbf{g}$ is parallel to $\mathbf{\Omega}$, and if conditions are statistically homogeneous over horizontal planes, then, using

$$\nabla.\mathbf{u} = \nabla_\perp.\mathbf{u}_\perp + \partial w/\partial z = 0, \tag{10.75}$$

where w is the vertical component of $\mathbf{u}$, we have from (10.74)

$$2\rho_0\Omega^2\langle\mathbf{u}_\perp.\nabla\wedge\mathbf{u}_\perp\rangle = -(\mathbf{\Omega}.\mathbf{g})\langle\rho'\,\partial w/\partial z\rangle, \tag{10.76}$$

a formula that provides a direct relationship between the pseudo-scalar $\mathbf{\Omega}.\mathbf{g}$ and at least one ingredient of the total mean helicity $\langle\mathbf{u}.(\nabla\wedge\mathbf{u})\rangle$. The phase relationship between ρ' and $\partial w/\partial z$ is evidently important in determining the magnitude and sign of this ingredient.

With $\mathbf{u} = \mathbf{u}_\perp + \mathbf{u}_\parallel$, the total helicity, averaged over planes $z = \text{cst.}$, is given by

$$\begin{aligned} \langle\mathbf{u}.\nabla\wedge\mathbf{u}\rangle &= \langle\mathbf{u}_\perp.\nabla\wedge\mathbf{u}_\perp\rangle + \langle\mathbf{u}_\parallel.\nabla\wedge\mathbf{u}_\perp\rangle + \langle\mathbf{u}_\perp.\nabla\wedge\mathbf{u}_\parallel\rangle \\ &= \langle\mathbf{u}_\perp.\nabla\wedge\mathbf{u}_\perp\rangle + 2\langle\mathbf{u}_\parallel.\nabla\wedge\mathbf{u}_\perp\rangle, \end{aligned} \tag{10.77}$$

using homogeneity with respect to x and y. Hence

$$\langle \mathbf{u} \,.\, \nabla \wedge \mathbf{u} \rangle = \langle \mathbf{u}_\perp \,.\, \nabla \wedge \mathbf{u}_\perp \rangle + 2\langle w\omega_z \rangle, \tag{10.78}$$

where ω_z is the component of vorticity parallel to $\mathbf{\Omega}$. The ingredient $2\langle w\omega_z \rangle$ is not determined (in terms of ρ') by the above argument; Hide maintains that this ingredient should be negligible compared with $\langle \mathbf{u}_\perp \,.\, \nabla \wedge \mathbf{u}_\perp \rangle$ when Ω is sufficiently strong. However, order of magnitude estimates of the two contributions in (10.77) suggest that

$$\frac{\langle \mathbf{u}_\parallel \,.\, \nabla \wedge \mathbf{u}_\perp \rangle}{\langle \mathbf{u}_\perp \,.\, \nabla \wedge \mathbf{u}_\perp \rangle} = O\left(\frac{U_\parallel/L_\perp}{U_\perp/L_\parallel}\right) = O\left(\frac{L_\parallel^2}{L_\perp^2}\right), \tag{10.79}$$

where $U_\parallel$, $U_\perp$, $L_\parallel$ and $L_\perp$ are velocity- and length-scales parallel and perpendicular to Ω, related (via $\nabla \,.\, \mathbf{u} = 0$) by $U_\parallel/L_\parallel = O(U_\perp/L_\perp)$. Since $L_\parallel/L_\perp$ is generally large in a rotation dominated system, it seems quite possible that the second contribution to (10.78) may in fact dominate the first. Estimates such as (10.79) do however depend critically on the phase relations between the velocity components, and the matter can really only be settled within the framework of more specific models such as that treated in the following section.

10.6. Excitation of magnetostrophic waves by unstable stratification

The following idealised problem (fig. 10.5), which has been widely studied in different forms (Braginskii, 1964*c*, 1967, 1970; Eltayeb, 1972, 1975; Roberts & Stewartson, 1974, 1975), provides a basis for the detailed consideration of the effects of unstable density stratification. Suppose that fluid is contained between two horizontal planes $z = \pm z_0$, on which the temperature Θ is prescribed as $\Theta = \Theta_0 \mp \beta z_0$ respectively. In the undisturbed state $-\beta\,(<0)$ is then the vertical temperature gradient, and instabilities are to be expected if β is sufficiently large for the rate of release of potential energy associated with overturning of the 'top-heavy' fluid to exceed the rate of dissipation of energy due to viscous and/or ohmic diffusion. The equation for $\Theta(\mathbf{x}, t)$ is the heat conduction equation

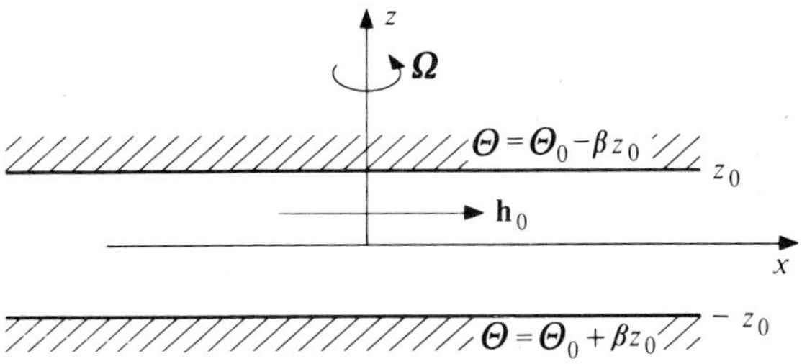

Fig. 10.5 Configuration considered in § 10.6. Fluid is contained between the planes $z = \pm z_0$, the lower plane being maintained at a higher temperature than the upper plane. The system rotates about the z-axis with angular velocity $\mathbf{\Omega}$, and a uniform magnetic field $(\mu_0\rho)^{1/2}\mathbf{h}_0$, with $\mathbf{h}_0$ parallel to the x-axis, is supposed present. The system is unstable when the applied temperature difference is sufficiently large.

(see § 3.3),

$$\partial\Theta/\partial t + \mathbf{u} \,.\, \nabla\Theta = \kappa\nabla^2\Theta, \tag{10.80}$$

where κ is the thermal diffusivity of the fluid, and writing $\Theta = \Theta_0 - \beta z + \theta(\mathbf{x}, t)$ where θ is a small perturbation, the linearised form of (10.80) (regarding $\mathbf{u} = (u, v, w)$ also as small) is

$$\partial\theta/\partial t - \beta w = \kappa\nabla^2\theta. \tag{10.81}$$

The perturbation θ induces a density perturbation $\rho' = \alpha\theta$ where α is the coefficient of thermal expansion, and the equation of motion, under the Boussinesq approximation, is

$$\partial\mathbf{u}/\partial t + \mathbf{u} \,.\, \nabla\mathbf{u} + 2\mathbf{\Omega} \wedge \mathbf{u} = -\nabla P + \mathbf{h} \,.\, \nabla\mathbf{h} + \alpha\theta\mathbf{g} + \nu\nabla^2\mathbf{u}, \tag{10.82}$$

where P includes a contribution due to the undisturbed density gradient. We shall restrict attention to the possibility of modes of instability having growth rate small compared with Ω; for these we have seen in § 10.2 that provided $\nu/\lambda \lesssim 1$, a legitimate approximation to (10.82) is the magnetostrophic equation

$$2\mathbf{\Omega} \wedge \mathbf{u} = -\nabla P + \mathbf{h} \,.\, \nabla\mathbf{h} + \alpha\theta\mathbf{g}. \tag{10.83}$$

We suppose that in the undisturbed situation the field $\mathbf{h}_0$ is uniform and horizontal. Putting $\mathbf{h} = \mathbf{h}_0 + \mathbf{h}_1$ where $|\mathbf{h}_1| \ll h_0$, the linearised form of (10.83) is

$$2\mathbf{\Omega}\wedge\mathbf{u}=-\nabla p+\mathbf{h}_0\,.\,\nabla\mathbf{h}_1+\alpha\theta\mathbf{g}, \tag{10.84}$$

where p is the perturbation in P, and the linearised induction equation still has the form

$$\partial\mathbf{h}_1/\partial t=\mathbf{h}_0\,.\,\nabla\mathbf{u}+\lambda\nabla^2\mathbf{h}_1. \tag{10.85}$$

Equations (10.81), (10.84) and (10.85) together with $\nabla\,.\,\mathbf{u}=\nabla\,.\,\mathbf{h}_1=0$ determine the evolution of small perturbations ($\mathbf{u}$, $\mathbf{h}_1$, θ, p). We must of course impose boundary conditions on $z=\pm z_0$; we have firstly that

$$\theta=0,\qquad u_z=0\quad\text{on}\quad z=\pm z_0. \tag{10.86}$$

Secondly, neglect of all viscous effects is consistent with the adoption of the 'stress-free' conditions

$$\partial u_x/\partial z=0,\qquad \partial u_y/\partial z=0\quad\text{on}\quad z=\pm z_0. \tag{10.87}$$

Finally, the simplest conditions on $\mathbf{h}_1$ result from the assumption that the regions $|z|>z_0$ are perfectly conducting and that both magnetic and electric fields are confined to the region $|z|<z_0$; it follows that (cf. 3.117 and 3.118)

$$h_{1z}=0,\qquad \partial h_{1x}/\partial z=0,\qquad \partial h_{1y}/\partial z=0\quad\text{on}\quad z=\pm z_0. \tag{10.88}$$

The above equations and boundary conditions admit solutions of the form

$$\begin{aligned}(u, v, h_{1x}, h_{1y}, p)&=(\hat{u}, \hat{v}, \hat{h}_{1x}, \hat{h}_{1y}, \hat{p})\cos nz\,.\,e^{i(lx+my-\omega t)},\\ (w, h_{1z}, \theta)&=(\hat{w}, \hat{h}_{1z}, \hat{\theta})\sin nz\,.\,e^{i(lx+my-\omega t)},\end{aligned} \tag{10.89}$$

where nz_0/π is an integer. Substituting in (10.81), (10.84) and (10.85), and eliminating $\hat{\mathbf{h}}_1$ gives, with $\mathbf{k}=(l, m, n)$,

$$-i\sigma\hat{\mathbf{u}}+2\mathbf{\Omega}\wedge\hat{\mathbf{u}}=-i(l, m, -in)\hat{p}+\alpha\hat{\theta}\mathbf{g}, \tag{10.90}$$

$$-i(\omega+\kappa k^2)\hat{\theta}-\beta\hat{w}=0, \tag{10.91}$$

and, from $\nabla\,.\,\mathbf{u}=0$, we have also

$$il\hat{u}+im\hat{v}+n\hat{w}=0. \tag{10.92}$$

Here, $\sigma = -(\omega + i\lambda k^2)(\mathbf{h}_0 \,.\, \mathbf{k})^2$ (cf. (10.30) in the magnetostrophic limit). We may solve (10.90) and (10.91) to give the velocity components in terms of $\hat{p}$ in the form

$$\frac{\hat{u}}{-l\sigma - 2i\Omega m} = \frac{\hat{v}}{-m\sigma + 2i\Omega l} = \frac{\hat{w}(\alpha\beta g + \sigma(\omega + i\kappa k^2))}{in(\omega + i\kappa k^2)(4\Omega^2 - \sigma^2)} = \frac{\hat{p}}{4\Omega^2 - \sigma^2}, \qquad (10.93)$$

and (10.92) then yields the dispersion relationship for ω, which may be simplified to the form

$$Y^2(-i\omega + \lambda k^2)^2(-i\omega + \kappa k^2) + (-i\omega + \kappa k^2) - Z(-i\omega + \lambda k^2) = 0, \qquad (10.94)$$

where

$$Y = 2n\Omega/h_0^2 m^2 k, \qquad Z = \alpha\beta g(l^2 + m^2)/m^2 k^2 h_0^2. \qquad (10.95)$$

We are interested in the possibility of unstable modes characterised by $\omega = \omega_r + i\omega_i$ with $\omega_i > 0$. If $\omega_r = 0$, these modes are non-oscillatory, while if $\omega_r \neq 0$ they have the character of propagating magnetostrophic waves of increasing amplitude. Suppose first that we neglect dissipative effects in (10.94), i.e. we put $\lambda = \kappa = 0$. This was the ideal situation to which Braginskii (1964*c*, 1967) restricted attention. The roots of the cubic (10.94) are then $\omega = 0$ and

$$\omega = \pm i(Z - 1)^{1/2}/Y, \qquad (10.96)$$

and the mode corresponding to the upper sign is clearly unstable whenever $\mathbf{k}$ is such that $Z > 1, 0 < Y < \infty$. Now in the geophysical or solar contexts, it is appropriate (as in § 9.6) to regard the y-direction as east, so that possible values of m are restricted by the requirement of periodicity in longitude to a discrete set Nm_1, ($N = 0, \pm 1, \pm 2, \ldots$). The 'axisymmetric' mode corresponding to $N = 0$ is of no interest since, for it, $Y = \infty$ and $\omega = 0$. The mode that is most prone to instability is that for which $N = \pm 1$, since (other things being equal) this gives the largest value of Z. This mode would correspond to $e^{\pm i\varphi}$ dependence on the azimuth angle φ in the corresponding spherical geometry. Note that the above description in terms of the magnetostrophic approximation can only detect low frequency modes ($|\omega| \ll \Omega$) and the approximation breaks down for modes not satisfying this condition.

Neglect of dissipative effects in this problem is a dangerous simplification, for reasons spelt out by Roberts & Stewartson (1974): if $g\alpha\beta/h_0^2$ exceeds m_1^2 by any amount, no matter how small, an infinite number of modes (corresponding to large values of l) apparently become unstable according to (10.96), since as $l \to \infty$, $Z \sim \alpha\beta g/h_0^2 m_1^2 > 1$; this result is however spurious, since $Y \to 0$ as $l \to \infty$ (m and n being fixed), and the condition $|\omega| \ll \Omega$ is not therefore satisfied when l is very large; i.e. the contribution of $\mathbf{Du}/\mathrm{D}t$ in the equation of motion undoubtedly becomes important when l is large. Also, more obviously, diffusion effects must become important for large wave-number disturbances, and presumably have a stabilising effect.

Weak diffusion also has the effect of shifting the degenerate root $\omega = 0$ of (10.94) away from the origin in the complex ω-plane. Indeed linearisation of (10.94) in the quantities λ, κ and ω (assumed small) gives for this root the expression

$$\omega = -\mathrm{i}\lambda k^2(Z-q)/(Z-1), \qquad q = \kappa/\lambda. \tag{10.97}$$

If

$$0 < q < Z < 1, \tag{10.98}$$

then the modes given by (10.96) are stable, whereas that given by (10.97) is clearly unstable with a slow growth rate determined by the weak diffusion effects. The manner in which the ratio of the small diffusivities λ and κ enters this criterion is noteworthy.

The behaviour of the roots of (10.94) for varying values of k, $q = \kappa/\lambda$, Y and Z was investigated by Eltayeb (1972), and the results have been summarised in § 3 of Roberts & Stewartson (1974). Of particular interest is the question of whether, for given values of q, Q and R_a where

$$Q = 2\Omega\lambda/h_0^2, \qquad R_a = g\alpha\beta z_0^2/\Omega\kappa, \tag{10.99}$$

the preferred mode of instability (i.e. that for which ω_i is maximum) is oscillatory ($\omega_r \neq 0$) or non-oscillatory ($\omega_r = 0$). Fig. 10.6 (from Roberts & Stewartson, 1974) shows the region of the (q, Q) plane in which oscillatory modes are preferred; in particular, non-oscillatory

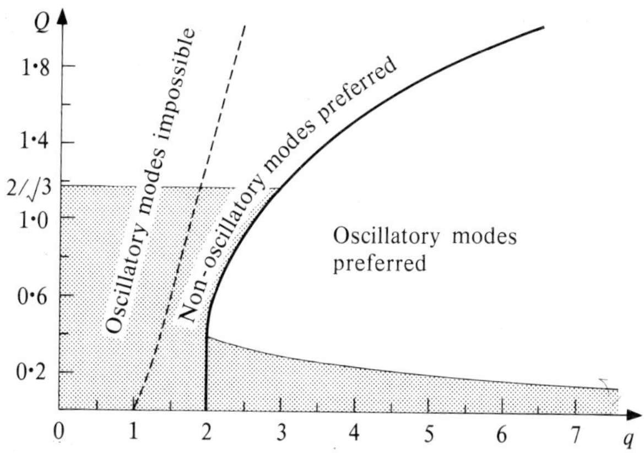

Fig. 10.6 Regions of the Q–q plane in which oscillatory modes are possible and preferred. In the shaded regions, the preferred modes are oblique to the applied field (i.e. $m \neq 0$ in (10.101)), while in the unshaded regions the preferred modes are transverse to $\mathbf{B}_0$ (i.e. $m = 0$). (From Roberts & Stewartson, 1974.)

modes are always preferred (when unstable) if

$$\text{either} \quad q < 2 \quad \text{or} \quad Q \gtrsim 3{\cdot}273, \tag{10.100}$$

and this result is independent of the value of R_a.

Let us now consider the helicity associated with unstable disturbances. With

$$\mathbf{u} = \text{Re}\,(\hat{u}\cos nz, \hat{v}\cos nz, \hat{w}\sin nz)\,e^{i(lx+my-\omega t)}, \tag{10.101}$$

it is straightforward to show that

$$\langle \mathbf{u}\,.\,\nabla \wedge \mathbf{u}\rangle = \text{Re}\,[\hat{w}^*(il\hat{v} - im\hat{u})]\cos nz \sin nz\, e^{2\omega_i t}, \tag{10.102}$$

the average being over horizontal planes. (Incidentally, the contribution $\langle \mathbf{u}_\perp\,.\,\nabla \wedge \mathbf{u}_\perp\rangle$ to the total helicity in the notation of § 10.5 is zero in this case!) Now if the disturbance is non-oscillatory[4], then ω and σ are both pure imaginary; putting

$$\omega = i\omega_i, \qquad \sigma = i\sigma_i, \qquad \sigma_i = (\omega_i + \lambda k^2)^{-1}(\mathbf{h}_0\,.\,\mathbf{k})^2, \tag{10.103}$$

[4] The statement in Moffatt (1976) that the helicity vanishes in this case is incorrect.

we have from (10.93) and (10.102)

$$\langle \mathbf{u} \, . \, \nabla \wedge \mathbf{u} \rangle = \frac{n\Omega(m^2+l^2)(\omega_i+\kappa k^2)|\hat{p}|^2}{(4\Omega^2+\sigma_i^2)[\alpha\beta g-\sigma_i(\omega_i+\kappa k^2)]} \sin 2nz \, e^{2\omega_i t}. \tag{10.104}$$

This is antisymmetric about the centre-plane $z=0$; in the case $nz_0=\pi$, the helicity is positive or negative in $0<z<z_0$ according as

$$\alpha\beta g(\omega_i+\lambda k^2) \gtrless (\mathbf{h}_0 \, . \, \mathbf{k})^2(\omega_i+\kappa k^2). \tag{10.105}$$

In the critically stable case ($\omega_i=0$), the crucial importance of the ratio $q=\kappa/\lambda$ is again apparent.

Further detailed consideration of this problem, and of variations involving different orientations of $\mathbf{\Omega}$ and $\mathbf{h}_0$ relative to the boundaries, and different boundary conditions, may be found in Eltayeb (1975).

10.7. Instability due to magnetic buoyancy

The concept of *magnetic buoyancy* was introduced by Parker (1955*a*) in a discussion of the process of formation of sunspots by instabilities of a subsurface solar toroidal magnetic field. Compressibility is an essential element in this type of instability, which is closely related to the instability that occurs in a stratified atmosphere when the (negative) temperature gradient is super-adiabatic.

Suppose that in an equilibrium situation (fig. 10.7) with gravity $\mathbf{g}=(0,0,-g)$, we have a density distribution $\rho_0(z)$ and a magnetic field distribution $\mathbf{B}_0=(0, B_0(z), 0)$. The equation of magnetostatic equilibrium is

$$0=-\frac{d}{dz}\left(p_0(z)+\frac{1}{2\mu_0}B_0^2\right)-\rho_0(z)g, \tag{10.106}$$

where $p_0(z)$ is the pressure distribution, which may be supposed a monotonic increasing function $f(\rho_0)$ of density $\rho_0(z)$ (in the particular case of an isothermal 'atmosphere', $p_0(z)=c^2\rho_0(z)$ where c^2 is constant).

Suppose now that a flux-tube of cross-sectional area A_1 at height $z=z_1$, with $B_0(z_1)=B_1$, $\rho(z_1)=\rho_1$, is displaced upwards to the level $z=z_2$, and that ohmic diffusion may be neglected. Then A_1, B_1 and ρ_1 will change to, say, $\tilde{A}_1$, $\tilde{B}_1$, $\tilde{\rho}_1$, and conservation of mass and of

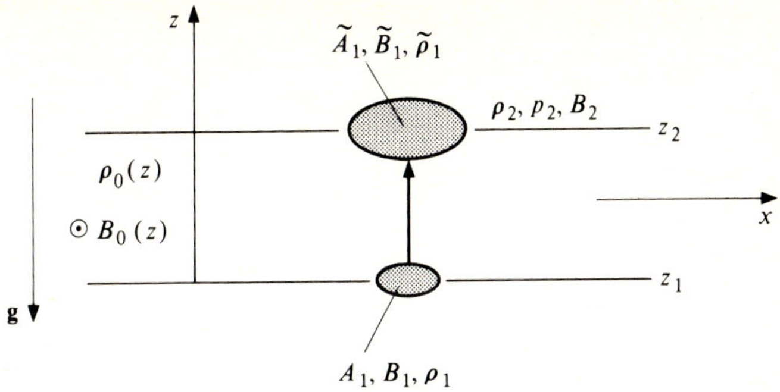

Fig. 10.7 Sketch illustrating the simplest instability due to magnetic buoyancy. The shaded flux-tube is displaced from the level $z = z_1$ to the level $z = z_2$; it will continue to rise if its new density $\tilde{\rho}_1$ is less than the ambient density ρ_2.

magnetic flux implies that

$$\rho_1 A_1 = \tilde{\rho}_1 \tilde{A}_1, \qquad B_1 A_1 = \tilde{B}_1 \tilde{A}_1, \tag{10.107}$$

and in particular

$$\tilde{B}_1/\tilde{\rho}_1 = B_1/\rho_1. \tag{10.108}$$

During the displacement of the tube, it will tend to expand (if $d\rho_1/dz < 0$) in such a way as to maintain pressure equilibrium with its new environment, and if the displacement is sufficiently slow, we may suppose that this equilibrium is maintained, i.e.

$$\tilde{p}_1 + \frac{1}{2\mu_0}\tilde{B}_1^2 = p_2 + \frac{1}{2\mu_0}B_2^2, \tag{10.109}$$

where

$$\tilde{p}_1 = f(\tilde{\rho}_1), \qquad p_2 = p(z_2), \qquad B_2 = B_0(z_2).$$

The tube will be in equilibrium in its new environment only if $\tilde{\rho}_1 = \rho_2$: clearly if $\tilde{\rho}_1 < \rho_2$, it will continue to rise, while if $\tilde{\rho}_1 > \rho_2$ it will tend to return to its original level. The neutral stability condition $\tilde{\rho}_1 = \rho_2$ implies $\tilde{p}_1 = p_2$, and so from (10.109), $\tilde{B}_1 = B_2$. Hence, under neutral stability conditions, (10.108) becomes

$$B_1/\rho_1 = B_2/\rho_2, \tag{10.110}$$

or since this must hold for every pair of levels (z_1, z_2), neutral stability requires that $\mathrm{d}(B_0/\rho_0)/\mathrm{d}z = 0$, and clearly the atmosphere is stable or unstable to the type of perturbations considered according as

$$\frac{\mathrm{d}}{\mathrm{d}z}\left(\frac{B_0}{\rho_0}\right) > \quad \text{or} \quad < 0. \tag{10.111}$$

Defining scale-heights

$$L_B = -B_0/(\mathrm{d}B_0/\mathrm{d}z), \qquad L_\rho = -\rho_0/(\mathrm{d}\rho_0/\mathrm{d}z), \tag{10.112}$$

the condition for instability may equally be written

$$L_B < L_\rho. \tag{10.113}$$

The above simple argument does not take any account of possible distortion of the field lines, and merely shows that the medium is prone to instabilities when the magnetic field strength decreases sufficiently rapidly with height. Distortion of the field lines is however of crucial importance in the problem of sunspot formation (§ 5.3) and it is essential to consider perturbations which do distort the field lines, and which are affected by Coriolis forces when account is taken of rotation. Different instability models have been studied by Gilman (1970) and by Parker (1971*f*, 1975). Parker studies instability modes for which the pressure perturbation δp may be neglected, i.e. modes whose time-scales are *short* compared with the time-scale associated with the passage of acoustic waves through the system. Gilman by contrast supposes that the thermal conductivity (due to radiative or other effects) is very large, and restricts attention to slow instability modes in which pressure and density remain in isothermal balance ($\delta p = c^2 \delta\rho$) and the perturbation in total pressure (fluid plus magnetic) is negligible (as in the qualitative discussion above). We shall here follow the analysis of Gilman, including the effects of rotation, but making the additional simplifying assumption that the instability growth rates are sufficiently small for the magnetostrophic approximation (neglect of $\mathrm{D}\mathbf{u}/\mathrm{D}t$ in the equation of motion) to be legitimate. This greatly simplifies the dispersion relationship, and allows some simple conclusions to be drawn.

The Gilman model

The equations describing small magnetostrophic perturbations about the equilibrium state of fig. 10.7 are

$$\left.\begin{aligned} 2\rho_0\boldsymbol{\Omega}\wedge\boldsymbol{u} &= -\nabla\chi+\mu_0^{-1}\mathbf{B}_0\,.\,\nabla\mathbf{b}+\mu_0^{-1}\mathbf{b}\,.\,\nabla\mathbf{B}_0-\rho g\mathbf{i}_z,\\ \partial\mathbf{b}/\partial t &= -\mathbf{u}\,.\,\nabla\mathbf{B}_0+\mathbf{B}_0\,.\,\nabla\mathbf{u}-\mathbf{B}_0(\nabla\,.\,\mathbf{u}),\\ \partial\rho/\partial t &= -\mathbf{u}\,.\,\nabla\rho_0-\rho_0\nabla\,.\,\mathbf{u},\\ \nabla\,.\,\mathbf{b}=0, &\qquad \chi=p+\mu_0^{-1}\mathbf{B}_0\,.\,\mathbf{b}, \end{aligned}\right\} \tag{10.114}$$

and, following Gilman (1970), we assume that the pressure and density perturbations p and ρ are related by

$$p=c^2\rho, \tag{10.115}$$

where c is the isothermal speed of sound. This is the situation if the effective diffusivity of heat (due to radiative transfer) is very large. In (10.114), viscous and ohmic diffusion effects are neglected, and also $\partial\mathbf{u}/\partial t$ is omitted[5], so that we focus attention on modes with growth rate small compared with Ω. Note also that the centrifugal contribution to χ is neglected, on the reasonable assumption (in the solar context) that gravitational acceleration greatly exceeds centrifugal acceleration. If *Ox*, *Oy*, *Oz* are interpreted as south, east and vertically upwards with origin *O* in the convective zone at colatitude θ, then the components of $\boldsymbol{\Omega}$ are given by

$$\boldsymbol{\Omega}=(-\Omega\sin\theta,\,0,\,\Omega\cos\theta). \tag{10.116}$$

Equations (10.114), (10.115) admit solutions proportional to $e^{i(lx+my-\omega t)}$. The analysis is greatly simplified in focussing attention on disturbances for which $|l|\gg|m|$; such disturbances are particularly relevant in the sunspot context – the longitudinal separation of the two members of a typical sunspot pair gives a measure of the spatial variation in the y-direction; variation in the x-direction is unconstrained by the magnetic field $\mathbf{B}_0$, and can have length-scales small compared with the longitudinal separation scale. If, formally,

[5] In this respect we depart from the treatment of Gilman (1970) who obtained a more general dispersion relationship than (10.120) describing both fast and slow modes of instability. When $\Omega=0$, Gilman found modes of instability whenever $|B_0(z)|$ decreases with height. Such modes are not helical in character, although there is little doubt that they become so when perturbed by rotation effects.

we let $l \to \infty$ in the Fourier transform of (10.114), the x-component of the equation of motion degenerates to $\hat{\chi} = 0$. Consequently, in modes for which $|l| \gg |m|$, we have equivalently

$$\chi = p + \mu_0^{-1}\mathbf{B}_0 \,.\, \mathbf{b} \approx 0. \qquad (10.117)$$

With $\mathbf{u} = (u, v, w)$, $\mathbf{b} = (b_x, b_y, b_z)$, we may then easily obtain from (10.114)–(10.117) three linear equations relating v, w and b_y, viz.

$$\begin{pmatrix} 0 & \omega q - mh_0^2K_B & -imh_0^2c^2 \\ \omega q & -im^2h_0^2 & gh_0^2 \\ -im & K_\rho - K_B & -i(c^2+h_0^2) \end{pmatrix} \begin{pmatrix} v \\ w \\ \omega b_y/B_0c^2 \end{pmatrix} = 0, \qquad (10.118)$$

where[6] $q = 2\Omega \sin\theta$, $h_0 = (\mu_0\rho_0)^{-1/2}B_0$ is the local Alfvén speed and

$$K_\rho = -\rho_0'/\rho_0 = L_\rho^{-1}, \qquad K_B = -B_0'/B_0 = L_B^{-1}. \qquad (10.119)$$

The vanishing of the determinant of coefficients in (10.118) gives a dispersion relationship in the form

$$\omega^2q^2(h_0^2+c^2) - \omega qmh_0^2(g + h_0^2K_B + c^2K_\rho) + m^2h_0^4(gK_B - m^2c^2) = 0. \qquad (10.120)$$

If the roots of this equation are complex, then we have instability; the condition for this is evidently

$$(g + h_0^2K_B + c^2K_\rho)^2 < 4(h_0^2 + c^2)(gK_B - m^2c^2). \qquad (10.121)$$

Since the equilibrium condition (10.106) may (in the case of an isothermal atmosphere with $c^2 = \text{cst.}$) be written

$$g = h_0^2K_B + c^2K_\rho, \qquad (10.122)$$

the conclusion from (10.121) is that the medium is unstable to perturbations whose wave-number in the y (or azimuth) direction satisfies

$$m < m_c = (g(K_B - K_\rho)(h_0^2 + c^2)^{-1})^{1/2}. \qquad (10.123)$$

As anticipated in the introductory comments, a necessary condition for this type of instability is $K_B > K_\rho$ (or equivalently $L_B < L_\rho$); when this condition is satisfied, (10.123) gives the minimum scale

[6] Note that only the component of $\mathbf{\Omega}$ perpendicular to $\mathbf{g}$ influences the perturbations when $|l| \gg |m|$.

$2\pi/m_c$ on which perturbations will grow. It is interesting that this scale does not depend on the rotation parameter q (although the growth rates given by (10.120) are proportional to q^{-1}).

With $\omega=\omega_r+i\omega_i$, when $m<m_c$, the unstable mode ($\omega_i>0$) is given from (10.120)–(10.123) by

$$\omega_r q = gmh_0^2(h_0^2+c^2)^{-1} \tag{10.124}$$

$$\omega_i q = h_0^2(h_0^2+c^2)^{-1/2}|m|c\,(m_c^2-m^2)^{1/2}. \tag{10.125}$$

Since $\omega_r m>0$, these instability waves propagate in the positive y-direction, i.e. towards the east.

The helicity distribution $\langle \mathbf{u}\,.\,\nabla\wedge\mathbf{u}\rangle$ associated with this type of instability is given (for large l) by

$$\mathcal{H}\sim\tfrac{1}{2}\,\mathrm{Re}\;il(vw^*-v^*w), \tag{10.126}$$

the ratio of v to w being given by (10.118). A straightforward calculation using (10.125) gives[7]

$$v/w=-(h_0^2+c^2)^{1/2}(m_c^2-m^2)^{1/2}/c|m|,$$

and since this is real, the leading order contribution (10.126) is in fact zero. Terms of order m/l would have to be retained in the analysis throughout in order to obtain the correct expression for $\mathcal{H}$. We can however directly derive an α-effect associated with the type of instability considered. Writing

$$\begin{aligned}(\mathbf{u}\wedge\mathbf{b})_y &= \tfrac{1}{2}\mathrm{Re}\,(ub_z^*-wb_x^*)\\ &= -(m/2l)\,\mathrm{Re}\,(vb_z^*-wb_y^*)=\alpha B_0,\end{aligned}$$

the coefficient α may be obtained in the form

$$\alpha=-\frac{m|m|(m_c^2-m^2)^{1/2}h_0^2[(h_0^2-c^2)K_B+c^2K_\rho]|w|^2}{2lqc\,(h_0^2+c^2)^{1/2}}. \tag{10.127}$$

The fact that this is in general non-zero is of course an indication of the lack of reflexional symmetry of the motion.

[7] I am indebted to Dr David Acheson and Mr M. Gibbons who pointed out errors in the first draft of this section, and who proved in particular that the helicity vanishes at leading order.

10.8. Helicity generation due to flow over a bumpy surface

As a final example of a mechanism whereby helicity (and an associated α-effect) may be generated, consider the problem depicted in fig. 10.8: an insulating solid is separated from a conduct-

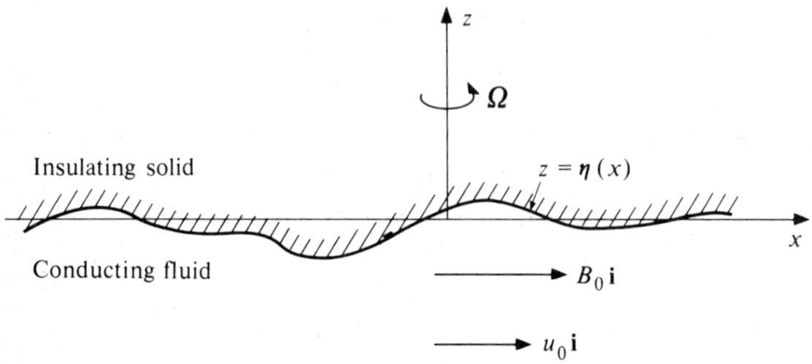

Fig. 10.8 Sketch of the configuration considered in § 10.8. This provides a crude model for the generation of magnetic fluctuations near the Earth's core–mantle boundary.

ing fluid by the 'bumpy' boundary $z=\eta(x)$, and as in § 10.6 we suppose that $\mathbf{\Omega}=(0, 0, \Omega)$, $\mathbf{B}_0=(B_0, 0, 0)$. Moreover we suppose that the fluid flows over the bumps, the velocity tending to the uniform value $(U_0, 0, 0)$ as $z\to-\infty$. This may be regarded as a crude model of flow near the core–mantle boundary in the terrestrial context, crude because (i) the spherical geometry is replaced by a Cartesian 'equivalent', (ii) the tangential magnetic field in the terrestrial context falls to a near zero value at the core–mantle boundary because of the low conductivity of the mantle, whereas here we suppose the field to be uniform, and (iii) the bumps are regarded as two-dimensional whereas in reality they are surely three-dimensional. The model in this crude form has been considered from different points of view by Anufriyev & Braginskii (1975) and by Moffatt & Dillon (1976). These studies are relevant (i) to the tangential stress (or equivalently angular momentum) transmitted from core to mantle and (ii) to the observed correlation between fluctuations in the Earth's gravitational and magnetic fields over its surface (Hide, 1970; Hide & Malin, 1970). Here we focus attention simply on the structure of the motion induced in the liquid and the associated helicity distribution, which is the most relevant aspect as far as large-scale dynamo effects are concerned.

Neglecting effects associated with variable density, and supposing that $|\eta'(x)| \ll 1$ so that all perturbations are small, the governing equations for the steady perturbations generated are (10.24) and (10.25) (with $\partial \mathbf{u}_1/\partial t = \partial \mathbf{h}_1/\partial t = 0$) together with $\nabla \cdot \mathbf{u}_1 = \nabla \cdot \mathbf{h}_1 = 0$. We shall also make the magnetostrophic approximation, which in this situation involves neglect of the convective acceleration term $\mathbf{u}_0 \cdot \nabla \mathbf{u}_1$ and of the viscous term $\nu \nabla^2 \mathbf{u}_1$ in (10.24). (The viscous term leads to a thin Ekman layer on the surface, and an associated small perturbation of the effective boundary condition (10.134*a*) below.) These equations admit solutions of the form

$$\mathbf{u}_1, \mathbf{h}_1, p_1 = 2 \operatorname{Re} \int_0^\infty (\hat{\mathbf{u}}, \hat{\mathbf{h}}, \hat{p})\, e^{i\mathbf{m}.\mathbf{x}}\, dk, \qquad (10.128)$$

where $m = k(1, 0, \gamma)$ and possible values of γ are to be determined; these must satisfy $\operatorname{Im} \gamma < 0$ since the perturbations must vanish as $z \to -\infty$. Substitution in the equations and straightforward elimination of the amplitudes $\hat{\mathbf{u}}, \hat{\mathbf{h}}, \hat{p}$ leads to a cubic equation for γ^2:

$$(1+\gamma^2)+\gamma^2[Q(1+\gamma^2)+2iA\kappa^{-1}]^2 = 0, \qquad (10.129)$$

where

$$Q = 2\Omega\lambda/h_0^2, \qquad A = U_0 h_0, \qquad \kappa = h_0 k/\Omega. \qquad (10.130)$$

Moreover, if the three solutions of (10.129) satisfying $\operatorname{Im} \gamma < 0$ are denoted by γ_n $(n = 1, 2, 3)$, then the ratios of the velocity and magnetic components are given in the corresponding modes by

$$\hat{\mathbf{u}}_n = a_n(k)(1, -\sigma_n^{-1}, -\gamma_n^{-1}), \qquad \hat{\mathbf{h}}_n = (i\sigma_n/\kappa)\hat{\mathbf{u}}_n, \qquad (10.131)$$

where

$$\sigma_n^2 = -4\gamma_n^2(1+\gamma_n^2)^{-1}. \qquad (10.132)$$

In the insulating region, the field $\mathbf{B}$ is harmonic, i.e. $\mathbf{B}-\mathbf{B}_0 = (\mu_0\rho_m)^{1/2}\nabla\Psi$ say, where ρ_m is the density of the solid ('mantle') region, and Ψ, being a potential function, has Fourier transform given by

$$\Psi(x, y) = 2 \operatorname{Re} \int_0^\infty \hat{\psi}_0(k)\, e^{-kz}\, e^{ikx}\, dk. \qquad (10.133)$$

The amplitudes $a_n(k)$, $\hat{\psi}_0(k)$ may be obtained in terms of the Fourier transform $\hat{\eta}(k)$ of $\eta(x)$ by applying the linearised boundary conditions

$$u_z = u_0 \partial\eta/\partial x, \qquad \mathbf{h} = -(\rho_m/\rho)^{1/2}\nabla\Psi \quad \text{on} \quad z = 0. \tag{10.134}$$

The first of these expresses the fact that the normal velocity on $z = \eta(x)$ is zero; the second expresses continuity of both normal and tangential components of $\mathbf{B}$.

The nature of the cubic equation (10.129) makes the detailed subsequent analysis rather complicated. In the terrestrial context however both Q and A are small, and asymptotic methods may be adopted. If $Q \ll 1$ and $A\kappa^{-1} \ll Q$, the three relevant roots of (10.129) are, to leading order,

$$\gamma_1 \sim -\mathrm{i}, \qquad \gamma_2 \sim -\tfrac{1}{2}(1+\mathrm{i})Q^{-1/2}, \qquad \gamma_3 \sim \tfrac{1}{2}(1-\mathrm{i})Q^{-1/2}. \tag{10.135}$$

This means that the mode corresponding to $n = 1$ has spatial dependence e^{kz+ikx}; the corresponding velocity and magnetic perturbations are irrotational (to leading order) and penetrate a distance $O(k^{-1})$ into the fluid. The helicity associated with this mode is zero. By contrast the modes corresponding to $n = 2$ and 3 have a boundary layer character, penetrating a distance $\delta_B = O(k^{-1}Q^{1/2})$ into the fluid. Moreover these modes are strongly helical: the helicity $\mathcal{H}_n$ ($n = 2$ or 3) associated with either mode is given (Moffatt & Dillon, 1976) by

$$\mathcal{H}_n(k) = \mathrm{Re}\,(\hat{\mathbf{u}}_n^* \,.\, \mathrm{i}\mathbf{k} \wedge \hat{\mathbf{u}}_n) = 2k|\hat{\mathbf{u}}_n^2|\,\mathrm{Re}\,(\mathrm{i}\gamma_n/\sigma_n) = \tfrac{1}{2}k|\hat{\mathbf{u}}_n|^2 Q^{-1/2}. \tag{10.136}$$

Since $k > 0$ in the representation (10.128) adopted, this helicity is positive for both modes, a fact that could be anticipated from the arguments of § 10.2: there is clearly a *downward* flux of energy from the boundary into the fluid and the associated helicity is therefore positive. The apparent contradiction between (10.136) and the result $|F(k)| \leqslant 2kE(k)$ bounding the helicity spectrum (§ 7.6) is accounted for by the fact that here we are dealing with a strongly anisotropic situation with severe attenuation of modes in the z-direction, and the argument leading to (7.55) simply does not apply.

The amplitude $|\hat{\mathbf{u}}_n|^2$ in (10.136) is proportional to $|\hat{\eta}|^2$ and also decays as $e^{kz/Q}$ as $z \to -\infty$.

In this context, Andrews & Hide (1975) have demonstrated that, in the non-dissipative limit ($\nu = \lambda = \kappa = 0$), free wave motions subject to the influence of buoyancy forces, Coriolis forces and Lorentz forces can in a similar way be trapped against a solid plane boundary provided the boundary is inclined to the horizontal and provided the basic field $\mathbf{B}_0$ is parallel to the wall; the trapped modes cease to exist if $\mathbf{B}_0$ has a component perpendicular to the wall. In the study described above, a component of $\mathbf{B}_0$ perpendicular to the wall may likewise be expected to exert a strong influence on the structure of the three modes.

CHAPTER 11

TURBULENCE WITH HELICITY AND ASSOCIATED DYNAMO ACTION

11.1. Effects of helicity on homogeneous turbulence

We have seen in chapters 7 and 8 that a lack of reflexional symmetry in a random 'background' velocity field $\mathbf{u}(\mathbf{x}, t)$, and in particular a non-zero value of the mean helicity $\langle \mathbf{u} \cdot \nabla \wedge \mathbf{u} \rangle$, is likely to be a crucial factor as far as the effect on large-scale magnetic field evolution is concerned. In these circumstances, it is appropriate to consider the general nature of the dynamics of a turbulent velocity field endowed with non-zero mean helicity. First it must be stated that turbulence exhibiting such a lack of reflexional symmetry has seldom been submitted to direct experimental investigation in the laboratory. Nearly all traditional studies of turbulence (e.g. grid turbulence, boundary layer turbulence, turbulence in wakes and jets, channel and pipe turbulence, etc.) have been undertaken in conditions that guarantee reflexional symmetry of the turbulence statistics. In order to study turbulence with helicity it is necessary to deliberately inject lack of reflexional symmetry through appropriate control of the source of energy for the flow. The natural way to do this, as indicated by the analysis of § 10.2, is to generate turbulence in a rotating fluid by some means that distinguishes between the directions $\pm\mathbf{\Omega}$, where $\mathbf{\Omega}$ is the rotation vector. For example, if a grid is rapidly drawn through a rotating fluid in the direction of $\mathbf{\Omega}$, the resulting random velocity field may be expected to lack reflexional symmetry. This situation has been realised in the laboratory by Ibbetson & Tritton (1975), who measured the decay of the mean square of the three velocity components; no technique however has as yet been realised for direct measurement of mean helicity[1].

[1] A possible method for making such a measurement is proposed by Léorat (1975, Appendix A3).

Energy cascade in non-helical turbulence

Before considering the effects of helicity on the dynamics of turbulence, let us first recall the essential features of high Reynolds number turbulence, as conceived by Kolmogorov (1941) and as expounded by Batchelor (1953). Consider a statistically steady state in which kinetic energy is generated on a length-scale l_0 at a rate ε per unit mass. Let $u_0 = \langle \mathbf{u}^2 \rangle^{1/2}$ and suppose that the Reynolds number $R_e = u_0 l_0 / \nu$ is large. Due to dynamical instability, the energy then 'cascades' through a sequence of decreasing length scales $l_n \ll l_{n-1}$ $(n = 1, 2, \ldots)$ until it reaches a scale l_ν say at which viscous dissipation is adequate to dissipate the energy at the rate ε. This scale is determined in order of magnitude by ε and ν and is therefore given (dimensionally) by

$$l_\nu \sim (\nu^3/\varepsilon)^{1/4}. \tag{11.1}$$

Moreover, on this picture, the energy level $\frac{1}{2}u_0^2$ depends on the rate of supply of energy ε to the system and on l_0, but not on ν, so that again on dimensional grounds

$$u_0^2 \sim (\varepsilon l_0)^{2/3}. \tag{11.2}$$

Eliminating ε between (11.1) and (11.2) gives

$$l_\nu \sim l_0 R_e^{-3/4}. \tag{11.3}$$

Equally, we may define characteristic wave-numbers $k_0 = l_0^{-1}$, $k_\nu = l_\nu^{-1}$, with

$$k_\nu \sim k_0 R_e^{3/4}. \tag{11.4}$$

In the range of wave-numbers $k_0 \ll k \ll k_\nu$ (described as the 'inertial range'), the energy spectrum tensor $\Phi_{ij}(\mathbf{k})$ (see § 7.8) is statistically decoupled from the energy source (which is confined to wave-numbers of order k_0), and may therefore be expected to be isotropic and to be determined solely by the parameter ε which represents the rate of flow of energy across any wave-number magnitude k in the inertial range. In the absence of any helicity effect, $\Phi_{ij}(\mathbf{k})$ is then given by

$$\Phi_{ij}(\mathbf{k}) = \frac{E(k)}{4\pi k^4}(k^2\delta_{ij} - k_i k_j), \tag{11.5}$$

and the energy spectrum function $E(k)$ is given, again on dimensional grounds, by

$$E(k)=C\varepsilon^{2/3}k^{-5/3} \quad (k_0 \ll k \ll k_\nu), \tag{11.6}$$

where C is a dimensionless constant of order unity. The associated vorticity spectrum function is given by

$$\Omega(k)=k^2E(k)=C\varepsilon^{2/3}k^{1/3} \quad (k_0 \ll k \ll k_\nu). \tag{11.7}$$

For $k \gtrsim k_\nu$, both $E(k)$ and $\Omega(k)$ experience a rapid (quasi-exponential) cut-off due to viscous dissipation. The mean-square vorticity $\langle \boldsymbol{\omega}^2 \rangle = \int_0^\infty \Omega(k)\,dk$ is clearly dominated by contributions from the neighbourhood of the viscous cut-off wave-number $k = k_\nu$, and increases as ν decreases; in fact, since the rate of dissipation of energy is $2\nu\langle \boldsymbol{\omega}^2 \rangle$, we have the exact result

$$\langle \boldsymbol{\omega}^2 \rangle = \varepsilon/2\nu. \tag{11.8}$$

From (11.2) and (11.8)

$$\langle \boldsymbol{\omega}^2 \rangle \sim R_e(u_0/l_0)^2, \tag{11.9}$$

or equivalently

$$l_c = u_0/\langle \boldsymbol{\omega}^2 \rangle^{1/2} \sim R_e^{-1/2} l_0. \tag{11.10}$$

Here l_c is an intermediate length-scale satisfying

$$l_0 \gg l_c \gg l_\nu. \tag{11.11}$$

One interpretation of the length l_c (the Taylor micro-scale) is that it provides a measure of the mean radius of curvature of an instantaneous streamline of the flow.

Effect of helicity on energy cascade

Suppose now that the source of energy on the scale l_0 is such as to impart non-zero helicity $\langle \mathbf{u} \cdot \boldsymbol{\omega} \rangle$ to the velocity field that is generated. A 'thought experiment' incorporating this behaviour has been described in § 7.6. In such a situation, we may talk of 'helicity injection' as well as 'energy injection' at wave-numbers of order k_0 (Brissaud *et al.*, 1973). However the level of helicity generated is limited by a Schwarz-type inequality: defining

$$\mathcal{H}_0 = \langle \mathbf{u} \cdot \boldsymbol{\omega} \rangle / \langle \mathbf{u}^2 \rangle^{1/2} \langle \boldsymbol{\omega}^2 \rangle^{1/2}, \tag{11.12}$$

we must clearly have

$$|\mathscr{H}_0| \leqslant 1. \tag{11.13}$$

We have already commented in § 3.2 on the fact that the total helicity of a localised disturbance in an inviscid fluid is a conserved quantity (like its total energy). In homogeneous turbulence, a similar result holds (as recognised by Betchov, 1961): the helicity density $\mathbf{u} \cdot \boldsymbol{\omega}$ satisfies the equation (cf. (3.7))

$$\frac{\mathrm{D}}{\mathrm{D}t}(\mathbf{u} \cdot \boldsymbol{\omega}) = -\nabla \cdot (\boldsymbol{\omega} p) \tag{11.14}$$

when $\nu = 0$, so that, taking a spatial average and using homogeneity, we have immediately

$$\mathrm{d}\langle \mathbf{u} \cdot \boldsymbol{\omega} \rangle / \mathrm{d}t = 0. \tag{11.15}$$

The helicity spectrum function $F(k)$, defined (cf. (7.52)) by

$$F(k) = \mathrm{i} \int_{S_k} \varepsilon_{ikl} k_k \Phi_{il}(\mathbf{k})\, \mathrm{d}S, \tag{11.16}$$

is therefore presumably controlled by a process of transfer of helicity from the 'source' at wave-numbers of order k_0 to the viscous sink at wave-numbers of order k_ν and greater. Injection of helicity may be thought of in terms of injection of large-scale linkages in the vortex lines of the flow. These linkages survive during the (essentially inviscid) cascade process through the inertial range, but they are eliminated by viscosity on the length-scale $l_\nu = k_\nu^{-1}$.

We have seen (cf. (7.55)) that $F(k)$ must satisfy a 'realisability' condition

$$|F(k)| \leqslant 2kE(k) \tag{11.17}$$

for all k. This condition is in general much stronger than the condition (11.13); the two conditions in fact coincide only when a single wave-number magnitude is represented in the velocity spectrum tensor. In normal turbulence with energy distributed over a wide range of length-scales, the condition (11.17) implies that $|\mathscr{H}_0| \ll 1$. To see this, suppose for example that we have maximal positive helicity at each wave-number, so that $F(k) = 2kE(k)$, and that $E(k)$ is given in an inertial range by the Kolmogorov spectrum

(11.6). Then $F(k) \propto k^{-2/3}$ in this inertial range, and so $\langle \mathbf{u} \cdot \boldsymbol{\omega} \rangle = \int_0^\infty F(k)\, dk$ is dominated by values of k near k_ν; this means that $\langle \mathbf{u} \cdot \boldsymbol{\omega} \rangle$ is determined by ε and ν, and so on dimensional grounds

$$\langle \mathbf{u} \cdot \boldsymbol{\omega} \rangle \sim \varepsilon^{5/4} \nu^{-7/4}. \tag{11.18}$$

Hence

$$|\mathcal{H}_0| = |\langle \mathbf{u} \cdot \boldsymbol{\omega} \rangle| / \langle \mathbf{u}^2 \rangle^{1/2} \langle \boldsymbol{\omega}^2 \rangle^{1/2} \sim R_e^{-1/4}. \tag{11.19}$$

Even (11.19) however is an overestimate of $|\mathcal{H}_0|$ in any real turbulent situation. Suppose for example that, at time $t = 0$, we impulsively generate a random velocity field on the scale $l_0 = k_0^{-1}$ with maximal helicity, i.e. $|\mathcal{H}_0| = O(1)$, and we allow this field to evolve under the Navier–Stokes equations. Viscosity has negligible effect until $\langle \boldsymbol{\omega}^2 \rangle$ has increased (by random stretching) by a factor $O(R_e)$; during this process $\langle \mathbf{u} \cdot \boldsymbol{\omega} \rangle$ and $\langle \mathbf{u}^2 \rangle$ remain essentially constant. Hence $|\mathcal{H}_0|$ decreases by a factor $O(R_e^{-1/2})$, i.e. at this stage

$$|\mathcal{H}_0| = O(R_e^{-1/2}), \tag{11.20}$$

and it will presumably remain of this order of magnitude (at most) during the subsequent decay process.

The difference between (11.19) and (11.20) provides an indication that a state of maximal helicity $F(k) = 2kE(k)$ is not in fact compatible with natural evolution under the Navier–Stokes equations. Direct evidence for this is provided by the following simple argument of Kraichnan (1973). Suppose that at time $t = 0$ we have a velocity field consisting of two pure 'helicity modes', $\mathbf{u} = \mathbf{u}_1 + \mathbf{u}_2$, where

$$\begin{aligned} \mathbf{u}_1 &= (\hat{\mathbf{u}}_1 + \mathrm{i}\hat{\mathbf{k}}_1 \wedge \hat{\mathbf{u}}_1)\, \mathrm{e}^{\mathrm{i}\mathbf{k}_1 \cdot \mathbf{x}} + c.c., \\ \mathbf{u}_2 &= (\hat{\mathbf{u}}_2 \pm \mathrm{i}\hat{\mathbf{k}}_2 \wedge \hat{\mathbf{u}}_2)\, \mathrm{e}^{\mathrm{i}\mathbf{k}_2 \cdot \mathbf{x}} + c.c., \end{aligned} \tag{11.21}$$

where $c.c.$ represents the complex conjugate, and $\hat{\mathbf{k}}_n = \mathbf{k}_n / k_n \, (n = 1, 2)$. These modes satisfy

$$\boldsymbol{\omega}_1 = \nabla \wedge \mathbf{u}_1 = k_1 \mathbf{u}_1, \qquad \boldsymbol{\omega}_2 = \nabla \wedge \mathbf{u}_2 = \pm k_2 \mathbf{u}_2. \tag{11.22}$$

The choice of sign in $\mathbf{u}_2$ (corresponding to right-handed or left-handed circular polarisation) is retained in order to shed light on the nature of the interactions between helical modes of like or opposite polarities. Writing the Navier–Stokes equation in the form

$$\left(\frac{\partial}{\partial t}-\nu\nabla^2\right)\mathbf{u}=-\nabla\tilde{p}+\mathbf{u}\wedge\boldsymbol{\omega}, \tag{11.23}$$

where $\tilde{p}=p/\rho+\frac{1}{2}\mathbf{u}^2$, it is evident that the strength of the interaction is given by the non-linear term

$$\mathbf{u}\wedge\boldsymbol{\omega}=\mathbf{u}_1\wedge\boldsymbol{\omega}_2+\mathbf{u}_2\wedge\boldsymbol{\omega}_1=(k_1\mp k_2)\mathbf{u}_2\wedge\mathbf{u}_1. \tag{11.24}$$

Since $k_1+k_2>|k_1-k_2|$, it is immediately apparent that modes of opposite polarity have a tendency to interact more strongly than modes of like polarity. In fact when $k_1=k_2$, $\mathbf{u}$ is parallel to $\boldsymbol{\omega}$ when the modes are of like polarity ($\mathcal{H}_0=1$) and there is no non-linear interaction whatsoever!

To see that maximal helicity is not conserved, it is necessary to obtain $(\partial\mathbf{u}/\partial t)_{t=0}$ from (11.23), and this involves elimination of $\tilde{p}$ using $\nabla\,.\,\mathbf{u}=0$. The initial interaction generates modes with wave-vectors $\mathbf{k}_1\pm\mathbf{k}_2$. Let $\mathbf{k}=\mathbf{k}_1+\mathbf{k}_2$ and choose axes (fig. 11.1) so that

$$\begin{aligned}\mathbf{k}_1&=k_1(\sin\gamma,0,-\cos\gamma),\\ \mathbf{k}_2&=k_2(-\sin\beta,0,-\cos\beta),\\ \hat{\mathbf{u}}_1&=u_0(0,\tfrac{1}{2},0),\\ \hat{\mathbf{u}}_2&=u_0(0,\tfrac{1}{2},0),\end{aligned} \tag{11.25}$$

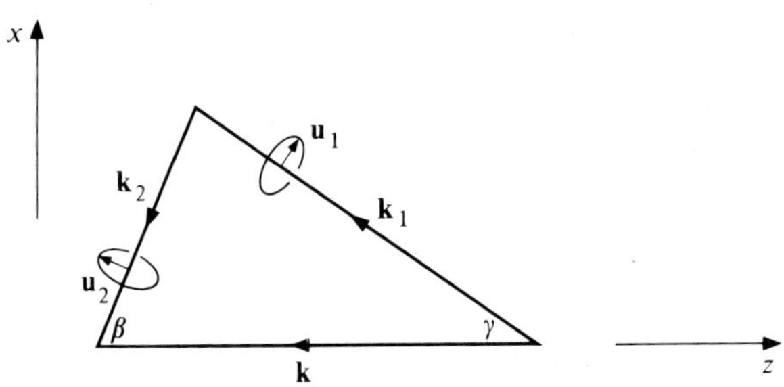

Fig. 11.1 Interaction of two helical modes; modes with wave-vectors $\mathbf{k}_1$ and $\mathbf{k}_2$ are each circularly polarised (cf. fig. 10.2) and of maximum positive helicity; these interact to generate a mode of wave-vector $\mathbf{k}=\mathbf{k}_1+\mathbf{k}_2$ which is not of maximum helicity. (After Kraichnan, 1973.)

where $k_1 \sin\gamma - k_2 \sin\beta = 0$. Then, as shown by Kraichnan (1973), the initial excitation of the mode $\mathbf{u}(\mathbf{k})\,e^{i\mathbf{k}.\mathbf{x}}$ is given (ignoring the viscous effect) by

$$\frac{\partial \mathbf{u}(\mathbf{k})}{\partial t} = -\tfrac{1}{4}ku_0^2[\mp i \sin(\beta-\gamma), -\sin\gamma \pm \sin\beta, 0], \quad (11.26)$$

so that, as anticipated above, $|\dot{\mathbf{u}}(\mathbf{k})|_{t=0}$ is greater when the interacting waves have opposite polarity than when they have the same polarity; and moreover since $i\mathbf{k} \wedge \dot{\mathbf{u}}(\mathbf{k}) \neq \pm k\dot{\mathbf{u}}(\mathbf{k})$ when $\gamma \neq \beta$, the condition of maximal helicity cannot be conserved under non-linear interactions.

The conclusion of these arguments is that, no matter how strong the level of helicity injection may be at wave-numbers of order k_0, the relative level of helicity as measured by the dimensionless ratio $F(k)/2kE(k)$ must grow progressively weaker with increasing k; and when k/k_0 is sufficiently large it may be conjectured (Brissaud *et al.*, 1973) that the helicity has negligible dynamical effect, and is itself convected and diffused in much the same way as a dynamically passive scalar contaminant (Batchelor, 1959). Suppose that the rate of injection of helicity (a pseudo-scalar) at wave-numbers of order k_0 is η. This is clearly bounded by an inequality of the form $|\eta| \lesssim k_0\varepsilon$ since helicity cannot be injected without simultaneous injection of energy. If helicity is injected at a maximal rate, then

$$|\eta| \sim k_0\varepsilon \sim u_0^3/l_0^2. \quad (11.27)$$

The helicity spectrum $F(k)$ must be proportional to η (the pseudo-scalar character of both quantities ensures this) and in the inertial range $k_0 \ll k \ll k_\nu$, the only other parameters that can serve to determine $F(k)$ are ε and k; hence on dimensional grounds[2]

$$F(k) = C_H \eta \varepsilon^{-1/3} k^{-5/3}, \quad (k_0 \ll k \ll k_\nu), \quad (11.28)$$

where C_H is a universal constant, analogous to the Kolmogorov constant C in (11.6). $E(k)$ is still given by (11.6), being unaffected

[2] The result (11.28) was presented as one of two possibilities by Brissaud *et al.* (1973). The other possibility conjectured involved a 'pure helicity cascade' without any energy casade; this possibility is incompatible with dynamical arguments, and has now been abandoned (André & Lesieur, 1977).

by helicity in the inertial range. Note that (11.27), together with (11.6) and (11.28), implies that

$$|F(k)| \ll 2kE(k), \quad (k_0 \ll k \ll k_\nu), \tag{11.29}$$

consistent with the discussion of the previous paragraphs.

From the point of view of dynamo theory, perhaps the most significant point in the foregoing discussion is that the mean helicity in a turbulent field is given by (11.20) rather than (11.19); equivalently it is a property of the 'energy-containing eddies' of the flow (scale $O(l_0)$) rather than of the dissipative eddies (scale $O(l_\nu)$). A crude maximal estimate of $|\langle \mathbf{u} \cdot \boldsymbol{\omega} \rangle|$ is therefore given by u_0^2/l_0, independent of the Reynolds number of the turbulence. This is in fact the estimate that we adopted in previous sections (see for example the argument leading to the estimate (7.90) for α), but it is reassuring now to have some retrospective dynamical justification.

A second general conclusion from the foregoing discussion is that the presence of helicity may be expected to exert a mild constraint on the energy cascade process, at any rate at wave numbers of order k_0 where the relative helicity level $|F(k)|/2kE(k)$ may be quite high. The study of interacting helicity modes (Kraichnan, 1973) indicates that non-linear interactions are weaker when the interacting modes have like polarity; if *all* the modes present at the initial instant have the same polarity (corresponding to a state of maximal helicity at each wave-number magnitude) the net energy transfer to higher wave-numbers may be expected to be inhibited, and so the decay of turbulence should be delayed. More transparently perhaps, since

$$(\mathbf{u} \cdot \boldsymbol{\omega})^2 + (\mathbf{u} \wedge \boldsymbol{\omega})^2 = \mathbf{u}^2 \boldsymbol{\omega}^2, \tag{11.30}$$

maximisation of $|\langle \mathbf{u} \cdot \boldsymbol{\omega} \rangle|$ may plausibly be associated with minimisation of $\langle (\mathbf{u} \wedge \boldsymbol{\omega})^2 \rangle$, and so with a weakening (on the average) of non-linear effects associated with the term $\mathbf{u} \wedge \boldsymbol{\omega}$ in (11.23) (Frisch *et al.*, 1975).

Finally, we note that André & Lesieur (1977) have numerically analysed the development of energy and helicity spectra on the basis of a closure of the infinite system of equations that gives the rate of change of nth order correlations in terms of correlations up to order $n+1$ (Batchelor, 1953); the particular closure scheme adopted by André & Lesieur is a variant of the 'Eddy-damped quasi-normal Markovian' closure (EDQNM) (Orszag, 1970, 1977) by which 4th order correlations are expressed in terms of 2nd order correlations in a way that guarantees that the realisability conditions $E(k,t)\geq 0$, $|F(k,t)|\leq 2kE(k,t)$ are satisfied at all times. Numerical integration of the resulting equations indicates (i) the development of $k^{-5/3}$ inertial ranges for both $E(k,t)$ and $F(k,t)$, and (ii) a significant delay in the process of energy dissipation when a high level of helicity is initially present. These results, although not conclusive (being based on a closure scheme that is to some extent arbitrary), do support the general description of the dynamics of turbulence with helicity, as outlined in this section.

11.2. Influence of magnetic helicity conservation in energy transfer processes

Suppose now that the turbulent fluid is permeated by an equally turbulent magnetic field $(\mu_0\rho)^{1/2}\mathbf{h}(\mathbf{x},t)$, the mean value of $\mathbf{h}$ being zero. What can be said about the joint evolution of the spectra of $\mathbf{u}$ and $\mathbf{h}$? We shall suppose in the following discussion that dissipation effects (both viscous and ohmic) are weak, or equivalently that both the Reynolds number $R_e=u_0l_0/\nu$ and the magnetic Reynolds number $R_m=u_0l_0/\lambda$ are large.

Consider first the new quadratic invariants that exist in the dissipationless limit $\lambda=\nu=0$. These are the total energy E_T, the magnetic helicity I_M, and the cross-helicity[3] I_C:

$$E_T=\tfrac{1}{2}\langle\mathbf{u}^2+\mathbf{h}^2\rangle,\qquad I_M=\langle\mathbf{a}\,.\,\mathbf{h}\rangle,\qquad I_C=\langle\mathbf{u}\,.\,\mathbf{h}\rangle,\quad (11.31)$$

(see §§ 3.1 and 10.1). Here $\mathbf{a}$ is a vector potential for $\mathbf{h}$, i.e. $\mathbf{h}=\nabla\wedge\mathbf{a}$. Note that the value of I_M is independent of the gauge of $\mathbf{a}$. The kinetic helicity $\langle\mathbf{u}\,.\,\boldsymbol{\omega}\rangle$ is no longer invariant in the presence of a magnetic field distribution, because, under the influence of the rotational Lorentz force, vortex lines are no longer frozen in the fluid.

The equations of magnetohydrodynamics, (10.9) and (10.10), are invariant under the transformation $\mathbf{h}(\mathbf{x},t)\rightarrow-\mathbf{h}(\mathbf{x},t)$ (i.e. to every

[3] In (11.31), factors involving the (uniform) density ρ and the constant μ_0 are omitted for simplicity. The term 'cross-helicity' was introduced by Frisch *et al.* (1975); see also Moffatt (1969).

solution ($\mathbf{u}(\mathbf{x}, t), \mathbf{h}(\mathbf{x}, t)$) (there corresponds another solution $\mathbf{u}(\mathbf{x}, t)$, $-\mathbf{h}(\mathbf{x}, t)$). If, therefore, at some initial instant $t=0$ the statistical properties of the $(\mathbf{u}, \mathbf{h})$ field are invariant under change of sign of $\mathbf{h}$, then they will remain so invariant for all $t>0$ (this holds for arbitrary values of λ and ν and also if there is an arbitrary forcing term in the equation of motion)[4]. Under this condition, which we may describe as the condition of *magnetic sign invariance*, any statistical function of the fields $\mathbf{u}$ and $\mathbf{h}$ which appears to change sign under replacement of $\mathbf{h}$ by $-\mathbf{h}$ must in fact be permanently zero. In particular, under the condition of magnetic sign invariance,

$$I_C = 0, \tag{11.32}$$

and similarly all moments of the form $\langle u_i u_j \ldots h_\alpha h_\beta \ldots \rangle$ involving an odd number of $\mathbf{h}$-factors will vanish for all t. We shall suppose, unless explicitly stated otherwise in the following discussion, that this condition of magnetic sign invariance is satisfied.

Consider now for simplicity the idealised situation in which the $\mathbf{u}$ and $\mathbf{h}$ fields are statistically invariant under rotations (but not necessarily under point reflexions). The spectrum tensors $\Phi_{ij}(\mathbf{k}, t)$, $\Gamma_{ij}(\mathbf{k}, t)$ of $\mathbf{u}$ and $\mathbf{h}$ are then given by

$$\Phi_{ij}(\mathbf{k}, t) = \frac{E(k, t)}{4\pi k^4}(k^2\delta_{ij} - k_i k_j) + \frac{\mathrm{i}F(k, t)}{8\pi k^4}\varepsilon_{ijk}k_k, \tag{11.33}$$

$$\Gamma_{ij}(\mathbf{k}, t) = \frac{M(k, t)}{4\pi k^4}(k^2\delta_{ij} - k_i k_j) + \frac{\mathrm{i}N(k, t)}{8\pi k^2}\varepsilon_{ijk}k_k, \tag{11.34}$$

where

$$\tfrac{1}{2}\langle \mathbf{u}^2 \rangle = \int_0^\infty E(k, t)\,\mathrm{d}k, \qquad \tfrac{1}{2}\langle \mathbf{h}^2 \rangle = \int_0^\infty M(k, t)\,\mathrm{d}k, \tag{11.35}$$

$$\langle \mathbf{u} \cdot \boldsymbol{\omega} \rangle = \int_0^\infty F(k, t)\,\mathrm{d}k, \qquad \langle \mathbf{a} \cdot \mathbf{h} \rangle = \int_0^\infty N(k, t)\,\mathrm{d}k. \tag{11.36}$$

[4] This result is mentioned by Pouquet *et al.* (1976).

Note that the spectrum tensor of the field **a** is (under the condition of isotropy) just $k^{-2}\Gamma_{ij}(\mathbf{u}, t)$, and that $N(k, t)$ is defined with respect to the invariant $\langle \mathbf{a} \cdot \mathbf{h} \rangle$ (rather than with respect to $\langle \mathbf{h} \cdot \nabla \wedge \mathbf{h} \rangle$); this leads to the factor k^{-2} in the antisymmetric term of (11.34), as contrasted with k^{-4} in the corresponding term of (11.33). The realisability conditions on $\Gamma_{ij}(\mathbf{k}, t)$ are simply

$$M(k, t) \geq 0, \qquad |N(k, t)| \leq 2k^{-1}M(k, t). \tag{11.37}$$

Let us now consider in a qualitative way how the system will respond to injection of kinetic energy and kinetic helicity at wave-numbers of order k_0 (a situation studied within the framework of the EDQNM closure scheme by Pouquet, *et al.* (1976). Suppose that an initially weak magnetic field is present with, in particular, ingredients on scales large compared with k_0^{-1}. On the scale k_0^{-1}, these ingredients will provide an almost uniform field $\mathbf{h}_0$ say. Helical motion on scales $\lesssim k_0^{-1}$ will generate a perturbation field with magnetic helicity related to the kinetic helicity. If we represent the action of motions on scale $\ll k_0^{-1}$ by an eddy diffusivity λ_e, then the magnetic fluctuations on scales $\sim k_0^{-1}$ are determined by the equation

$$\lambda_e \nabla^2 \mathbf{h} \approx -(\mathbf{h}_0 \cdot \nabla)\mathbf{u}, \tag{11.38}$$

with Fourier transform

$$\lambda_e k^2 \tilde{\mathbf{h}} \approx \mathrm{i}(\mathbf{h}_0 \cdot \mathbf{k})\tilde{\mathbf{u}}. \tag{11.39}$$

The corresponding relation between the spectra of **h** and **u** is

$$\Gamma_{ij}(\mathbf{k}, t) \approx \frac{(\mathbf{h}_0 \cdot \mathbf{k})^2}{\lambda_e^2 k^4} \Phi_{ij}(\mathbf{k}, t), \qquad (k \approx k_0), \tag{11.40}$$

(a result obtained in a related context by Golitsyn, 1960). The result (11.40) gives an anisotropic form for Γ_{ij} when Φ_{ij} is isotropic, due to the preferred direction of $\mathbf{h}_0$. However, if we now take account of the fact that the large-scale field $\mathbf{h}_0$ is non-uniform, all directions being equally likely, we may average over these directions to obtain the isotropic relationship

$$\Gamma_{ij}(\mathbf{k}, t) \approx \frac{\langle \mathbf{h}_0^2 \rangle}{2\lambda_e^2 k^2} \Phi_{ij}(\mathbf{k}, t), \tag{11.41}$$

and hence, from (11.33) and (11.34),

$$M(k,t)\approx\frac{\langle\mathbf{h}_0^2\rangle}{2\lambda_e^2k^2}E(k,t),$$
$$N(k,t)\approx\frac{\langle\mathbf{h}_0^2\rangle}{2\lambda_e^2k^4}F(k,t). \quad (11.42)$$

The magnetic helicity generated at wave-numbers of order k_0 is therefore of the same sign as the kinetic helicity.

Consider now the development of the large-scale field $\mathbf{h}_0$. We know from the general considerations of chapter 7 that positive kinetic helicity gives rise to a negative α-effect. There is of course no *a priori* justification for a two-scale approach when the spectrum of $\mathbf{h}(\mathbf{x}, t)$ is continuous; but *if* a two-scale approach is adopted, the evolution equation for $\mathbf{h}_0$ is

$$\frac{\partial \mathbf{h}_0}{\partial t}=\alpha\nabla\wedge\mathbf{h}_0+\lambda\nabla^2\mathbf{h}_0. \quad (11.43)$$

Writing $\mathbf{h}_0=\nabla\wedge\mathbf{a}_0$, we have equivalently

$$\frac{\partial \mathbf{a}_0}{\partial t}=\alpha\mathbf{h}_0-\nabla\phi_0+\lambda\nabla^2\mathbf{a}_0. \quad (11.44)$$

for some scalar ϕ_0, and from (11.43) and (11.44), the development of the large-scale magnetic helicity is given by

$$\frac{\partial}{\partial t}\langle\mathbf{a}_0\,.\,\mathbf{h}_0\rangle=\alpha\langle\mathbf{a}_0\,.\,\nabla\wedge\mathbf{h}_0\rangle+\alpha\langle\mathbf{h}_0^2\rangle+\lambda\langle\mathbf{a}_0\,.\,\nabla^2\mathbf{h}_0+\mathbf{h}_0\,.\,\nabla^2\mathbf{a}_0\rangle$$
$$=2\alpha\langle\mathbf{h}_0^2\rangle-2\lambda\langle(\partial a_j/\partial x_i)(\partial h_j/\partial x_i)\rangle, \quad (11.45)$$

using the property of homogeneity. Assuming $\langle\mathbf{u}\,.\,\boldsymbol{\omega}\rangle>0$, so that $\alpha<0$, it is evident that the large-scale magnetic helicity generated by the α-effect will be negative (the effects of diffusion being assumed weak).

We know, moreover, from the results of § 9.2 that, when dissipation is negligible, the growth of magnetic Fourier components on the scale k^{-1} due to the α-effect has a time-scale of order $(|\alpha|k)^{-1}$; hence for a given level of helicity maintained by 'injection' at wave-numbers of order k_0, it is to be expected that magnetic energy (and associated magnetic helicity) will develop on progressively increasing length-scales of order $|\alpha|t$ (or equivalently at wave-numbers of order $(|\alpha|t)^{-1}$) as t increases.

These qualitative considerations receive striking support from the work of Pouquet *et al* (1976), who have numerically integrated the four equations describing the evolution of the functions $E(k, t)$, $F(k, t)$, $M(k, t)$, $N(k, t)$, closed on the basis of the EDQNM scheme. Fig. 11.2(*a*), (*b*) show the development of $M(k, t)$, $N(k, t)$ as

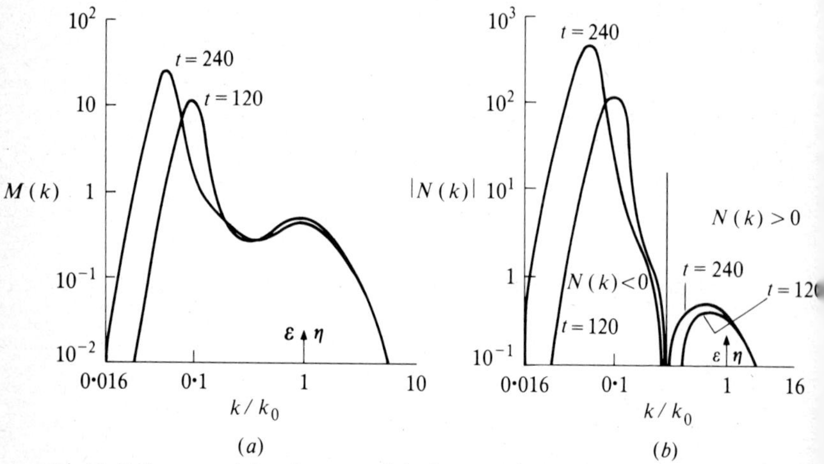

Fig. 11.2 Computed development (*a*) of magnetic energy spectrum and (*b*) of magnetic helicity spectrum on the basis of the EDQNM closure. Kinetic energy and helicity are injected at rates ε and η; the injection spectra are given by $F_\varepsilon(\kappa) = \kappa^{-1}F_\eta(\kappa) = C\varepsilon\kappa^4\,e^{-2\kappa^2}$ with C chosen such that $\int_0^\infty F_\varepsilon(\kappa)\,d\kappa = \varepsilon$ and $\kappa = k/k_0$. At time $t = 0$, the (normalised) kinetic and magnetic energy and helicity spectrum functions are given by

$$E(\kappa) = 10M(\kappa) = C\kappa^4\,e^{-2\kappa^2}, \qquad F(\kappa) = N(\kappa) = 0.$$

The minimum and maximum wave-numbers retained in the computation were $\kappa_{\min} = 2^{-6}$, $\kappa_{\max} = 2^4$. The Reynolds number based on the initial *rms* velocity and the length-scale k_0^{-1} was 30, and the magnetic Prandtl number ν/λ was unity. (From Pouquet, Frisch & Léorat 1976).

functions of k and for $t = 120, 140$ (the unit for time being $(k_0u_0)^{-1}$). The system is excited by injection of kinetic energy and helicity at wave-numbers of order k_0. The figures show (i) the excitation of magnetic energy at progressively larger scales as t increases, and

(ii) the fact that magnetic helicity has the same sign as the injected kinetic helicity when $k/k_0 = O(1)$, but the opposite sign when $k/k_0 \ll 1$. It is evident from both figures that a double-scale structure emerges in the magnetic field spectrum, the separation in the two spectral peaks becoming more marked as t increases. It must be emphasised that these results emerge from a fully dynamic model in which the back-reaction of the Lorentz force distribution on the velocity field is fully incorporated.

Pouquet *et al.* (1976) have described the above excitation of magnetic modes on ever-increasing length-scales in terms of an 'inverse cascade' of magnetic energy and magnetic helicity. It may be merely a matter of semantics, but this terminology could be just a little misleading in the present context. The word 'cascade' suggests 'successive excitation' due to non-linear mode interactions, and 'inverse cascade' suggests successive excitations that may be represented diagrammatically in the form

$$k_0 \to 2^{-1}k_0 \to 2^{-2}k_0 \to \dots \to 2^{-n}k_0 \to \dots. \qquad (11.46)$$

In fact in the present context, interaction of **u** and **h** fluctuations on the scale k_0^{-1} *simultaneously* generates large-scale Fourier components on all scales $k^{-1} \gg k_0^{-1}$; this is not a step-by-step process, but rather a long-range spectral process; the fact that larger scales take longer to excite does not in itself justify the use of the term 'cascade'.

A further numerical study of direct relevance to the present discussion has been carried out by Pouquet & Patterson (1977). In this work, the Fourier transforms of the coupled momentum and induction equations, appropriately truncated, are directly integrated numerically, subject to initial conditions for the Fourier components of **u** and **h** chosen from Gaussian distributions of random numbers. No attempt was made to average over different realisations of the turbulent field, and in fact the behaviour of the spectra (defined by averaging over spherical shells in wave-number space) showed significant variation between different realisations. Nevertheless in all cases Pouquet & Patterson found a substantial net transfer of energy from kinetic to magnetic modes. They interpret this transfer in the following terms: (i) cascade of kinetic energy towards higher wave-numbers due to conventional non-linear interactions; (ii) sharing of energy between kinetic and magnetic modes at small scales due to excitation of Alfvén waves propagating on the large-scale magnetic field; (iii) intensification of the large-scale field due to what is again described as an 'inverse cascade' of magnetic energy, an effect that is most noticeable in the numerical solutions when magnetic helicity (rather than kinetic helicity) is initially present. These calculations (and the EDQNM calculations of Pouquet *et al.*, 1976) were carried out for a magnetic Prandtl number λ/ν equal to unity; further calculations with $\lambda/\nu > 1$ (or even $\gg 1$) would perhaps be more relevant in the solar context and will be awaited with great interest.

11.3. Modification of inertial range due to large-scale magnetic field

The presence of a strong large-scale magnetic field profoundly modifies the energy transfer process in small scales, and in particular in the inertial range, as pointed out by Kraichnan (1965). In the inertial range, the total energy content is small, relative to the total energy in scales $\gtrsim k_0^{-1}$, and dissipative processes are negligible. The large-scale velocity field $\mathbf{u}_0$ simply convects eddies on the scale k^{-1} (where $k \gg k_0$) without significant distortion, and can in effect be 'eliminated' by Galilean transformation. The large-scale magnetic field $\mathbf{h}_0$ cannot be eliminated by Galilean transformation, and in fact provides an important coupling between $\mathbf{u}$ and $\mathbf{h}$ fields on scales $k^{-1} \ll k_0^{-1}$ through the Alfvén wave mechanism discussed in § 10.1. The inertial range may be pictured as a random sea of localised disturbances on scales k^{-1} ($k \gg k_0$), propagating along the local mean field $\mathbf{h}_0$ with velocity $\pm\mathbf{h}_0$, energy transfer to smaller scales occurring due to the collision of oppositely directed disturbances. In each such disturbance magnetic and kinetic energy are equal; hence $E(k) = M(k)$ in the inertial range. Energy transfer will be maximised when there are as many disturbances travelling in the direction $+\mathbf{h}_0$ as in the direction $-\mathbf{h}_0$, i.e. when the cross-helicity I_C is zero. It is relevant to note here that Pouquet & Patterson (1977) have found that non-zero values of I_C leads to a decrease in energy transfer to small-scales in numerical simulations.

The time-scale for the interaction of blobs on the scale k^{-1} is $t_k = O(h_0 k)^{-1}$, and, as argued by Kraichnan (1965), the rate of energy transfer ε through the inertial range may be expected to be proportional to t_k. Since there is no dissipation in the inertial range, ε must be independent of k; moreover, if the energy cascade is local in k-space, the only other parameters that can serve to determine ε are $E(k)$ and k itself; dimensional analysis then gives

$$\varepsilon \propto h_0^{-1}(E(k))^2 k^3, \tag{11.47}$$

or equivalently

$$E(k) = M(k) = A(\varepsilon h_0)^{1/2} k^{-3/2} \qquad (k_0 \ll k \ll k_d), \tag{11.48}$$

where A is a universal constant of order unity, and k_d an upper wave-number for the inertial range at which dissipative effects (ohmic and/or viscous) become important. This cut-off wave-number is presumably determined (in order of magnitude) by equating the time-scale of interaction of blobs $(h_0 k)^{-1}$ with the time-scale of ohmic or viscous dissipation $(\lambda k^2)^{-1}$ or $(\nu k^2)^{-1}$, whichever is smaller. If, as is more usual, $\lambda \gg \nu$, then evidently

$$k_d \sim h_0/\lambda. \tag{11.49}$$

In (11.48) and (11.49), h_0 must be interpreted as $\langle \mathbf{h}^2 \rangle^{1/2}$, and the condition that there should exist an inertial range of the form (11.48) is $k_0 \gg k_d$, or

$$\langle \mathbf{h}^2 \rangle^{1/2} / k_0 \lambda \gg 1. \tag{11.50}$$

11.4. Non-helical turbulent dynamo action

The generation of magnetic fields on scales of order k_0^{-1} and greater, as discussed in § 11.2, is directly attributable to the presence of helicity in the smaller scale velocity and/or magnetic fields. Let us now briefly consider the problem that presents itself when both fields are assumed reflexionally symmetric, so that the α-effect mechanism for the generation of large-scale fields is no longer present.

It has long been recognised that random stretching of magnetic lines of force leads to exponential increase of magnetic energy density, as long as ohmic diffusion effects can be neglected. This random stretching is however associated with a systematic decrease in scale of the magnetic field (as in the differential rotation problem studied in § 3.8), and ultimately ohmic diffusion must be taken into account. The question then arises as to whether magnetic energy generation by random stretching can still win over degradation of magnetic energy due to ohmic dissipation.

Batchelor (1950) invoked the analogy between the vorticity equation in a non-conducting fluid

$$\frac{\partial \boldsymbol{\omega}}{\partial t} = \nabla \wedge (\mathbf{u} \wedge \boldsymbol{\omega}) + \nu \nabla^2 \boldsymbol{\omega}, \tag{11.51}$$

and the induction equation in a conducting fluid

$$\frac{\partial \mathbf{h}}{\partial t} = \nabla \wedge (\mathbf{u} \wedge \mathbf{h}) + \lambda \nabla^2 \mathbf{h}, \tag{11.52}$$

(see § 3.2). If, in either case, $\mathbf{u}(\mathbf{x}, t)$ is a homogeneous turbulent velocity field, we may deduce equations for $\langle \boldsymbol{\omega}^2 \rangle$ and $\langle \mathbf{h}^2 \rangle$:

$$\frac{\mathrm{d}}{\mathrm{d}t} \langle \boldsymbol{\omega}^2 \rangle = \left\langle \omega_i \omega_j \frac{\partial u_i}{\partial x_j} \right\rangle - \nu \left\langle \left(\frac{\partial \omega_i}{\partial x_j} \right)^2 \right\rangle, \tag{11.53}$$

$$\frac{\mathrm{d}}{\mathrm{d}t} \langle \mathbf{h}^2 \rangle = \left\langle h_i h_j \frac{\partial u_i}{\partial x_j} \right\rangle - \lambda \left\langle \left(\frac{\partial h_i}{\partial x_j} \right)^2 \right\rangle. \tag{11.54}$$

Batchelor now argued that $\langle \boldsymbol{\omega}^2 \rangle$, being a 'small-scale' property of the turbulence, is in statistical equilibrium on any time-scale (e.g. l_0/u_0) associated with the energy-containing eddies. Moreover the random stretching of magnetic lines of force may be expected to generate a similar statistical structure in the **h**-field as the random stretching of vortex lines generates in the $\boldsymbol{\omega}$-field. Hence if $\lambda = \nu$, $\langle \mathbf{h}^2 \rangle$ may also be expected to be in statistical equilibrium, and therefore to remain constant on time-scales of order l_0/u_0; if $\lambda < \nu$, $\langle \mathbf{h}^2 \rangle$ may be expected to grow (and Batchelor argued that this growth would continue until arrested by the back-reaction of Lorentz forces); and if $\lambda > \nu$, $\langle \mathbf{h}^2 \rangle$ may be expected to decay.

The analogy with vorticity is of course not perfect, as has been pointed out in § 3.2. A particular difficulty in its application to the turbulent dynamo problem is that we are really interested in the possibility of magnetic field maintenance on time-scales at least as large as the ohmic time-scale l_0^2/λ, and this is *large* compared with l_0/u_0, when $R_m = u_0 l_0/\lambda \gg 1$. In the absence of 'random stirring devices', $\langle \boldsymbol{\omega}^2 \rangle$ decays on the time-scale l_0/u_0, and if $\lambda = \nu$, $\langle \mathbf{h}^2 \rangle$ may be expected to do likewise, i.e. we do not have a dynamo in the usual sense. If $\lambda < \nu$, $\langle \mathbf{h}^2 \rangle$ may at first increase by a factor of order ν/λ (as $\langle \boldsymbol{\omega}^2 \rangle$ would do if ν in (11.53) were suddenly decreased to a value $\nu_1 = \lambda < \nu$), but may again be expected to decay in a time of order l_0/u_0. We can of course prevent the long-term decay of $\langle \boldsymbol{\omega}^2 \rangle$ in (11.53) by invoking a random torque distribution in (11.51); but to maintain the analogy we should then introduce the curl of a random electromotive force in (11.52) – and we are then no longer considering the problem of a self-excited dynamo.

While the above considerations might suggest that reflexionally symmetric turbulence cannot provide sustained dynamo action, there is also the consideration (more prominent in the context of turbulence with helicity) that since the field $\mathbf{h}$ is not subject to the constraint $\boldsymbol{\omega} = \nabla \wedge \mathbf{u}$, modes of excitation may be available to $\mathbf{h}$ that are simply not available to $\boldsymbol{\omega}$ and so $\langle \mathbf{h}^2 \rangle$ may grow (until Lorentz forces intervene) even when $\lambda > \nu$.

It seems likely that the questions raised by these considerations will be fully resolved only by extended numerical experimentation by methods similar to those of Pouquet & Patterson (1977). Computer capacities are however not yet adequate to yield reliable asymptotic laws in integrations over large times. Fortunately, as mentioned in the introductory chapter, the reflexionally symmetric problem, although profoundly challenging, has been largely bypassed in terrestrial and astrophysical contexts by the realisation that lack of reflexional symmetry will generally be present in turbulence generated in a rotating system, and that this lack of reflexional symmetry swamps all other purely turbulent effects as far as magnetic field generation is concerned.

CHAPTER 12

DYNAMICALLY CONSISTENT DYNAMOS

12.1. The Taylor constraint and torsional oscillations

We shall now consider some general aspects of the dynamics of a rotating fluid within a spherical boundary subject to a combination of Lorentz, Coriolis and buoyancy forces. Under the Boussinesq approximation the equation of motion is

$$\begin{aligned}\partial \mathbf{U}/\partial t + \mathbf{U}\,.\,\nabla\mathbf{U} + 2\boldsymbol{\Omega}\wedge\mathbf{U} \\ = -\nabla P + \rho^{-1}\mathbf{J}\wedge\mathbf{B} + \alpha\Theta\mathbf{g} + \nu\nabla^2\mathbf{U},\end{aligned} \qquad (12.1)$$

where ρ is the mean density, and $\mathbf{g}$ the radial gravitational acceleration; Θ is the temperature field, its evolution being determined by (10.80). $\mathbf{U}$ is solenoidal, and satisfies $\mathbf{U}=0$ on the sphere $r=R$.

The order of magnitude of $\mathbf{U}\,.\,\nabla\mathbf{U}$ and $\nu\nabla^2\mathbf{U}$ relative to $2\boldsymbol{\Omega}\wedge\mathbf{U}$ in (12.1) is given by the global estimates

$$\frac{|\mathbf{U}\,.\,\nabla\mathbf{U}|}{|\boldsymbol{\Omega}\wedge\mathbf{U}|} = O\left(\frac{U}{\Omega L}\right), \qquad \frac{|\nu\nabla^2\mathbf{U}|}{|\boldsymbol{\Omega}\wedge\mathbf{U}|} = O\left(\frac{\nu}{\Omega L^2}\right), \qquad (12.2)$$

where $L(=O(R))$ is the scale of variation of $\mathbf{U}$. We shall suppose that

$$R_o = U/\Omega L \ll 1 \quad \text{and} \quad E = \nu/\Omega L^2 \ll 1, \qquad (12.3)$$

and shall neglect the convective acceleration and the viscous term in (12.1). The boundary condition for $\mathbf{U}$ is then simply

$$\mathbf{U}\,.\,\mathbf{n} = 0 \quad \text{on} \quad r = R, \qquad (12.4)$$

the no-slip condition being accommodated by an Ekman layer, thickness $O(E^{1/2})$ on the surface (see e.g. Greenspan, 1968). Equation (12.1) may now be written in the form

$$\partial\mathbf{U}/\partial \mathrm{t} + 2\boldsymbol{\Omega}\wedge\mathbf{U} = -\nabla P + \mathbf{F}, \qquad (12.5)$$

where

$$\mathbf{F} = \rho^{-1}\mathbf{J}\wedge\mathbf{B} + \alpha\Theta\mathbf{g}. \qquad (12.6)$$

Necessary condition for a steady solution $\mathbf{U}(\mathbf{x})$

The φ-component of (12.5) is, in cylindrical polars (s, φ, z),

$$\partial U_\varphi/\partial t + 2\Omega U_s = -\partial P/\partial\varphi + F_\varphi. \qquad (12.7)$$

The fact that P must be single-valued places an important constraint (Taylor, 1963) on functions $F_\varphi(s, \varphi, z)$ for which steady solenoidal solutions of (12.5) exist. Let $C(s_0)$ denote the cylindrical surface $s = s_0$, $|z| < (R^2 - s^2)^{1/2}$ (fig. 12.1). Since $\nabla \cdot \mathbf{U} = 0$, and $\mathbf{n} \cdot \mathbf{U} = 0$ on $r = R$, it is evident that the flux of $\mathbf{U}$ across $C(s_0)$ must vanish, i.e.

$$\iint_{C(s_0)} U_s(s, \varphi, z) s \, d\varphi \, dz = 0, \quad \text{for all } s_0. \qquad (12.8)$$

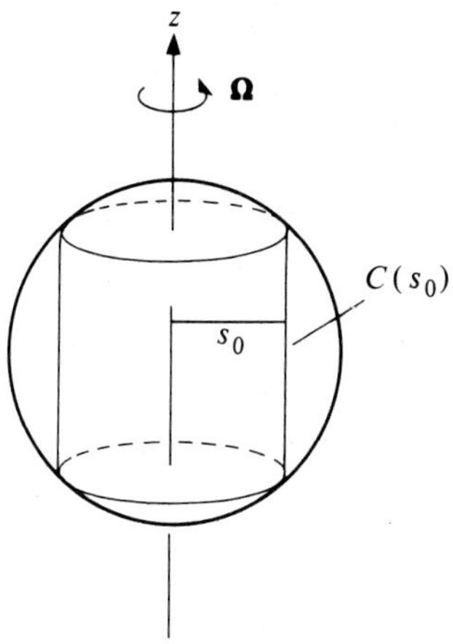

Fig. 12.1 The surface $C(s_0)$: $s = s_0$, $|z| < (R^2 - s^2)^{1/2}$.

Hence from (12.7) we have directly that if $\partial U_\varphi/\partial t = 0$ then

$$\mathcal{T}(s_0) \equiv \iint_{C(s_0)} F_\varphi(s, \varphi, z) s \, d\varphi \, dz = 0, \quad \text{for all } s_0. \qquad (12.9)$$

$\rho\mathcal{T}(s_0)\,\delta s_0$ is evidently the torque exerted by the Lorentz force $\mathbf{J} \wedge \mathbf{B}$ on the annular cylinder of fluid $s_0 < s < s_0 + \delta s_0$; unless this torque is identically zero, angular acceleration must result. Since $g_\varphi = 0$, the condition (12.9) may equally be written (under steady conditions $\partial\mathbf{U}/\partial t = 0$)

$$\mathcal{T}(s_0) \equiv \rho^{-1} \iint_{C(s_0)} (\mathbf{J} \wedge \mathbf{B})_\varphi s \, d\varphi \, dz \equiv 0. \qquad (12.10)$$

In this form, the condition is generally known as the *Taylor constraint.*

Sufficiency of the Taylor constraint for the existence of a steady solution **U**(**x**)

It is a little more difficult to establish that the condition $\mathcal{T}(s)\equiv 0$ is also sufficient to ensure that (12.5) can be solved for **U**(**x**). The following discussion is based on Taylor (1963), although differing in points of detail. The curl of (12.5) (with $\partial\mathbf{U}/\partial t=0$) gives simply

$$\partial\mathbf{U}/\partial z=(2\Omega)^{-1}\nabla\wedge\mathbf{F}=\mathbf{A}(\mathbf{x}),\text{ say.}\tag{12.11}$$

Note at once that

$$\iint_{C(s)} A_z s\,\mathrm{d}\varphi\,\mathrm{d}z=\frac{1}{2\Omega s}\frac{\partial}{\partial s}(s\mathcal{T}(s)),\tag{12.12}$$

using the definition (12.9) of $\mathcal{T}(s)$. Hence if $\mathcal{T}(s)\equiv 0$, then also

$$\iint_{C(s)} A_z s\,\mathrm{d}\varphi\,\mathrm{d}z\equiv 0.\tag{12.13}$$

We now suppose that **A**(**x**) is a given solenoidal function satisfying (12.13), and we attempt to solve (12.11) for **U**(**x**). Let $z^-=-(R^2-s^2)^{1/2}$, $z^+=+(R^2-s^2)^{1/2}$, and for any function $\psi(s,\varphi,z)$, let

$$\bar{\psi}(s,\varphi)=\frac{1}{z^+-z^-}\int_{z^-}^{z+}\psi(s,\varphi,z)\,\mathrm{d}z.\tag{12.14}$$

Also let

$$\mathbf{U}_0(s,\varphi,z)=\int_{z^-}^{z}\mathbf{A}(s,\varphi,\zeta)\,\mathrm{d}\zeta.\tag{12.15}$$

Equation (12.11) then integrates to give

$$\mathbf{U}(s,\varphi,z)=\mathbf{U}_0(s,\varphi,z)+\mathbf{V}(s,\varphi),\tag{12.16}$$

where **V** is to be found; and evidently, integrating with respect to z,

$$\mathbf{V}(s,\varphi)=\bar{\mathbf{U}}(s,\varphi)-\bar{\mathbf{U}}_0(s,\varphi).\tag{12.17}$$

The s and z components of $\mathbf{V}$ are determined by satisfying $\mathbf{n}.\,\mathbf{U}=0$ on $r=R$, i.e. at $z=z^-, z^+$;

$$\begin{aligned} sV_s+zV_z&=0 \quad \text{at } z=z^-,\\ s(V_s+U_{0s})+z(V_z+U_{0z})&=0 \quad \text{at } z=z^+. \end{aligned} \tag{12.18}$$

These equations uniquely determine $V_s(s,\varphi)$ and $V_z(s,\varphi)$ in terms of the known function $\mathbf{U}_0$. It remains therefore to determine $V_\varphi(s,\varphi)$. Now from $\nabla\,.\,\mathbf{U}=0$,

$$\frac{1}{s}\frac{\partial}{\partial s}(sU_s)+\frac{1}{s}\frac{\partial U_\varphi}{\partial\varphi}=-\frac{\partial U_z}{\partial z}=-A_z \quad \text{from (12.11)}, \tag{12.19}$$

and so

$$\frac{1}{s}\frac{\partial \bar{U}_\varphi}{\partial\varphi}=-\bar{A}_z-\frac{1}{s}\frac{\partial}{\partial s}(s\bar{U}_s), \tag{12.20}$$

and hence

$$\bar{U}_\varphi(s,\varphi)=-\int_0^\varphi\left(\bar{A}_z+\frac{1}{s}\frac{\partial}{\partial s}s\bar{U}_s\right)s\,\mathrm{d}\varphi+v(s), \tag{12.21}$$

where $v(s)$ is an arbitrary function of integration. By virtue of (12.8) and (12.13), the function defined by (12.21) is single-valued, and so $V_\varphi(s,\varphi)$ as given by (12.17) is single-valued also. This completes the proof that the condition $\mathcal{T}(s)\equiv 0$ is sufficient for the solvability of (12.11) for steady $\mathbf{U}(\mathbf{x})$.

Torsional oscillations when the Taylor constraint is violated

If $\mathcal{T}(s)\not\equiv 0$, then integration of (12.7) over the cylinder $C(s)$ gives

$$\frac{\partial}{\partial t}\iint_{C(s)} U_\varphi(s,\varphi,z)s\,\mathrm{d}\varphi\,\mathrm{d}z=\mathcal{T}(s), \tag{12.22}$$

i.e. angular acceleration is inevitable. Defining

$$U_G(s)=\frac{1}{A(s)}\iint_{C(s)} U_\varphi(s,\varphi,z)s\,\mathrm{d}\varphi\,\mathrm{d}z, \tag{12.23}$$

where $A(s)=4\pi s(R^2-s^2)^{1/2}$ is the area of $C(s)$, (12.22) becomes

$$\frac{\partial}{\partial t}A(s)U_G(s,t)=\mathcal{T}(s,t), \tag{12.24}$$

where we now take explicit account of time dependence. $U_G(s, t)$ is the geostrophic ingredient of the total velocity field, and it is unaffected by the rotation $\mathbf{\Omega}$. Determination of U_G is clearly equivalent to the determination of v in (12.21).

Again following Taylor (1963) we can obtain a further equation by differentiating (12.10) with respect to t and using the induction equation. This gives

$$\frac{\partial \mathcal{T}(s,t)}{\partial t}=\frac{1}{\mu_0\rho}\iint_{C(s)}\{\nabla\wedge[\nabla\wedge(\mathbf{U}\wedge\mathbf{B})+\lambda\nabla^2\mathbf{B}]\wedge\mathbf{B}$$
$$+(\nabla\wedge\mathbf{B})\wedge[\nabla\wedge(\mathbf{U}\wedge\mathbf{B})+\lambda\nabla^2\mathbf{B}]\}_\varphi\ s\,d\varphi\,dz. \qquad (12.25)$$

In this complicated integral, let us write

$$\mathbf{U}=\mathbf{U}_1(s,\varphi,z,t)+U_G(s,t)\mathbf{i}_\varphi, \qquad (12.26)$$

and regard $U_G(s, t)$ as the unknown ingredient. We know, from § 3.11, that

$$\nabla\wedge(U_G\mathbf{i}_\varphi\wedge\mathbf{B})=sB_s\,\partial(U_G/s)/\partial s\,\mathbf{i}_\varphi, \qquad (12.27)$$

and so

$$[\nabla\wedge[\nabla\wedge(U_G\mathbf{i}_\varphi\wedge\mathbf{B})]\wedge\mathbf{B}]_\varphi=s^{-1}B_s\frac{\partial}{\partial s}\left(s^2B_s\frac{\partial}{\partial s}\frac{U_G}{s}\right). \qquad (12.28)$$

Hence (12.25) reduces to the form

$$\frac{\partial \mathcal{T}(s,t)}{\partial t}=a(s)\frac{\partial^2}{\partial s^2}\left(\frac{U_G}{s}\right)+b(s)\frac{\partial}{\partial s}\left(\frac{U_G}{s}\right)+c(s), \qquad (12.29)$$

where

$$a(s)=\frac{1}{\mu_0\rho}\iint_{C(s)}s^2B_s^2\,d\varphi\,dz,$$
$$b(s)=\frac{1}{\mu_0\rho}\iint_{C(s)}B_s\frac{\partial}{\partial s}(s^2B_s)\,d\varphi\,dz, \qquad (12.30)$$

and $c(s)$ contains all the other contributions to (12.25), which do not involve U_G.

Equations (12.24) and (12.29) clearly constitute a hyperbolic system, which may be expected to admit oscillatory solutions about the steady state for which $\mathscr{T} \equiv 0$. In such oscillatory solutions, each cylinder $C(s)\,ds$ of fluid rotates about its axis, coupling between the cylinders being provided by the radial field B_s via the coefficients $a(s)$ and $b(s)$. Such torsional oscillations have been studied by Braginskii (1970); the damping of the oscillations, due to Ekman layer effects on $r = R-$, has been discussed by Roberts & Soward (1972, § 3). The time-scale associated with the oscillations is determined essentially by the mean value of $h_s^2 = B_s^2/\mu_0\rho$; it is therefore of order $R/\langle h_s^2\rangle^{1/2}$, the time for an Alfvén wave to propagate on the radial field a distance of the order of the radius of the sphere.

In a situation in which the poloidal field $\mathbf{B}_P$ develops on a very long time-scale ($t_\lambda = O(R^2/\lambda)$) by dynamo action (associated with, say, an α-effect), then, provided the damping time t_d for the above torsional oscillations is small compared with t_λ, a quasi-static situation in which $\mathscr{T} \equiv 0$ and U_G/s is the resulting steady solution of (12.29) may be anticipated. If t_d is not small compared with t_λ then torsional oscillations may be expected to persist as long as the development of the poloidal field continues.

12.2. Dynamo action incorporating mean flow effects

We have seen in § 9.5 that dynamo action in a sphere can result from the α-effect alone; and that when, for example,

$$\alpha = -\alpha_0 \cos\theta, \tag{12.31}$$

the critical value of α_0 at which dipole modes are excited is given (numerically) by

$$R_\alpha = |\alpha_0| R/\lambda = 7{\cdot}64 = R_{\alpha c}, \quad \text{say.} \tag{12.32}$$

If $R_\alpha > R_{\alpha c}$ then a mode of dipole symmetry will grow exponentially until the effects of the Lorentz force become important. We have seen in § 11.2 that one manifestation of the increasing influence of the Lorentz force will be an ultimate reduction in the level of the α-effect. There is another mechanism of equilibration that may however intervene and stop the growth at an earlier stage if R_α is just a little larger than $R_{\alpha c}$, viz. the effect of the mean velocity distribution that will be driven by the large-scale Lorentz force distribution; this mean velocity will modify the structure of the growing **B**-field and in general may be expected to modify it in such a way that the rate of ohmic dissipation increases. The total energy in the **B**-field may then be expected to saturate at a low level, which tends to zero as $R_\alpha - R_{\alpha c} \to 0$.

A formalism for this problem has been developed by Malkus & Proctor (1975), by developing all the fields as power series in $(R_\alpha - R_{\alpha c})$ and seeking conditions for steady-state finite-amplitude magnetic field distributions. It is important to recognise that in this work the α-effect is supposed *given* by a formula such as (12.31) and unaffected by the magnetic field. Attention is focussed on the development of large-scale or 'macro' fields, and all the usual difficulties associated with the small-scale turbulent dynamics are bypassed. The equations studied are

$$\partial \mathbf{U}/\partial t + \mathbf{U} \,.\, \nabla \mathbf{U} + 2\mathbf{\Omega} \wedge \mathbf{U} = -\nabla P + \rho^{-1} \mathbf{J} \wedge \mathbf{B} + \nu \nabla^2 \mathbf{U}, \tag{12.34}$$

$$\partial \mathbf{B}/\partial t = \nabla \wedge (\alpha \mathbf{B}) + \nabla \wedge (\mathbf{U} \wedge \mathbf{B}) + \lambda \nabla^2 \mathbf{B}, \tag{12.35}$$

with $\alpha(\mathbf{x})$ prescribed and with initial conditions

$$\mathbf{U}(\mathbf{x}, 0) = 0, \qquad \mathbf{B}(\mathbf{x}, 0) = \mathbf{B}_0(\mathbf{x}), \tag{12.36}$$

where $\mathbf{B}_0(\mathbf{x})$ is the eigenfunction of the kinematic problem (12.35) when $\mathbf{U} = 0$ and $R_\alpha = R_{\alpha c}$ (as discussed in § 9.5). The field **B** is of course as usual matched to an irrotational field of dipole symmetry in the region $r > R$.

When $R_\alpha = R_{\alpha c}(1+\varepsilon)$ with $0 < \varepsilon \ll 1$, the magnetic field initially grows exponentially, and the Lorentz force (which is certainly non-zero by the general result of § 2.4) generates a velocity field which develops according to (12.34). This velocity field will continue to grow until it has a significant effect in (12.35); and comparing the terms $\nabla \wedge (\mathbf{U} \wedge \mathbf{B})$ and $\lambda \nabla^2 \mathbf{B}$, it is evident that the relevant scale for **U** at this stage[1] is $U_0 = \lambda/R$.

The problem is characterised by three dimensionless numbers,

$$E = \nu/\Omega R^2, \qquad E_m = \lambda/\Omega R^2 = U_0/\Omega R, \qquad R_\alpha = |\alpha|_{\max} R/\lambda, \tag{12.37}$$

[1] Strictly $U_0 = (\lambda/R) f(\varepsilon)$ where $f(\varepsilon) \to 0$ as $\varepsilon \to 0$.

and interest centres on the geophysically relevant situation

$$E \ll 1, \qquad E_m \ll 1. \tag{12.38}$$

Note that E_m (the 'magnetic Ekman number') in fact plays the role of a Rossby number here, providing a measure of the importance of inertia forces relative to the Coriolis force in (12.34). In the limit (12.38), the ultimate level of magnetic energy is determined by magnetostrophic balance in which Lorentz forces and Coriolis forces are of the same order of magnitude, i.e. the relevant scale for **B** is B_0 where

$$(\mu_0\rho)^{-1}B_0^2 = \Omega U_0 R = \Omega\lambda, \tag{12.39}$$

and the magnetic energy may be expected to level off at a value of order $\Omega\lambda$ (again multiplied by a function of ε which vanishes with ε).

Solutions of the above problem have been computed by Proctor (1977*a*), for the particular case when α is given by (12.31), and values of R_α in the range between the critical value 7·64 and 10·0. In each of the cases studied, Proctor found that the growth of magnetic energy was arrested by the mean flow effect, the level of equilibrium magnetic energy increasing with $R_\alpha - R_{\alpha c}$; fig. 12.2 shows the ultimate level of magnetic energy associated with the toroidal field for two cases of particular interest, (i) $E_m = 0{\cdot}04$, $E = 0{\cdot}01$, and (ii) $E_m = 0{\cdot}0025$, $E = 0{\cdot}005$. In case (i), the magnetic energy settled down to its ultimate level after some damped oscillations about this level. In case (ii), which was as near to the geostrophic limit ($E_m = 0$, $E = 0$) as the numerical scheme would permit, there were again oscillations of small amplitude about the ultimate mean level of magnetic energy, and these oscillations showed no tendency to decay, with increasing time. The frequency of these oscillations was $O(\Omega E_m^{1/2}) = O((\Omega\lambda)^{1/2}/R)$, and Proctor identified them with the torsional oscillations described in § 12.1. The suggestion here is that when E_m and E are sufficiently small, the steady state in which the Taylor constraint is operative is in fact unattainable, and that torsional oscillations about this state are inevitable. Further investigation at different values of E_m and E in order to explore this phenomenon further would be of great interest.

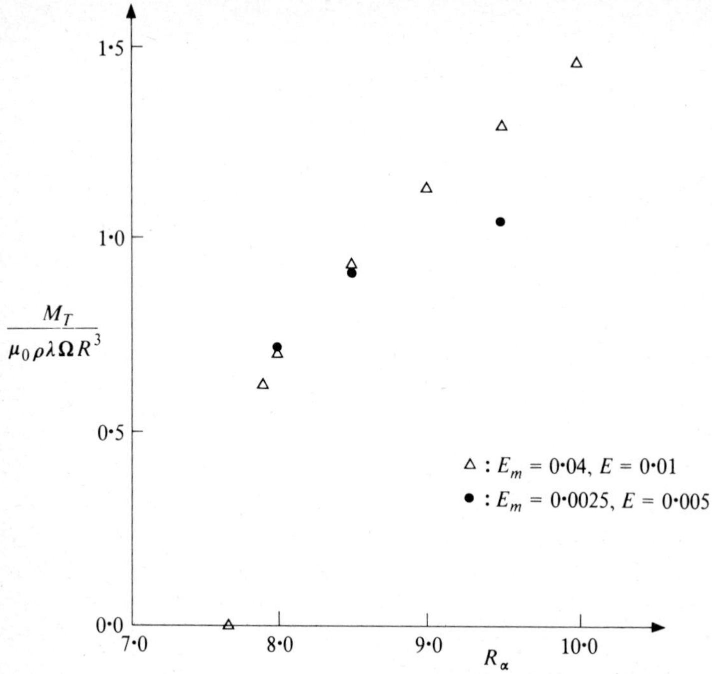

Fig. 12.2 Equilibrium level of magnetic energy M_T of the toroidal field as a function of R_α for two choices of (E_m, E). Note that when viscosity is decreased (i.e. when E is decreased) the equilibrium level is lower because the Lorentz force more easily drives the mean flow which provides the equilibration mechanism. (From Proctor 1977*a*.)

It would be interesting also to extend the approach to cover dynamos of $\alpha\omega$-type. Here it would be necessary to specify not only the distribution of $\alpha(s, z)$ but also the distribution $\omega(s, z)$, and since the Lorentz force tends to change the toroidal (as well as the poloidal) velocity, the only consistent way to achieve this would be through the assumption of a prescribed toroidal force field $F_\varphi(s, z)$ which is unaffected by Lorentz forces.

Greenspan (1974) has studied a non-linear eigenvalue problem that presents itself in the magnetostrophic limit $E = E_m = 0$, viz. suppose that $U = v(s)\mathbf{i}_\varphi$, that $\alpha(s, z)$ is prescribed, that

$$\nabla \wedge (\mathbf{U} \wedge \mathbf{B}) + \nabla \wedge (\alpha \mathbf{B}) + \lambda \nabla^2 \mathbf{B} = 0 \quad (r < R), \qquad (12.40)$$

and finally that

$$\mathscr{T}(s)=(\mu_0\rho)^{-1}\iint_{C(s)}(\nabla\wedge\mathbf{B})\wedge\mathbf{B}s\,\mathrm{d}\varphi\,\mathrm{d}z=0\quad(r<R).\qquad(12.41)$$

The matching conditions to a current-free field in $r>R$ are understood. The problem is to determine the values of the parameter $|\alpha|_{\max}R/\lambda$ for which solutions $\{\mathbf{B}(\mathbf{x}), v(s)\}$ exist, and to determine the form of these solutions. Greenspan finds a formal solution to this problem in terms of infinite series when $\alpha(s, z)$ is non-zero only in a thin layer on $(s^2+z^2)^{1/2}=R-$. The physical structure and significance of this solution remain to be determined.

A similar formulation to that of Malkus & Proctor (1975) has also been studied by Braginskii (1975) who however puts forward the possibility that $|B_s|$ may be much weaker than $|B_z|$ in the liquid core of the earth, and that in consequence the coupling of cylindrical shells due to the 'threading' of the field B_s may be so weak that the Taylor function $\mathscr{T}(s)$, though small, need not vanish identically in the steady state. Solutions based on this supposition have not yet however been worked out in detail and it remains to be seen whether this approach will give a class of dynamos that is genuinely distinct from that of Malkus & Proctor.

12.3. Dynamos driven by buoyancy forces

As discussed in § 4.4, the question of whether the liquid core of the Earth is stably or unstably stratified is not yet fully resolved, and buoyancy forces of thermal or non-thermal origin remain as one of the most likely sources of energy for core motions. We have discussed in § 10.6 the stability problem for a plane rotating layer heated from below, and have shown there that when the system is unstable, the motions that ensue may generally be expected to have non-zero mean helicity, and therefore at least to have the potential to act as a dynamo. The key question that now arises is to what level the magnetic energy will rise when such dynamo action does occur.

If the Rayleigh number R_a (see (12.42b) below) describing the state of the system is just above the critical value R_{ac} for the onset of cellular convection, then in the absence of any magnetic effects, the amplitude of the motions may on general grounds be expected to settle down to a value of order $\varepsilon = (R_a - R_{ac})^{1/2}$ (Malkus & Veronis, 1958). If this motion is strongly helical, then there will be an associated α-effect, in general strongly anisotropic due to the influence of rotation, but with $\alpha_0 = O(\varepsilon^2)$, where α_0 is a typical ingredient of the pseudo-tensor α_{ij}. If now the system is sufficiently large (in at least one of its dimensions) for the growth of a large-scale field (due essentially to an α^2-type process), then, as in the discussion of § 10.3, this growth may be expected to continue until the Lorentz force modifies the cellular convection pattern in such a way as to reduce the α-effect. If ε decreases (keeping all other parameters fixed), then we must reach a critical value, ε_1 say, at which dynamo action *just* occurs. If now ε is just greater than ε_1 (i.e. $\eta = (\varepsilon - \varepsilon_1)/\varepsilon_1 \ll 1$), then there *may* exist a stable state near to the original state of steady convection with a typical field amplitude B_0 of order η (or some positive power of η depending on the detailed equilibration mechanism), and if η is sufficiently small the magnetic energy in this new equilibrium situation will be small compared with the kinetic energy of the cellular motion. In this situation, a perturbation approach may lead to the determination of the equilibrium field structure and amplitude. It is not however certain *a priori* that such a neighbouring state exists, since, as indicated in the earlier discussion of § 10.3, it is always possible that a suitably oriented magnetic field may release the constraint of strong rotation, permitting more vigorous convection in which case a runaway situation would result; the field would then presumably grow to a large amplitude, the Lorentz force becoming at least comparable with the Coriolis force, i.e. a state of magnetostrophic equilibrium might be anticipated.

The perturbation approach based on the above idea of equilibrium at weak field strength has been developed by Soward (1974, see also Childress & Soward, 1972) for the plane rotating layer problem, and by Busse (1975*b*) for an annular geometry. The detailed analysis of these papers is complicated, and it is not practicable to give the details here. It will be sufficient to describe the essential approximations adopted and the nature of the conclusions in the two cases.

The Soward convection-driven dynamo[2]

Soward (1974) studies the configuration sketched in fig. 12.3(*a*). The planes $z = 0, z_0$ are supposed stress-free, perfectly conducting and isothermal, conditions that (as in § 10.6) lead to the simplest combination of boundary conditions for analytical solution of the stability problem. (If the stress-free condition were replaced by a no-slip condition, then, as Soward observes, Ekman layers on the two boundaries can play an important part in the long-term finite amplitude behaviour of convection cells; the system may be expected to be equally sensitive to the electrical and thermal boundary conditions.) In the absence of any magnetic field, the system is characterised by the Ekman, Rayleigh and Prandtl numbers,

$$E = \nu/\Omega z_0^2, \qquad R_a = \alpha g z_0^3 \Delta\Theta/\nu\kappa, \qquad \sigma = \nu/\kappa, \qquad (12.42)$$

where $\Delta\Theta$ is the temperature difference between the plates, and κ the thermal diffusivity, and it is supposed that

$$E \ll 1, \qquad \sigma = O(1). \qquad (12.43)$$

The critical Rayleigh number (Chandrasekhar, 1961) is in these circumstances

$$R_{ac} = O(E^{-4/3}), \qquad (12.44)$$

and the horizontal wave-number for which R_{ac} is minimal is

$$k_c = O(E^{-1/3}) z_0^{-1}. \qquad (12.45)$$

Soward supposes that the layer is just unstable, in the sense that

$$\frac{R_a - R_{ac}}{R_{ac}} = O(E^{1/3}), \qquad (12.46)$$

so that only modes with horizontal wave-vectors $\mathbf{k} = (k_1, k_2, 0)$ such that $|\mathbf{k}| \approx k_c$ are excited. He supposes moreover that $\sigma > 1$, under which condition (Chandrasekhar, 1961) the motion at the onset of convection is steady, rather than time-periodic. Under the

[2] Our notation in this subsection differs slightly from Soward's, in the interest of retaining consistency with the notation of earlier chapters.

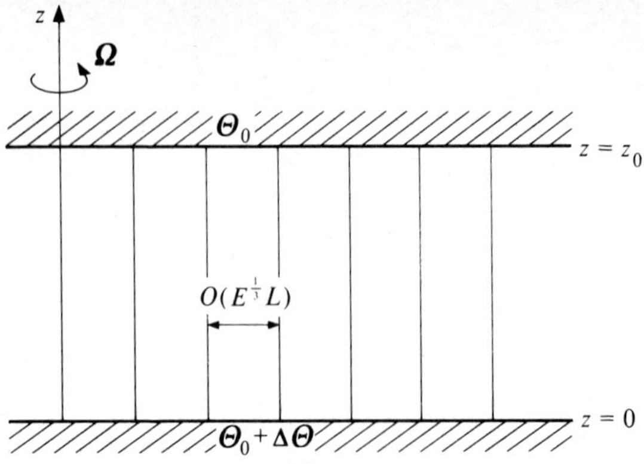

(*a*)

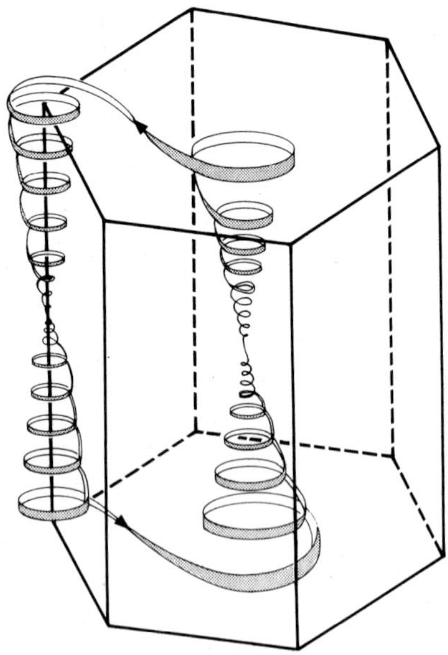

[From Veronis 1959]

(*b*)

Fig. 12.3 (*a*) Sketch of the configuration studied by Soward (1974); the vertical lines indicate the boundaries of convection cells when the Rayleigh number exceeds the critical value; when $E \ll 1$, the horizontal scale is $O(E^{1/3}z_0)$ as indicated. (*b*) A perspective sketch of a fluid particle path which passes through the centre of a cell when the cell planform is hexagonal; the helicity associated with this type of motion is evidently antisymmetric about the centre plane $z = \frac{1}{2}z_0$, and the α-effect (given by (12.51) and (12.52)) is likewise antisymmetric about $z = \frac{1}{2}z_0$.

condition (12.46), the amplitude of the velocity in the convection cells excited has order of magnitude (Malkus & Veronis, 1958)

$$u_0 = O(E^{1/6})\nu/z_0. \tag{12.47}$$

This indicates that an expansion of all mean and fluctuating field variables as power series in the small parameter $E^{1/6}$ is appropriate, and this is the basis of Soward's perturbation procedure.

Motions periodic in the x and y directions are determined uniquely by the vertical velocity distribution

$$w(\mathbf{x}, t) = \mathrm{Re} \sum_{\mathbf{k}} \hat{w}(\mathbf{k}, t) \sin \frac{\pi z}{z_0} \mathrm{e}^{i\mathbf{k}.\mathbf{x}}. \tag{12.48}$$

The fact that the horizontal scale $l = O(E^{1/3})z_0$ is small compared with z_0 now permits the use of the methods of mean-field electrodynamics, means being defined over the horizontal plane. The magnetic Reynolds number based on u_0 and l is

$$u_0 l/\lambda = O(E^{1/6})\nu/\lambda, \tag{12.49}$$

and this is small provided[3] $\nu/\lambda \lesssim O(1)$. Hence in calculating the pseudo-tensor α_{ij} (cf. § 7.8) the magnetic field perturbation $\mathbf{b}$ is effectively determined (at leading order in $E^{1/6}$) by

$$\lambda \nabla^2 \mathbf{b} = -\mathbf{B} \,.\, \nabla \mathbf{u}, \tag{12.50}$$

and calculation of

$$\mathscr{E}_i = \langle \mathbf{u} \wedge \mathbf{b} \rangle_i = \alpha_{ij} B_j \tag{12.51}$$

is a relatively straightforward matter, with the result

$$\alpha_{ij} = \frac{\pi}{E^{1/2}\lambda z_0} \sum_{\mathbf{k}} \frac{k_i k_j}{k^6} q(\mathbf{k}, t) \sin\left(\frac{2\pi z}{z_0}\right), \tag{12.52}$$

where $q(\mathbf{k}, t) = |\hat{w}(\mathbf{k}, t)|^2$. This calculation of course requires knowledge of the phase relationships between horizontal and vertical

[3] Soward assumes $\nu/\lambda = O(1)$; the fact that $\nu/\lambda \ll 1$ in the core of the Earth should perhaps be incorporated in the expansion procedure to make the theory more relevant in the terrestrial context.

velocity components, as determined by the linear stability analysis. In the expression (12.51), the mean field $\mathbf{B}(z, t)$ is necessarily horizontal, and it evolves according to the (now) well-known equation

$$\frac{\partial B_i}{\partial t} = \varepsilon_{ijk} \frac{\partial}{\partial x_j} (\alpha_{kl}(z, t) B_l) + \lambda \frac{\partial^2 B_i}{\partial z^2}. \quad (12.53)$$

The next stage of the calculation is to derive equations for the $q(\mathbf{k}, t)$ appearing in (12.52) by continuation of the systematic perturbation procedure to the order $(O(E^{1/2}))$ at which Lorentz forces have a significant effect on the dynamics. These equations, as derived by Soward, take the form

$$\partial q(\mathbf{k}, t)/\partial t + 2 \sum_{\mathbf{k}'} A(\mathbf{k}, \mathbf{k}') q(\mathbf{k}', t) q(\mathbf{k}, t)$$
$$- \left[m(\mathbf{k}, t) - \tfrac{1}{2} \sum_{\mathbf{k}'} m(\mathbf{k}', t) q(\mathbf{k}', t) \right] q(\mathbf{k}, t) = 0, \quad (12.54)$$

provided all the wave-vectors $\mathbf{k}$ in the velocity spectrum have equal magnitude. Here $A(\mathbf{k}, \mathbf{k}')$ is a coupling coefficient representing non-linear interactions between the instability modes; $m(\mathbf{k}, t)$ is a weighted average of $(\mathbf{B} \cdot \mathbf{k})^2$ over $0 < z < z_0$, and the terms involving m represent the effect of the mean magnetic field on the small-scale motions.

In order to integrate (12.53) and (12.54), it is necessary to adopt initial conditions, and in particular to specify the horizontal structure of the cellular convection pattern at time $t = 0$: if only one wave-vector $\mathbf{k}_1$ is represented (i.e. $q(\mathbf{k}, 0) = 0$ unless $\mathbf{k} = \pm\mathbf{k}_1$) then the motion has the form of cylindrical rolls with axes perpendicular to $\mathbf{k}_1$; it is almost self-evident that this motion has too simple a structure to provide dynamo maintenance of $\mathbf{B}$. If two wave-vectors $\mathbf{k}_1$, $\mathbf{k}_2$ are equally represented ($q(\mathbf{k}_1, 0) = q(\mathbf{k}_2, 0)$ with $|\mathbf{k}_1| = |\mathbf{k}_2|$ and $\mathbf{k}_1 \cdot \mathbf{k}_2 = 0$) and if the amplitudes of the various modes are equal, then the convection cells have square boundaries[4]. If three wave-

[4] Cf. the velocity field (3.114) in § 3.12, although in that case, four horizontal wave-vectors are represented, viz. $(k_1, 0, 0)$, $(0, k_1, 0)$, $2^{-1/2}(k_1, k_1, 0)$, $2^{-1/2}(k_1, -k_1, 0)$, with $k_1 = \pi/z_0$.

vectors $\mathbf{k}_1$, $\mathbf{k}_2$, $\mathbf{k}_3$, are equally represented, and if

$$|\mathbf{k}_1|^2 = |\mathbf{k}_2|^2 = |\mathbf{k}_3|^2 = -\frac{\sqrt{3}}{2}\mathbf{k}_2 \,.\, \mathbf{k}_3$$

$$= -\frac{\sqrt{3}}{2}\mathbf{k}_3 \,.\, \mathbf{k}_1 = -\frac{\sqrt{3}}{2}\mathbf{k}_1 \,.\, \mathbf{k}_2, \qquad (12.55)$$

then the cell boundaries are regular hexagons; a typical particle path in this case is sketched in fig. 12.3(*b*) (Veronis, 1959).

Numerical integration of (12.53) and (12.54) for the case of square cell boundaries indicates that both the magnetic energy

$$M(t) = \frac{1}{2\mu_0}\int_0^L \mathbf{B}^2\,\mathrm{d}z, \qquad (12.56)$$

and the quantity $v(t) = q(\mathbf{k}, t) - q(\mathbf{k}_2, t)$ settle down to a time-periodic behaviour when $\lambda t/L^2$ becomes large. The ratio of mean magnetic energy to mean kinetic energy in this asymptotic periodic state is of order $E^{1/3}$ (i.e. small as required for validity of the 'weak field' approach) and the fluctuations in magnetic energy are of order $\pm 3\%$ about the mean value. Soward presented these results in the form of an approach to a limit cycle in the plane of the variables $v(t), M(t)$ (see fig. 12.4).

The results for hexagonal cell boundaries are naturally more complicated since now the kinetic energy is shared among three modes rather than two; nevertheless Soward's numerical integrations again indicate asymptotic weak fluctuations of the magnetic energy about its ultimate mean level, while the kinetic energy (or rather a substantial fraction of it) appears to 'flow' cyclically among the three modes.

It is a characteristic feature of these dynamos that both velocity and magnetic field distributions are unsteady (though ultimately periodic in time with period of order z_0^2/λ).

Busse's (1975b) model of the geodynamo

One of the difficulties in analysing the problem of thermal convection in a rotating spherical annulus (as a model of convection between the solid inner core of the Earth and the mantle) is that the radial gravity vector **g** does not make a constant angle with the

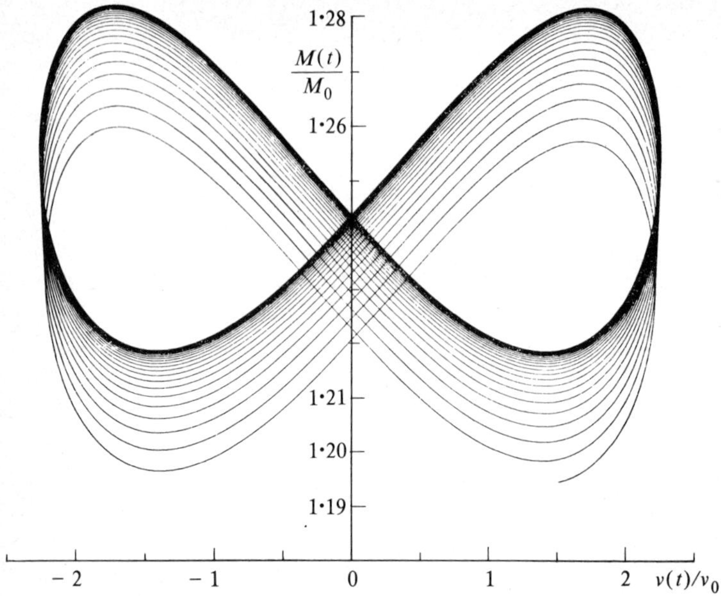

Fig. 12.4 Phase plane evolution of magnetic energy $M(t)$ and the quantity $v(t)$ which represents twice the difference between the kinetic energies in the modes corresponding to wave-vectors $\mathbf{k}_1$ and $\mathbf{k}_2$; the normalising constants M_0 and v_0 depend on the precise initial conditions of the problem. The figure shows the evolution, over 16 000 time-steps in the integration procedure, towards a limit cycle in which $M(t)$ and $v(t)$ vary periodically with time. (From Soward, 1974.)

rotation vector $\mathbf{\Omega}$. However convection at small Ekman number (and when Lorentz forces are sufficiently weak) is characterised by long thin convection cells aligned with the direction of $\mathbf{\Omega}$ (fig. 12.5(a)). Thermal instabilities of this kind have been investigated by Busse (1970). Only the component $g \sin \theta$ of $\mathbf{g}$ perpendicular to $\mathbf{\Omega}$ is effective for cells of this structure. Since the centrifugal force $\Omega^2 r \sin \theta$ has the same θ-dependence, the effects of radial gravity can be simulated in laboratory experiments by the centrifugal force in a spherical annulus rotated about a vertical axis; in such experiments, unstable stratification is provided by heating the outer sphere and cooling the inner sphere (a deliberate reversal to meet the fact that the centrifugal force is outwards, whereas the gravitational force in the terrestrial context is inwards). Such experiments (Busse & Carrigan, 1974) confirm the appearance of convection

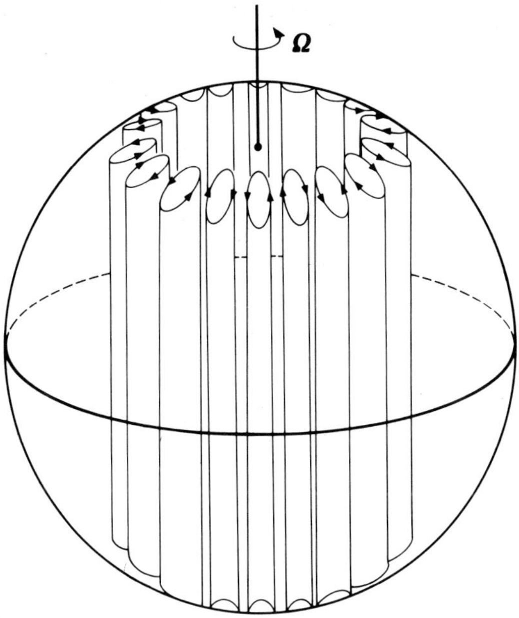

(a)

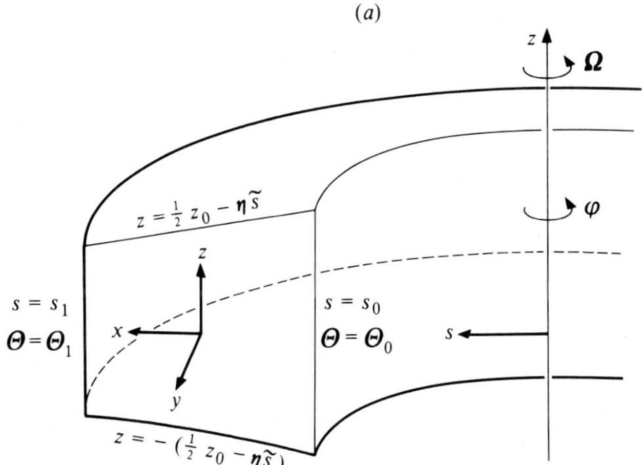

(b)

Fig. 12.5 (a) Sketch of the marginally unstable convective motions in a sphere of rapidly rotating fluid with a uniform distribution of heat sources. (From Busse, 1970.) (b) Section of the cylindical geometry which incorporates the essential features of the spherical system; the outer cylinder is heated ($\Theta_1 > \Theta_0$); the (locally) Cartesian coordinates *Oxyz* used in the analysis are as indicated. (From Busse, 1975b.)

columns in the region outside the cylinder $C(s_0)$ where s_0 is the radius of the inner sphere.

These considerations led Busse (1975*b*) to consider the possibility of dynamo action due to thermal convection in the simpler cylindrical geometry of fig. 12.5(*b*). The cylindrical annulus

$$s_0 < s < s_1, \quad |z| < \tfrac{1}{2}z_0 - \eta\tilde{s}, \quad \tilde{s} = s - \tfrac{1}{2}(s_0 + s_1), \qquad (12.57)$$

has approximately the same form as the region occupied by convection cells in fig. 12.5(*a*). The parameter η represents the slope of the upper and lower surfaces of the annulus and is assumed small in the theory[5]. Moreover it is assumed that $z_0 \ll s_1 - s_0 \ll s_1$ so that (as in earlier studies) the cylindrical polar coordinates (s, φ, z) may be replaced by Cartesian coordinates (x, y, z). These latter assumptions are of course *not* reflected in the spherical annulus problem, but the hope is that the qualitative behaviour of the latter problem will nevertheless be adequately represented by the quasi-cartesian model.

It is clear that simple motions of the kind represented by fig. 12.5(*a*) have zero helicity and will not in themselves be sufficient to provide an α-effect. What is needed is a superposed motion within each cell, roughly aligned with the axis of the cell, and correlated with the sense of rotation in the cell. This ingredient of the motion in Busse's theory is provided by the Ekman suction effect (Greenspan, 1968) associated with the Ekman layers on the upper and lower boundaries; it is of order $E^{1/2}$ relative to the primary cellular motion, and provides a mean helicity linear and antisymmetric in z, and of order $E^{1/2}z_0^{-1}$ relative to the mean kinetic energy. This need for an ingredient of flow parallel to two-dimensional convection cells, if dynamo action is to occur, was recognised in an earlier study of Busse (1973).

Busse's calculation proceeds in three stages. (i) The first stage consists in calculation of the critical stability conditions when there is no magnetic field present, of the structure of the critically stable disturbances, and of their amplitude when conditions are slightly supercritical; this involves the procedure of Malkus & Veronis

[5] The perturbation procedures adopted also require that $E^{1/4} \ll \eta \ll z_0/(s_1 - s_0)$. A non-zero value for η is of crucial importance in determining the stability properties of the system.

(1958), but with the additional feature here that the small slope of the upper and lower surfaces generates weak unsteadiness in the instability modes which propagate slowly in the azimuth direction. (ii) The second stage involves solution of the kinematic dynamo problem with the velocity field determined by stage (i); here, as in Soward's (1974) work, the methods of mean-field electrodynamics are applied, averages being defined over the x and y variables. The dynamo mechanism is again of α^2-type, the effective value of α being linear in z (like the helicity discussed above). As in all such dynamos, the toroidal and poloidal fields are of the same order of magnitude. Busse restricts attention to fields of dipole symmetry[6] about $z = 0$, and obtains a criterion for growth of the field. Since the α-effect is proportional in intensity to $E^{1/2}u_0^2$ where $u_0^2 = \langle \mathbf{u}^2 \rangle$, this criterion in effect puts a lower bound on $E^{1/2}u_0^2$ in terms of other dimensionless parameters of the problem, for dynamo action to occur. (iii) In the third stage, Busse calculates the small modification of the convection pattern due to the presence of the field excited (when $E^{1/2}u_0^2$ is just large enough) as calculated in stage (ii). The field has an as yet undetermined amplitude B_0 which is assumed small. The amplitude u_0 of the modified motion and the amplitude B_0 are simultaneously determined by the condition that the fundamental magnetic mode should have zero growth rate under the α^2-action of the slightly modified convection cells. Since increase of B_0 leads to a decrease in u_0 (other things being constant), an equilibrium in which the magnetic energy density is small compared with the kinetic energy density is attained.

It is this latter fact that most distinguishes the Busse and Soward dynamos from the type of dynamo conceived by Malkus & Proctor (1975), which is characterised by magnetostrophic force balance and consequently a magnetic energy density that is the more relevant as far as the Earth's liquid core is concerned. Extensions of the Busse and Soward models to strong (rather than weak) field

[6] Busse does not actually match the field to a field in the current-free region outside the annulus. He claims that the condition of dipole symmetry 'allows for the continuation of the meridional field towards infinity in such a way that it will decay at least as fast as a dipolar field'. The claim is plausible, but it is desirable that the actual matching should be explicitly carried out, since in the dynamo context the presence of a 'source at infinity', which might also generate a field *of dipole symmetry* about $z = 0$, must be carefully excluded.

situations (as outlined by Childress & Soward, 1972) are doubtless possible and will be eagerly awaited. Busse (1976) has already made a move in this direction.

12.4. Reversals of the Earth's field, as modelled by coupled disc dynamos

A simple model for reversals of the Earth's magnetic field was proposed by Rikitake (1958) and has been subsequently studied by Allan (1962) and Cook & Roberts (1970). This is the coupled disc dynamo model sketched in fig. 12.6(*a*), an elaboration of the single

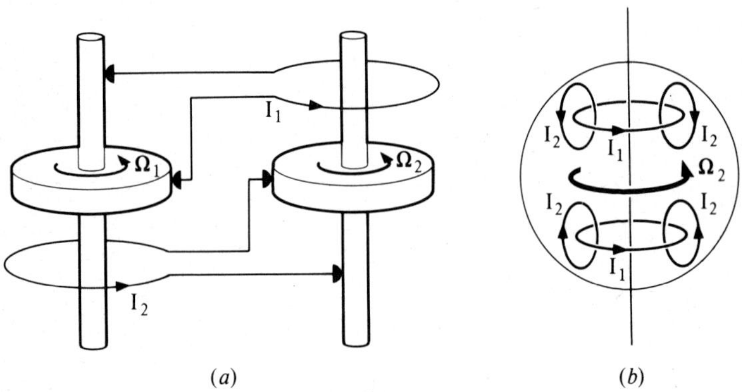

Fig. 12.6 (*a*) Coupled disc dynamo (after Rikitake, 1958); the discs rotate (under applied torques) with angular velocities $\Omega_1(t)$, $\Omega_2(t)$ and drive currents $I_1(t)$, $I_2(t)$ through the wires which make sliding contact with rims and axles as indicated. (*b*) The corresponding gross currents and rotation in the Earth's core; Ω_1 (not shown in the figure) is associated with the α-effect (see (12.61)) and Ω_2 is associated with differential rotation (see (12.60)) about the axis of mean rotation.

homopolar disc dynamo discussed in chapter 1. For the coupled system, there are two degrees of freedom represented by the angular velocities $\Omega_1(t)$ and $\Omega_2(t)$ of the discs, and two 'electrical degrees of freedom' represented by the currents $I_1(t)$ and $I_2(t)$ in the wires which must both be twisted in the same sense relative to the rotation vectors if field regeneration is to occur. Under symmetric conditions, the equations describing the evolution of this system, when a constant torque G is applied to each disc, are

$$\begin{aligned} L\,\mathrm{d}I_1/\mathrm{d}t + RI_1 &= M\Omega_1 I_2 \\ L\,\mathrm{d}I_2/\mathrm{d}t + RI_2 &= M\Omega_2 I_1, \\ C\,\mathrm{d}\Omega_1/\mathrm{d}t &= G - MI_1 I_2, \\ C\,\mathrm{d}\Omega_2/\mathrm{d}t &= G - MI_1 I_2, \end{aligned} \tag{12.58}$$

where L and R are the self-inductance and resistance of each circuit, C is the moment of inertia of each disc about its axis, and $2\pi M$ is the mutual inductance between the circuits. The terms $M\Omega_1 I_2$ and $M\Omega_2 I_1$ represent the electromotive forces arising from the rotations Ω_1 and Ω_2, while the term $-MI_1I_2$ represents the torque associated with the Lorentz force distribution in each disc.

In what sense may the simple system (12.58) be regarded as a model for the (literally) infinitely more complicated situation in the Earth's liquid core? First (fig. 12.6(b)) we can regard $I_1(t)$ and $I_2(t)$ as providing measures of the total toroidal current (integrated across a meridian plane) and 'total meridional current' which may be defined by

$$I_2(t) = \tfrac{1}{2} \max_z \iint_{z=\text{cst.}} |J_z(x, y, z, t)|\,\mathrm{d}x\,\mathrm{d}y. \tag{12.59}$$

$I_1(t)$ is then associated with mean poloidal field, and $I_2(t)$ with the mean toroidal field (the mean being over the azimuth angle φ). The term $M\Omega_2 I_1$ in (12.58b) then represents production of toroidal field, and $\Omega_2(t)$ is then best interpreted as a measure of the mean differential rotation in the core, say

$$\Omega_2(t) = \frac{R_C}{V_C} \iiint_{\text{core}} |\nabla\omega|\,\mathrm{d}V, \tag{12.60}$$

where V_C is the volume of the liquid core, and R_C the core radius. Similarly the term $M\Omega_1 I_2$ represents production of poloidal field, and $\Omega_1(t)$ is best interpreted as a measure of the mean intensity of the α-effect, say

$$\Omega_1(t) = \frac{1}{R_C V_C} \iiint_{\text{core}} |\alpha|\,\mathrm{d}V. \tag{12.61}$$

The term $-MI_1I_2$ in (12.58c) then represents the reduction of the α-effect as described in § 10.3, while in (12.58d) it represents the modification of the mean velocity distribution (purely toroidal in this idealisation) as described in § 12.2. The term G in (12.58c, d) represents on the one hand the driving forces (of thermal or other origin) which generate an α-effect, and on the other hand the source of differential rotation (due to transfer of angular momentum by small-scale mixing associated with the α-effect).

It would of course be nice to provide a *formal derivation* of a system of ordinary differential equations similar to (12.58), by appropriate averaging of the full magnetohydrodynamic equations, but the non-linearity of these equations makes this a difficult (if not impossible) task[7], and all that can be said at present is that the equations (12.58) provide a plausible model, the behaviour of which is at the least suggestive in the terrestrial context. The model can of course be varied (e.g. by introducing different coupling constants M_1, M_2, different coefficients of inertia C_1, C_2, and different torques G_1, G_2, or by increasing the number of degrees of freedom through the introduction of further discs), but the simple system (12.58) already contains enough parameters to permit a wide range of behaviour, and we therefore restrict attention to this system in the following discussion, which closely follows the treatment of Cook & Roberts (1970).

With the definition of dimensionless variables

$$\tau = (GM/CL)^{1/2}t, \qquad X, Y = (M/G)^{1/2}I_{1,2},$$
$$Z, V = (CM/GL)^{1/2}\Omega_{1,2}, \tag{12.62}$$

the equations (12.58) take the simpler form

$$\dot{X} + \mu X = ZY,$$
$$\dot{Y} + \mu Y = VX, \tag{12.63}$$
$$\dot{Z} = \dot{V} = 1 - XY,$$

where $\mu = (C/GLM)^{1/2}R$. These equations have a trivial first integral

$$Z - V = A, \tag{12.64}$$

where A is a constant which may be assumed non-negative. Note also that

$$\frac{\mathrm{d}}{\mathrm{d}\tau}(X^2 + Y^2 + Z^2 + V^2) = (Z + V) - \mu(X^2 + Y^2), \tag{12.65}$$

[7] An approach based on truncation of the system of moment equations derivable from (12.34) and (12.35) has been explored by Robbins (1976).

an energy equation for the system: the first bracketed term on the right represents the rate of working of the applied torques, while the second represents the ohmic dissipation in the two circuits; under steady conditions these two terms are in balance. The non-linear terms in (12.63) make no contribution to the total energy budget, and correspond simply to transfer of energy between magnetic and kinetic 'reservoirs'.

Steady state solutions of (12.63) satisfy

$$XY = 1, \qquad ZV = \mu^2. \tag{12.66}$$

Introducing a constant $K \geqslant 1$ such that

$$A = \mu(K^2 - K^{-2}), \tag{12.67}$$

these steady state solutions are given by

$$X = \pm K, \qquad Y = \pm K^{-1}, \qquad Z = \mu K^2, \qquad V = \mu K^{-2}. \tag{12.68}$$

In these two states, which may be (arbitrarily) described as the normal state S_+ and the reversed state S_-, the currents are either both positive (i.e. in the sense indicated in fig. 12.6(*a*)) or both negative. It is of course no surprise that two complementary states of this kind exist; we have previously (§ 11.2) referred to the general invariance of the equations of magneto-hydrodynamics under change of sign of **B**, a property shared by the system (12.63) under the transformation $(X, Y) \to (-X, -Y)$. The interesting question now is whether transitions from a neighbourhood of one stationary state to a neighbourhood of the other are possible, a question that naturally involves the stability of the two states.

To examine the stability of the state S_+ (the state S_- being entirely similar) we put

$$X = K + \xi, \qquad Y = K^{-1} + \eta, \qquad Z - \mu K^2 = V - \mu K^{-2} = \zeta, \tag{12.69}$$

substitute in (12.63) and linearise in (ξ, η, ζ); this gives three linear homogeneous equations in (ξ, η, ζ) which admit solutions proportional to $e^{p\tau}$ possible values of p being given by the vanishing of a

3×3 determinant. This gives a cubic for p with the three roots

$$p_1=-2\mu, \qquad p_2=\mathrm{i}(K^2+K^{-2})^{1/2}, \qquad p_3=-\mathrm{i}(K^2+K^{-2})^{1/2}. \tag{12.70}$$

The pure imaginary value of p_2 and p_3 suggest neutral stability. However in this situation the terms non-linear in the small perturbations ζ, η and ζ may have a long-term destabilising influence. Cook & Roberts examined this possibility using the method of Liapounov (1947), and showed that when $K>1$ the S_+ state is indeed unstable (as is the S_- state also) as a result of the cumulative effect of these non-linear terms over times τ large compared with $|K^2+K^{-2}|^{-1/2}$.

A solution of the system (12.63) may be represented by the motion of a point $P(\tau)$ with Cartesian coordinates $(X(\tau),\ Y(\tau),\ Z(\tau))$, the orbit in this phase space naturally depending on the initial position $P(0)$. At each point of this phase space is defined a 'velocity field'

$$\mathbf{u}=(\dot{X},\ \dot{Y},\ \dot{Z})=(-\mu X+ZY,\ -\mu Y+(Z-A)X,\ 1-XY), \tag{12.71}$$

for which

$$\nabla\,.\,\mathbf{u}=\frac{\partial u_1}{\partial X}+\frac{\partial u_2}{\partial Y}+\frac{\partial u_3}{\partial Z}=-2\mu. \tag{12.72}$$

The associated 'density field' $\rho(X,\ Y,\ Z)$ satisfies the conservation equation (in Lagrangian form)

$$\frac{\mathrm{D}\rho}{\mathrm{D}\tau}=-\rho\nabla\,.\,\mathbf{u}=2\mu\rho, \tag{12.73}$$

with Lagrangian solution

$$\rho=\rho_0\,\mathrm{e}^{2\mu\tau}. \tag{12.74}$$

Equivalently the volume $\mathrm{d}V$ occupied by any element of 'phase fluid' tends to zero in a time of order μ^{-1}, suggesting that, for nearly all possible initial positions $P(0)$, the trajectory of $P(\tau)$ will ultimately lie arbitrarily close to a limit surface $F(X,\ Y,\ Z)=0$, with the property that

$$(-\mu X+ZY)\frac{\partial F}{\partial X}+(\mu Y+(Z-A)X)\frac{\partial F}{\partial Y}+(1-XY)\frac{\partial F}{\partial Z}=0. \tag{12.75}$$

This type of behaviour was confirmed by Cook & Roberts who computed solution trajectories of (12.71), and who succeeded in describing the topological character of the limit surfaces for particular values of the parameters μ and A. Fig. 12.7 shows a typical evolution of $X(\tau)$, and a projection of a typical trajectory on the (Y, Z) plane, in both cases for $\mu = 1, K = 2$. The stationary states $S_\pm$ for this case are given by

$$X = \pm 2, \qquad Y = \pm\tfrac{1}{2}. \qquad Z = 4, \qquad V = \tfrac{1}{4}. \qquad (12.76)$$

The transitions from a neighbourhood of S_+ to a neighbourhood of S_- are evident in both figures, although what is most striking is the fact that the oscillations about these states are typically of *large* amplitude; even if $P(0)$ is very near to either S_+ or S_-, $P(\tau)$ ultimately follows a trajectory which orbits alternatively about S_+ and S_- *without ever again* approaching very near to either state.

This type of behaviour is at least qualitatively comparable with the behaviour of the north–south component $\mu_z^{(1)}(t)$ of the Earth's dipole moment over time-scales of the order of millions of years, as described in § 4.3. The existence of two time-scales (one characteristic of the mean time between reversals, the other of the duration of reversals), the oscillations about the present 'normal' state on the shorter time-scale, and the occasional appearance of short 'bursts' of one state between much longer periods of the other, are all reflected in the behaviour of the solution curve of fig. 12.7(a). This provides a strong indication that reversals of the Earth's field are indeed the result of coupled non-linear dynamic and magnetic oscillations. The work of Proctor (1977a) and Soward (1974) as described in §§ 12.2 and 12.3 indicates that for some dynamo models, even when a steady dynamically equilibrated state exists, the system may well prefer to seek a state involving finite amplitude oscillations of magnetic energy. Whether such models, based on systematic deduction from the equations of magnetohydrodynamics in a rotating fluid, can also yield genuine field reversals in a spherical geometry is as yet unknown; this problem continues to present an exciting challenge in an important area of mathematical geophysics.

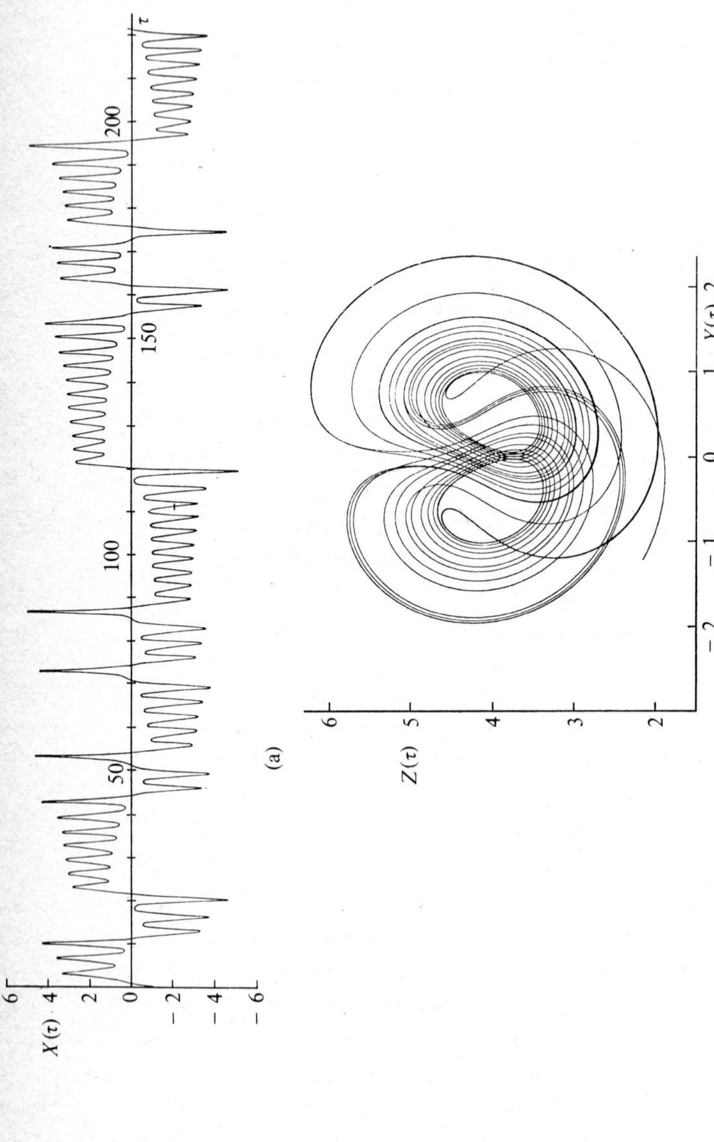

Fig. 12.7 Computed solution of the system (12.63) with $\mu = 1$, $K = 2$. (*a*) Note how $X(\tau)$ exhibits large amplitude oscillations about the stationary state value ± 2 and reversals of sign at irregular intervals. (*b*) The projection of a typical trajectory in phase space on the (Y, Z) plane; note that $Y(\tau)$ (like $X(\tau)$)

REFERENCES

Acheson, D. J. & Hide, R. (1973) Hydromagnetics of rotating fluids. *Rep. Prog. Phys.* **36**, 159–211.

Acuna, M. H. & Ness, N. F. (1976) The main magnetic field of Jupiter. *J. Geophys. Res.* **81**, 2917–22.

Allan, D. W. (1962) On the behaviour of systems of coupled dynamos. *Proc. Camb. Phil. Soc.* **58**, 671–93.

Alfvén, H. (1942) On the existence of electromagnetic-hydrodynamic waves. *Arkiv. f. Mat. Astron. Fysik* **29**B, no. 2. (7 pp.).

Altschuler, M. D., Trotter, D. E., Newkirk, G. Jr. & Howard, R. (1974) The large-scale solar magnetic field. *Solar Phys.* **39**, 3–17.

Andrews, D. G. & Hide, R. (1975) Hydromagnetic edge waves in a rotating stratified fluid. *J. Fluid Mech.* **72**, 593–603.

André, J. D. & Lesieur, M. (1977) Evolution of high Reynolds number isotropic three-dimensional turbulence; influence of helicity. *J. Fluid Mech.* **81**, 187–208.

Angel, J. R. P. (1975) Strong magnetic fields in White Dwarfs. *Ann. New York Acad. Sci.* **257**, 80–1.

Anufriev, A. P. & Braginskii, S. I. (1975) Effect of a magnetic field on the stream of a liquid rotating at a rough surface. *Mag. Gidrod.* **4**, 62–8. [English translation: *Magnetohydrodynamics* **11**, 461–7.]

Babcock, H. D. (1959) The Sun's polar magnetic field. *Astrophys. J.* **130**, 364–80.

Babcock, H. W. (1947) Zeeman effect in stellar spectra. *Astrophys. J.* **105**, 105–19.

Babcock, H. W. & Babcock, H. D. (1953) Mapping the magnetic fields of the Sun. *In* The Sun, ed. Kuiper, G. R., Univ. of Chicago Press, pp. 704–10.

Babcock, H. W. & Babcock, H. D. (1955) The Sun's magnetic field 1952–1954. *Astrophys. J.* **121**, 349–66.

Backus, G. E. (1957) The axisymmetric self-excited fluid dynamo *Astrophys. J.* **125**, 500–24.

Backus, G. E. (1958) A class of self-sustaining dissipative spherical dynamos. *Ann. Phys.* **4**, 372–447.

Backus, G. E. & Chandrasekhar, S. (1956) On Cowling's theorem on the impossibility of self-maintained axisymmetric homogeneous dynamos. *Proc. Nat. Acad. Sci.* **42**, 105–9.

Barraclough, D. R., Harwood, J. M., Leaton, B. R. & Malin, S. C. R. (1975) A model of the geomagnetic field at epoch 1975. *Geoph. J. Roy. Astr. Soc.* **43**, 645–59.

Batchelor, G. K. (1950) On the spontaneous magnetic field in a conducting liquid in turbulent motion. *Proc. Roy. Soc.* A**201**, 405–16.

Batchelor, G. K. (1952) The effect of homogeneous turbulence on material lines and surfaces. *Proc. Roy. Soc.* A**213**, 349–66.

Batchelor, G. K. (1953) The theory of homogeneous turbulence. Cambridge Univ. Press.

Batchelor, G. K. (1956) Steady laminar flow with closed streamlines at large Reynolds numbers. *J. Fluid Mech.* **1**, 177–90.

Batchelor, G. K. (1959) Small-scale variation of convected quantities like temperature in turbulent fluid, I. General discussion and the case of small conductivity. *J. Fluid Mech.* **5**, 113–33.

Betchov, R. (1961) Semi-isotropic turbulence and helicoidal flows. *Phys. Fluids* **4**, 925–6.

Bolt, B. A. & Qamar, A. (1970) Upper bound to the density jump at the boundary of the Earth's inner core. *Nature* **228**, 148–50.

Bondi, H. & Gold, T. (1950) On the generation of magnetism by fluid motion. *Mon. Not. Roy. Astr. Soc.* **110**, 607–11.

Braginskii, S. I. (1964*a*) Self excitation of a magnetic field during the motion of a highly conducting fluid. *Sov. Phys. JETP* **20**, 726–35.

Braginskii, S. I. (1964*b*) Theory of the hydromagnetic dynamo. *Sov. Phys. JETP* **20**, 1462–71.

Braginskii, S. I. (1964*c*) Kinematic models of the Earth's hydromagnetic dynamo. *Geomagn. Aeron.* **4**, 572–83.

Braginskii, S. I. (1964*d*) Magnetohydrodynamics of the Earth's core. *Geomagn. Aeron.* **4**, 698–711.

Braginskii, S. I. (1967) Magnetic waves in the Earth's core. *Geomagn. Aeron.* **7**, 851–9.

Braginskii, S. I. (1970) Torsional magnetohydrodynamic vibrations in the Earth's core and variations in day length. *Geomagn. Aeron.* **10**, 1–8.

Braginskii, S. I. (1975) An almost axially symmetric model of the hydromagnetic dynamo of the Earth, I. *Geomagn. Aeron.* **15**, 149–56.

Braginskii, S. I. & Nikolaychik, V. V. (1973) Estimation of the electrical conductivity of the Earth's lower mantle from the lag of an electromagnetic signal. *Phys. Solid Earth* **9**, 601–2.

Brissaud, A., Frisch, U., Léorat, J., Lesieur, M. & Mazure, A. (1973) Helicity cascades in fully developed isotropic turbulence. *Phys. Fluids* **16**, 1366–7.

Brown, L. W. (1975) Saturn radio emission near 1 MHz. *Astrophys. J.* **198**, L89–92.

Bullard, E. C. (1955) The stability of a homopolar dynamo. *Proc. Camb. Phil. Soc.* **51**, 744–60.

Bullard, E. C. (1968) Reversals of the Earth's magnetic field. *Phil. Trans. Roy. Soc.* A**263**, 481–524.

Bullard, E. C., Freedman, C., Gellman, H. & Nixon, J. (1950) The westward drift of the Earth's magnetic field. *Phil. Trans. Roy. Soc.* A**243**, 67–92.

Bullard, E. C. & Gellman, H. (1954) Homogeneous dynamos and terrestrial magnetism. *Phil. Trans. Roy. Soc.* A**247**, 213–78.

Bullard, E. C. & Gubbins, D. (1977) Generation of magnetic fields by fluid motions of global scale. *Geophys. Fluid Dyn.* **8**, 43–56.

Bumba, V. & Kleczek, J. (1976) Proceedings of the IAU Symposium 71, Basic Mechanisms of Solar Activity. Prague.

Busse, F. H. (1968) Steady fluid flow in a precessing spheroidal shell. *J. Fluid Mech.* **33**, 739–51.

Busse, F. H. (1970) Thermal instabilities in rapidly rotating systems. *J. Fluid Mech.* **44**, 441–60.

Busse, F. H. (1973) Generation of magnetic fields by convection. *J. Fluid Mech.* **57**, 529–44.

Busse, F. H. (1975*a*) A necessary condition for the geodynamo. *J. Geophys. Res.* **80**, 278–80.

Busse, F. H. (1975*b*) A model of the geodynamo. *Geophys. J. Roy. Astr. Soc.* **42**, 437–59.

Busse, F. H. (1976) Generation of planetary magnetism by convection. *Phys. Earth Planet Inter.* **12**, 350–8.

Busse, F. H. & Carrigan, C. R. (1974) Convection induced by centrifugal buoyancy. *J. Fluid Mech.* **62**, 579–92.

Chandrasekhar, S. (1956) On force-free magnetic fields. *Proc. Nat. Acad. Sci.* **42**, 1–5.

Chandrasekhar, S. (1961) Hydrodynamic and hydromagnetic stability. Clarendon Press, Oxford.

Chapman, S. & Bartels, J. (1940) Geomagnetism: vol. I, Geomagnetic and related phenomena; vol. II, Analysis of the data and physical theories. Clarendon Press, Oxford.

Childress, S. (1970) New solutions of the kinematic dynamo problem. *J. Math. Phys.* **11**, 3063–76.

Childress, S. & Soward, A. M. (1972) Convection-driven hydromagnetic dynamo, *Phys. Rev. Lett.* **29**, 837–9.

Clarke, A. Jr. (1964) Production and dissipation of magnetic energy by differential fluid motion. *Phys. Fluids* **7**, 1299–305.

Clarke, A. Jr. (1965) Some exact solutions in magnetohydrodynamics with astrophysical applications. *Phys. Fluids* **8**, 644–9.

Cook, A. E. & Roberts, P. H. (1970) The Rikitake two-disc dynamo system. *Proc. Camb. Phil. Soc.* **68**, 547–69.

Cowling, T. G. (1934) The magnetic field of sunspots. *Mon. Not. Roy. Astr. Soc.* **94**, 39–48.

Cowling, T. G. (1957*a*) Magnetohydrodynamics, Interscience, New York.
Cowling, T. G. (1957*b*) The dynamo maintenance of steady magnetic fields. *Q. J. Mech. Appl. Math.* **10**, 129–36.
Cowling, T. G. (1975*a*) Magnetohydrodynamics. Adam Hilger Ltd.
Cowling, T. G. (1975*b*) Sunspots and the solar cycle. *Nature* **255**, 189–90.
Cox, A. (1969) Geomagnetic reversals. *Science* **163**, 237–45.
Cox, A. (1972) Geomagnetic reversals: characteristic time constants and stochastic processes. *In* Report on Int. Cong. on Core–Mantle interface. *Trans. Am. Geophys.* Un. **53**, 613.
Deinzer, W., von Kusserow, H. U. & Stix, M. (1974) Steady and oscillatory $\alpha\omega$-dynamos. *Astron. Astrophys.* **36**, 69–78.
Dolginov, Sh., Yeroshenko, Ye. & Zhuzgov, L. (1973) Magnetic field in the very close neighbourhood of Mars according to data from the Mars 2 and Mars 3 spacecraft. *J. Geophys. Res.* **78**, 4779–86.
Drobyshevski, E. M. & Yuferev, V. S. (1974) Topological pumping of magnetic flux by three-dimensional convection. *J. Fluid Mech.* **65**, 33–44.
Durney, B. R. (1974) On the Sun's differential rotation: its maintenance by large-scale meridional motions in the convection zone. *Astrophys. J.* **190**, 211–21.
Durney, B. R. (1976) On theories of solar rotation. Proc. IAU Symp., 71, Basic Mechanisms of Solar Activity, ed. Bumba & Kleczek, Prague, 1975.
Eddy, J. A. (1976) The Maunder minimum. *Science* **192**, 1189–202.
Eddy, J. A., Gilman, P. A. & Trotter, D. E. (1976) Solar radiation during the Maunder minimum. *Solar Phys.* **46**, 3–14.
Elsasser, W. M. (1946) Induction effects in terrestrial magnetism, I. Theory. *Phys. Rev.* **69**, 106–16.
Eltayeb, I. A. (1972) Hydromagnetic convection in a rapidly rotating fluid layer. *Proc. Roy. Soc.* A**326**, 229–54.
Eltayeb, I. A. (1975) Overstable hydromagnetic convection in a rotating fluid layer. *J. Fluid Mech.* **71**, 161–79.
Ferraro, V. C. A. (1937) The non-uniform rotation of the Sun and its magnetic field. *Mon. Not. Roy. Astr. Soc.* **97**, 458–72.
Frisch, U., Pouquet, A., Léorat, J. & Mazure, A. (1975) Possibility of an inverse cascade of magnetic helicity in magnetohydrodynamic turbulence. *J. Fluid Mech.* **68**, 769–78.
Gailitis, A. (1970) Self-excitation of a magnetic field by a pair of annular vortices. *Mag. Gidrod.* **6**, 19–22. [English Translation: *Magnetohydrodynamics* **6**, 14–17.]
Gibson, R. D. (1968*a*) The Herzenberg dynamo, I. *Q. J. Mech. Appl. Math.* **21**, 243–55.
Gibson, R. D. (1968*b*) The Herzenberg dynamo, II. *Q. J. Mech. Appl. Math.* **21**, 257–87.

Gibson, R. D. & Roberts, P. H. (1967) Some comments on the theory of homogeneous dynamos. *In* Magnetism and the Cosmos, ed. Hindmarsh, W. R., Lowes, F. J., Roberts, P. R. & Runcorn, S. K., Oliver & Boyd Ltd., Edinburgh, pp. 108–20.

Gibson, R. D. & Roberts, P. H. (1969) The Bullard–Gellman dynamo. *In* The Application of Modern Physics to the Earth and Planetary Interiors, ed. Runcorn, S. K., Wiley, Interscience, pp. 577–601.

Gilman, P. A. (1970) Instability of magnetohydrostatic stellar interiors from magnetic buoyancy, I. *Astrophys. J.* **162**, 1019–29.

Golitsyn, G. S. (1960) Fluctuations of the magnetic field and current density in a turbulent flow of a weakly conducting fluid. *Sov. Phys. Dokl.* **5**, 536–9.

Gough, D. O. & Weiss, N. O. (1976) The calibration of stellar convection theories. *Mon. Not. Roy. Astr. Aoc.* **176**, 589–607.

Greenspan, H. (1968) The theory of rotating fluids. Cambridge Univ. Press.

Greenspan, H. (1974) On α-dynamos. *Studies in Appl. Math.* **13**, 35–43.

Gubbins, D. (1973) Numerical solutions of the kinematic dynamo problem. *Phil. Trans. Roy. Soc.* A**274**, 493–521.

Hale, G. E. (1908) On the probable existence of a magnetic field in sunspots. *Astrophys. J.* **28**, 315–43.

Herzenberg, A. & Lowes, F. J. (1957) Electromagnetic induction in rotating conductors. *Phil. Trans. Roy. Soc.* A**249**, 507–84.

Hide, R. (1956) The hydrodynamics of the Earth's core. *In* Physics and Chemistry of the Earth **1**, ed. Ahrens, L. H., Rankama, K. & Runcorn, S. K., Pergamon Press, London, pp. 94–137.

Hide, R. (1956) The hydrodynamics of the Earth's core. *In* Physics and Chemistry of the Earth **1**, ed. Ahrens, L. H., Rankama, K. & Runcorn, S. K., Pergamon Press, London, pp. 94–137.

Hide, R. (1969) Interaction between the Earth's liquid core and solid mantle. *Nature* **222**, 1055–6.

Hide, R. (1970) On the Earth's core–mantle interface. *Q. J. Roy. Met. Soc.* **96**, 579–90.

Hide, R. (1974) Jupiter and Saturn. *Proc. Roy. Soc.* A**336**, 63–84.

Hide, R. (1976) A note on helicity. *Geophys. Fluid Dyn.* **7**, 157–61.

Hide, R. & Malin, S. R. C. (1970) Novel correlations between global features of the Earth's gravitational and magnetic fields. *Nature* **225**, 605–9.

Hide, R. & Malin, S. R. C. (1971) Novel correlations between global features of the Earth's gravitational and magnetic fields: further statistical considerations. *Nature* **230**, 63.

Hide, R. & Roberts, P. H. (1961) The origin of the main geomagnetic field. *In* Physics and Chemistry of the Earth **4**, ed. Ahrens, L. H., Rankama, K. & Runcorn, S. K., Pergamon Press, London, pp. 25–98.

Higgins, G. H. & Kennedy, G. C. (1971) The adiabatic gradient and the melting point gradient in the core of the Earth. *J. Geophys. Res.* **76**, 1870–8.

Howard, R. (1974) Studies of solar magnetic fields, I. The average field strengths. *Solar Phys.* **38**, 283–99.

Howard, R. & Harvey, J. (1970) Spectroscopic determinations of solar rotation. *Solar Phys.* **12**, 23–51.

Ibbetson, A. & Tritton, D. J. (1975) Experiments on turbulence in a rotating fluid. *J. Fluid Mech.* **68**, 639–72.

Irving, E. (1964) Paleomagnetism. Wiley, New York.

Jacobs, J. A. (1975) The Earth's Core. Academic Press.

Jacobs, J. A. (1976) Reversals of the Earth's magnetic field. *Phys. Rep.* (Sec. C of Phys. Letters) **26**, 183–225.

Jepps, S. A. (1975) Numerical models of hydromagnetic dynamos. *J. Fluid Mech.* **67**, 625–46.

Kennedy, G. C. & Higgins, G. H. (1973) The core paradox. *J. Geophys. Res.* **78**, 900–4.

Khan, M. A. (1971) Correlation function in geophysics. *Nature* **230**, 57–61.

Kolmogorov, A. N. (1941) The local structure of turbulence in incompressible viscous fluid for very large Reynolds numbers. *Dokl. AN SSSR* **30**, 299–303.

Kovasznay, L. S. G. (1960) Plasma turbulence. *Rev. Mod. Phys.* **32**, 815–22.

Kraichnan, R. H. (1965) Inertial-range spectrum of hydromagnetic turbulence. *Phys. Fluids* **8**, 1385–87.

Kraichnan, R. H. (1973) Helical turbulence and absolute equilibrium. *J. Fluid Mech.* **59**, 745–52.

Kraichnan, R. H. (1976*a*) Diffusion of weak magnetic fields by isotropic turbulence. *J. Fluid Mech.* **75**, 657–76.

Kraichnan, R. H. (1976*b*) Diffusion of passive-scalar and magnetic fields by helical turbulence. *J. Fluid Mech.* **77**, 753–68.

Kraichnan, R. H. & Nagarajan, S. (1967) Growth of turbulent magnetic fields. *Phys. Fluids* **10**, 859–70.

Krause, F. & Steenbeck, M. (1967) Some simple models of magnetic field regeneration by non-mirror-symmetric turbulence. *Z. Naturforsch.* **22**a, 671–5. See Roberts & Stix (1971), pp. 81–95.

Kuiper, G. P. (ed.) (1953) The Sun, Univ. of Chicago Press, Chicago, Illinois.

Kumar, S. & Roberts, P. H. (1975) A three-dimensional kinematic dynamo. *Proc. Roy. Soc.* A**344**, 235–58.

Landstreet, J. D. & Angel, J. R. P. (1974) The polarisation spectrum and magnetic field strength of the White Dwarf Grw + 70°8247 *Astrophys. J.* **196**, 819–25.

Larmor, J. (1919) How could a rotating body such as the Sun become a magnet? *Rep. Brit. Assoc. Adv. Sci. 1919*, 159–60.

Lehnert, B. (1954) Magnetohydrodynamic waves under the action of the Coriolis force. *Astrophys. J.* **119**, 647–54.

Léorat, J. (1975) La turbulence magnétohydrodynamique hélicitaire et la géneration des champs magnétiques a grande échelle. Thèse doctoral, Univ. de Paris VII.

Levy, E. H. (1972) Effectiveness of cyclonic convection for producing the geomagnetic field. *Astrophys. J.* **171**, 621–33.

Liapounov, A. M. (1947) Problème generale de la stabilité du mouvement. *Ann. Math. Studies* **17**, Princeton Univ. Press.

Liepmann, H. (1952) Aspects of the turbulence problem. *Zeit. Ang. Math. Phys.* **3**, 321–42.

Lighthill, M. J. (1959) Introduction to Fourier analysis and generalised functions. Cambridge Univ. Press.

Lilley, F. E. M. (1970) On kinematic dynamos. *Proc. Roy. Soc.* A**316**, 153–67.

Loper, D. E. (1975) Torque balance and energy budget for the precessionally driven dynamo. *Phys. Earth Planet. Inter.* **11**, 43–60.

Lortz, D. (1968) Impossibility of steady dynamos with certain symmetries. *Phys. Fluids* **11**, 913–15.

Lowes, F. J. (1971) Significance of the correlation between spherical harmonic fields. *Nature* **230**, 61–2.

Lowes, F. J. (1974) Spatial power spectrum of the main geomagnetic field, and extrapolation to the core. *Geophys. J. R. Astr. Soc.* **36**, 717–30.

Lowes, F. J. & Wilkinson, I. (1963) Geomagnetic dynamo: a laboratory model. *Nature* **198**, 1158–60.

Lowes, F. J. & Wilkinson, I. (1968) Geomagnetic dynamo: an improved laboratory model. *Nature* **219**, 717–18.

Lüst, R. & Schlüter, A. (1954) Kraftfreie Magnetfelder. *Z. Astrophys.* **34**, 263–82.

Malkus, W. V. R. (1963) Precessional torques as the cause of geomagnetism. *J. Geophys. Res.* **68**, 2871–86.

Malkus, W. V. R. (1968) Precession of the Earth as the cause of geomagnetism, *Science* **160**, 259–64.

Malkus, W. V. R. & Proctor, M. R. E. (1975) The macrodynamics of α-effect dynamos in rotating fluids. *J. Fluid Mech.* **67**, 417–44.

Malkus, W. V. R. & Veronis, G. (1958) Finite amplitude convection. *J. Fluid Mech.* **4**, 225–60.

Maunder, E. W. (1904) Note on the distribution of sunspots in heliographic latitude, 1874–1902. *Mon. Not. Roy. Astr. Soc.* **64**, 747–61.

Maunder, E. W. (1913) Distribution of sunspots in heliographic latitude, 1874–1913. *Mon. Not. Roy. Astr. Soc.* **74**, 112–16.

Mestel, L. (1967) Stellar magnetism. *In* Plasma Astrophysics, ed. Sturrock, P. Acad. Press, London, pp. 185–228.

Moffatt, H. K. (1969) The degree of knottedness of tangled vortex lines. *J. Fluid Mech.* **35**, 117–29.

Moffatt, H. K. (1970*a*) Turbulent dynamo action at low magnetic Reynolds number. *J. Fluid Mech.* **41**, 435–52.

Moffatt, H. K. (1970*b*) Dynamo action associated with random inertial waves in a rotating conducting fluid. *J. Fluid. Mech.* **44**, 705–19.

Moffatt, H. K. (1972) An approach to a dynamic theory of dynamo action in a rotating conducting fluid. *J. Fluid Mech.* **53**, 385–99.

Moffatt, H. K. (ed.) (1973) Report on the Nato Advanced Study Institute on magnetohydrodynamic phenomena in rotating fluids. *J. Fluid Mech.* **57**, 625–49.

Moffatt, H. K. (1974) The mean electromotive force generated by turbulence in the limit of perfect conductivity. *J. Fluid Mech.* **65**, 1–10.

Moffatt, H. K. (1976) Generation of magnetic fields by fluid motion. *Adv. Appl. Mech.* **16**, 119–81.

Moffatt, H. K. & Dillon, R. F. (1976) The correlation between gravitational and geomagnetic fields caused by interaction of the core fluid motion with a bumpy core–mantle interface. *Phys. Earth Planet. Inter.* **13**, 67–78.

Munk, W. H. & MacDonald, G. J. F. (1960) The rotation of the Earth, Cambridge Univ. Press.

Murthy, V. R. & Hall, H. T. (1970) The chemical composition of the Earth's core: possibility of sulphur in the core. *Phys. Earth Planet. Inter.* **2**, 276–82.

Ness, N. F., Behannon, K. W., Lepping, R. P. & Whang, Y. C. (1975) The magnetic field of Mercury, I. *J. Geophys. Res.* **80**, 2708–16.

Orszag, S. A. (1970) Analytical theories of turbulence. *J. Fluid Mech.* **41**, 363–86.

Orszag, S. A. (1977) Lectures on the statistical theory of turbulence. *In* Fluid Dynamics, Les Houches 1973, ed. Balian, R. & Peube, J. L. Gordon and Breach, pp. 235–374.

Parker, E. N. (1955*a*) The formation of sunspots from the solar toroidal field. *Astrophys. J.* **121**, 491–507.

Parker, E. N. (1955*b*) Hydromagnetic dynamo models. *Astrophys. J.* **122**, 293–314.

Parker, E. N. (1963) Kinematical hydromagnetic theory and its applications to the low solar photosphere. *Astrophys. J.* **138**, 552–75.

Parker, E. N. (1970) The generation of magnetic fields in astrophysical bodies, I. The dynamo equations. *Astrophys. J.* **162**, 665–73.

Parker, E. N. (1971*a*) The generation of magnetic fields in astrophysical bodies, II. The galactic field. *Astrophys, J.* **163**, 255–78.

Parker, E. N. (1971*b*) The generation of magnetic fields in astrophysical bodies, III. Turbulent diffusion of fields and efficient dynamos. *Astrophys. J.* **163**, 279–85.

Parker, E. N. (1971*c*) The generation of magnetic fields in astrophysical bodies, IV. The solar and terrestrial dynamos. *Astrophys. J.* **164**, 491–509.

Parker, E. N. (1971*d*) The generation of magnetic fields in astrophysical bodies, V. Behaviour at large dynamo numbers. *Astrophys. J.* **165**, 139–46.

Parker, E. N. (1971*e*) The generation of magnetic fields in astrophysical bodies, VI. Periodic modes of the galactic field. *Astrophys. J.* **166**, 295–300.

Parker, E. N. (1971*f*) The generation of magnetic fields in astrophysical bodies, VIII. Dynamical considerations. *Astrophys. J.* **168**, 239–49.

Parker, E. N. (1975) The generation of magnetic fields in astrophysical bodies, X. Magnetic buoyancy and the solar dynamo. *Astrophys. J.* **198**, 205–9.

Parker, R. L. (1966) Reconnexion of lines of force in rotating spheres and cylinders. *Proc. Roy. Soc.* A**291**, 60–72.

Pekeris, C. L. (1972) Stationary spherical vortices in a perfect fluid. *Proc. Nat. Acad. Sci.* **69**, 2460–2. (Corrigendum, p. 3849).

Pekeris, C. L., Accad, Y. & Shkoller, B. (1973) Kinematic dynamos and the Earth's magnetic field. *Phil. Trans. Roy. Soc.* A**275**, 425–61.

Pichackchi, L. D. (1966) Theory of the hydromagnetic dynamo. *Sov. Phys. JETP* **23**, 542–3.

Piddington, J. H. (1972*a*) The origin and form of the galactic magnetic field, I. Parker's dynamo model. *Cosmic Electrodynamics* **3**, 60–70.

Piddington, J. H. (1972*b*) The origin and form of the galactic magnetic field, II. The primordial-field model. *Cosmic Electrodynamics* **3**, 129–46.

Pouquet, A., Frisch, U., & Léorat J. (1976) Strong MHD turbulence and the nonlinear dynamo effect. *J. Fluid Mech.* **77**, 321–54.

Pouquet, A. & Patterson, G. S. (1977) Numerical simulation of helical magnetohydrodynamic turbulence. *J. Fluid Mech.* (to appear).

Preston, G. W. (1967) Studies of stellar magnetism – past, present and future. *In* Proc. AAS-NASA Symp. The Magnetic and Related Stars, ed. Cameron, R.C., Mono Book Corp., Baltimore Md., pp. 3–28.

Proctor, M. R. E. (1975) Non-linear mean field dynamo models and related topics. Ph.D. thesis, Cambridge University.

Proctor, M. R. E. (1977*a*) Numerical solutions of the nonlinear α-effect dynamo equations, *J. Fluid Mech.* **80**, 769–84.

Proctor, M. R. E. (1977*b*) On the eigenvalues of kinematic α-effect dynamos. *Astron. Nachrichten* **298**, 19–25.

Proctor, M. R. E. (1977*c*) The role of mean circulation in parity selection by planetary magnetic fields. *Geoph. Astrophys. Fluid Dyn.* (to appear).

Rädler, K.-H. (1969*a*) On the electrodynamics of turbulent fluids under the influence of Coriolis forces. *Monats. Dt. Akad. Wiss.* **11**, 194–201. [English translation: Roberts & Stix (1971) pp. 291–9.]

Rädler, K.-H. (1969*b*) A new turbulent dynamo, I. *Monats. Dt. Akad. Wiss.* **11**, 272–9. [English translation: Roberts & Stix (1971) pp. 301–8.]

Rikitake, T. (1958) Oscillations of a system of disk dynamos. *Proc. Camb. Phil. Soc.* **54**, 89–105.

Robbins, K. A. (1976) A moment equation description of magnetic reversals in the Earth. *Proc. Nat. Acad. Sci. USA* **73**, 4297–301.

Roberts, G. O. (1970) Spatially periodic dynamos. *Phil. Trans. Roy. Soc.* A**266**, 535–58.

Roberts, G. O. (1972) Dynamo action of fluid motions with two-dimensional periodicity. *Phil. Trans. Roy. Soc.* A**271**, 411–54.

Roberts, P. H. (1971) Dynamo theory. *In* Mathematical problems in the Geophysical Sciences, ed. Reid, W. H., Am. Math. Soc., Prov., R.I., pp. 129–206.

Roberts, P. H. (1972*a*) Kinematic dynamo models. *Phil. Trans. Roy. Soc.* A**272**, 663–98.

Roberts, P. H. (1972*b*) Electromagnetic core–mantle coupling. *J. Geomagn. Geoelectr.* **24**, 231–59.

Roberts, P. H. & Soward, A. M. (1972) Magnetohydrodynamics of the Earth's core. *Ann. Rev. Fluid Mech.* **4**, 117–54.

Roberts, P. H. & Stewartson, K. (1974) On finite amplitude convection in a rotating magnetic system. *Phil. Trans. Roy. Soc.* A**277**, 287–315.

Roberts, P. H. & Stewartson, K. (1975) On double roll convection in a rotating magnetic system. *J. Fluid Mech.* **68**, 447–66.

Roberts, P. H. & Stix, M. (1971) The turbulent dynamo: a translation of a series of paper by F. Krause, K.-H. Rädler and M. Steenbeck. Tech. Note 60, NCAR, Boulder, Colorado.

Roberts, P. H. & Stix, M. (1972) α-effect dynamos, by the Bullard–Gellman formalism. *Astron. Astrophys.* **18**, 453–66.

Rochester, M. G. (1962) Geomagnetic core–mantle coupling. *J. Geophys. Res.* **67**, 4833–6.

Rochester, M. G., Jacobs, J. A., Smylie, D. E. & Chong, K. F. (1975) Can precession power the geomagnetic dynamo? *Geophys. J. Roy. Astr. Soc.* **43**, 661–78.

Saffman, P. G. (1963) On the fine-scale structure of vector fields convected by a turbulent fluid. *J. Fluid Mech.* **16**, 545–72.

Schlüter, A. & Biermann, L. (1950) Interstellare Magnetfelder. *Z. Naturforsch.* **5**a, 237–51.

Smith, E., Davis, L., Jones, D. E., Colburn, D. S., Colemann, P. J., Dyal, P. L. & Sonnett, C. P. (1974) Magnetic field of Jupiter and its interaction with the solar wind. *Science* **183**, 305–6.

Smoluchowski, R. (1971) Metallic interiors and magnetic fields of Jupiter and Saturn. *Astrophys. J.* **166**, 435–9.

Soward, A. M. (1972) A kinematic theory of large magnetic Reynolds number dynamos. *Phil. Trans. Roy. Soc.* A**272**, 431–62.

Soward, A. M. (1974) A convection-driven dynamo I. The weak field case. *Phil. Trans. Roy. Soc.* A**275**, 611–51.

Soward, A. M. (1975) Random waves and dynamo action. *J. Fluid Mech.* **69**, 145–77.

Steenbeck, M., Kirko, I. M., Gailitis, A., Klawina, A. P., Krause, F., Laumanis, I. J. & Lielausis, O. A. (1967) An experimental verification of the α-effect. *Monats. Dt. Akad. Wiss.* **9**, 716–9. [English translation: Roberts & Stix (1971) pp. 97–102.]

Steenbeck, M., Krause, F. & Rädler, K.-H. (1966) A calculation of the mean electromotive force in an electrically conducting fluid in turbulent motion, under the influence of Coriolis forces. *Z. Naturforsch.* **21**a, 369–76. [English translation: Roberts & Stix (1971) pp. 29–47.]

Steenbeck, M. & Krause, F. (1966). The generation of stellar and planetary magnetic fields by turbulent dynamo action. *Z. Naturforsch.* **21**a, 1285–96. [English translation: Roberts & Stix (1971) pp. 49–79.]

Steenbeck, M. & Krause, F. (1969*a*) On the dynamo theory of stellar and planetary magnetic fields, I. A. C. dynamos of solar type. *Astron. Nachr.* **291**, 49–84 [English translation: Roberts & Stix (1971) pp. 147–220.]

Steenbeck, M. & Krause, F. (1969*b*) On the dynamo theory of stellar and planetary magnetic fields, II. D.C. dynamos of planetary type. *Astron. Nachr.* **291**, 271–86. [English translation: Roberts & Stix (1971) pp. 221–45.]

Stenflo, J. O. (1976) Small-scale solar magnetic fields. Proc. IAU Symp. 71, Basic Mechanisms of Solar Activity, Ed. Bumba & Kleczek, Prague 1975.

Stewartson, K. & Roberts, P. H. (1963) On the motion of a liquid in a spheroidal cavity of a precessing rigid body. *J. Fluid Mech.* **17**, 1–20.

Stix, M. (1972). Non-linear dynamo waves. *Astron. Astrophys.* **20**, 9–12.

Stix, M. (1973). Spherical $\alpha\omega$–dynamos, by a variational method. *Astron. Astrophys.* **24**, 275–81.

Stix, M. (1975) The galactic dynamo. *Astron. Astrophys.* **42**, 85–9.

Stix, M. (1976) Differential rotation and the solar dynamo. *Astron. Astrophys.* **47**, 243–54.

Taylor, G. I. (1921) Diffusion by continuous movements. *Proc. Lond. Math. Soc.* A**20**, 196–211. (Collected papers, vol. II, ed. Batchelor, G. K., Cambridge Univ. Press, 1960, pp. 172–84).

Taylor, J. B. (1963) The magnetohydrodynamics of a rotating fluid and the Earth's dynamo problem. *Proc. Roy. Soc.* A**274**, 274–83.

Toomre, A. (1966) On the coupling of the Earth's core and mantle during the 26000-year precession. *In* The Earth–Moon system, ed. Marsden, B. G. & Cameron, A. G. W., Plenum, New York, pp. 33–45.

Tough, J. G. (1967) Nearly symmetric dynamos. *Geophys. J. Roy. Astr. Soc.* **13**, 393–6. See also Corrigendum, *Geophys. J. Roy. Astr. Soc.* **15**, 343.

Tough, J. G. & Gibson, R. D. (1969) The Braginskii dynamo. *In* The application of modern physics to the Earth and planetary interiors, ed. Runcorn, S. K., Wiley-Interscience, London, New York, pp. 555–69.

Townsend, A. A. (1975) Turbulent shear flow (2nd edn.). Cambridge Univ. Press.

Urey, H. C. (1952) The Planets. Yale Univ. Press.

Vainshtein, S. I. & Zel'dovich Ya. B. (1972) Origin of magnetic fields in astrophysics. *Sov. Phys. Usp.* **15**, 159–72.

Veronis, G. (1959) Cellular convection with finite amplitude in a rotating fluid. *J. Fluid Mech.* **5**, 401–35.

Warwick, J. (1963) Dynamic spectra of Jupiter's decametric emission. *Astrophys. J.* **137**, 41–60.

Weir, A. D. (1976) Axisymmetric convection in a rotating sphere, I. Stress-free surface. *J. Fluid Mech.* **75**, 49–79.

Weiss, N. O. (1966) The expulsion of magnetic flux by eddies. *Proc. Roy. Soc.* A**293**, 310–28.

Weiss, N. O. (1971) The dynamo problem. *Q. J. Roy. Astr. Soc.* **12**, 432–46.

Weiss, N. O. (1976) The pattern of convection in the Sun. Proc. IAU Symp. 71, Basic Mechanisms of Solar Activity, ed. Bumba & Kleczek, Prague 1975.

Woltjer, L. (1958) A theorem on force-free magnetic fields. *Proc. Nat. Acad. Sci.* **44**, 489–91.

Woltjer, L. (1975) Astrophysical evidence for strong magnetic fields. *Ann. New York Acad. Sci.* **257**, 76–9.

Wrubel, M. H. (1952) On the decay of a primeval stellar magnetic field. *Astrophys. J.* **116**, 291–8.

Yoshimura, H. (1975) A model of the solar cycle driven by the dynamo action of the global convection in the solar convection zone. *Astrophys. J. Suppl.* **29**, 467.

Zel'dovich, Ya. B. (1957). The magnetic field in the two-dimensional motion of a conducting turbulent fluid. *Sov. Phys. JETP* **4**, 460–2.

INDEX